AF553670

Sustaining Soil and Water Quality in Changing Climate

Abouth the Editors

Dr. Savita Jangde earned her Ph.D. from the Institute of Agricultural Sciences, Banaras Hindu University, Varanasi, Uttar Pradesh, India. Currently, she is an Assistant Professor of Plant Physiology at Banaras Hindu University. Her primary research focuses on plant stress physiology, post-harvest physiology, nutriophysiology, and the impact of climate change on crop physiology. Dr. Jangde has supervised two Ph.D. students and numerous postgraduate students in their thesis work. Additionally, she has authored several research papers and articles published in both national and international journals. Her contributions extend to writing book chapters and books in her field of expertise.

Dr. Niranjan B.N. is presently working as an Extension officer in Dakshina Kannada Milk Union (KMF), Mangalore-575005, Karnataka, India. He has 1 year and 8 months of work experience. His research area includes Soil Science and Agricultural Chemistry, Plant Nutrition, Soil Fertility, and problematic soils. He has published 5 Research Papers. Qualified for the National Eligibility Test conducted by ASRB on 27th October 2021 in the discipline of Soil Sciences for eligibility for Assistant Professorship. Awarded national fellowship by GOI under the Ministry of Social Justice and Empowerment for a Doctorate program. He attended one conference and one workshop on good laboratory techniques. He was well-versed in sports activities and represented the university for 4 times. Being a multifaceted personality, he presented 2 posters of his research work on various platforms.

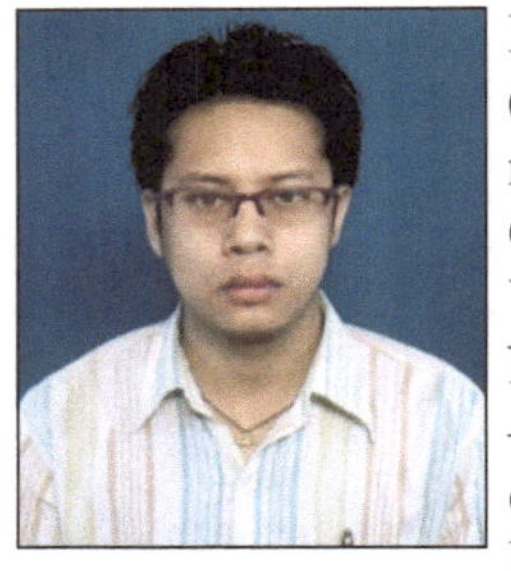

Dr. Kh. Chandrakumar Singh graduated in Agriculture (2009) from HNBGU, Uttarakhand, and achieved both master's (2011) and doctoral (2015) degree in Agricultural Chemistry and Soil Science from Bidhan Chandra Krish Viswavidyalaya, Mohanpur. Dr. Singh qualified ICAR NET conducted by ASRB in 2013. Dr. Singh served KV – North Sikkim as a Subject Matter Specialist (SMS) nd currently serving as an Assistant Professor at VCSGUUHF Bharsar, Uttarakhand. Dr. Singh is actively involved in teaching and research in the field of Soil Science. So far Dr. Singh has guided 13 students in their master's degree programs.

Er. Jeetendra Kumar obtained his M. Tech. (Agricultural Engineering) with specialization in Soil & Water Engineering from the College of Agricultural Engineering & Technology, Dr. Rajendra Prasad Central Agricultural University, Pusa, Samastipur, Bihar, India. Presently, he is working as a Subject Matter Specialist (Agricultural Engineering) at Krishi Vigyan Kendra, Jehanabad under Bihar Agricultural University, Sabour, Bhagalpur, Bihar, India and serving the farming community mostly for extension works. He has more than two decades of experience in extension, research and teaching. His research area focuses on soil and water conservation engineering, irrigation and drainage engineering, and soil water water-plant relationships in climate change scenarios. He has to his credit 26 research papers published in reputed journals (national and international), 2 Invited Lectures/Lead Papers, two books, 11 book chapters, 16 Booklet /Extension Manuals, 32 full-length Papers/abstracts in conference/ seminar/ symposium Proceedings, 45 Popular Articles/Extension Bulletins, 45 Success stories. He has been the editor of the Quarterly Newsletter, Krishak Samachar and the monthly booklet/magazine Krishak Sandesh of Krishi Vigyan Kendra, BAU, Sabour. Life Member of 6 Scientific/professional Societies including the Indian Association of Soil & Water Conservationists (IASWC), Dehradun and the Indian Society of Agricultural Engineers, New Delhi. He is a recipient of 05 awards from different scientific and professional societies for his contributions in the area of Soil & Water Conservation Engineering. Er. Kumar has been CO-PI in 3 projects namely National Innovations on Climate Resilient Agriculture (NICRA), Climate Resilient Agriculture Programme (CRA) and the IRRI Project. He has Co- Chaired the Technical Sessions of International Conferences conducted 552 Trainings for the farming community, and rural youths as well as extension functionaries, delivered 15 TV/Radio Talks, participated in 32 Conferences/Seminar/Workshop/Training programmes and organized 6 Conferences/Seminars as Co-Organizing/Joint Organizing Secretary/ Co-Convener.

Dr. Wajid Hasan, PhD PDF, is devoted to research and extension activities in Agriculture Entomology. Dr Hasan is a Subject Matter Specialist in Entomology at KVK Jehanabad, Bihar Agricultural University in Bihar. He was awarded Dr DS Kothari UGC-Post Doctoral Fellow (PDF) in 2010, offered by the University Grant Commission of India. He holds a doctorate in Agricultural Entomology from GB Pant University of Agri & Tech, Pantnagar, UK, India. He has good terms with farmers, and he serves them as an adviser for plant protection measures. Dr Hasan has received eleven awards from several societies for his outstanding contribution to the relevant discipline. He has 65 research papers,

35 books, 67 book chapters, 42 popular articles, 140 success stories, 3 Practical Manuals, 16 success stories and 48 delivered lectures in different seminars/ symposia/conferences/workshops and seven radio talks. Dr Hasan stayed as the chief organizer for eleven international conferences with more than three thousand participants each and managed a team of about three hundred co-organizers each. He has also organized 551 training programs for farmers, rural youth and agricultural professionals with 20450 participants. In addition, he has conducted three skill development training programs of 200 hours each and 49 On-Farm Trials (OFT). Dr Hasan has a lifetime membership in 6 scientific societies. He is the Editor-in-Chief of the International Journal of Agricultural and Applied Sciences (IJAAS) and an Editorial board member for 5 scientific International Journals.

Dr. Sheetanshu Gupta is an Assistant Professor at the Department of Biotechnology, Institute of Technology and Management, BKT, Lucknow, U.P. He has been working in Biochemistry, Molecular Biology & Biotechnology, and emphasizing Environmental Health and Climate Change for the last 12 years, an alumnus of G.B. Pant University of Agriculture & Technology, Pantnagar, Uttarakhand, India. He qualified CSIR and ICAR ARS NET in Life Sciences and Plant Biochemistry & Plant Physiology respectively. Dr. Sheetanshu Gupta's academic qualifications, research experience, and expertise in biochemistry and biotechnology make him a valuable asset. With his passion for teaching and research, dedication to innovative farming techniques, and student advocacy, Dr. Gupta continues to contribute significantly to advancing agricultural and biochemical sciences.

Sustaining Soil and Water Quality in Changing Climate

Savita Jangde
Department of Plant Physiology
Banaras Hindu University Varansi
U.P.,India

Niranjan B N
Dakshina Kannada Milk Union (KMF)
Mangalore-575005, Karnataka, India

Kh. Chandrakumar Singh
Department of Natural Resource Management
V.C.S.G Uttarakhand University of Horticulture and Forestry
Bharsar, Uttarakhand, India

Jeetendra Kumar
Krishi Vigyan Kendra, Jehanabad
Bihar Agricultural University
Bihar, India

Wajid Hasan
Krishi Vigyan Kendra, Jehanabad
Bihar Agricultural University
Bihar, India

Sheetanshu Gupta
Department of Biotechnology
Institute of Technology and Management
BKT, Lucknow, U.P., India

FU 75, Pitam Pura
New Delhi 110034
publishmywork23@gmail.com
Mob. 91-8447075807

ISBN: 978-93-61345-77-7

Preface

In the ever-evolving landscape of agriculture, where the impacts of climate change and environmental degradation are increasingly evident, the need for sustainable practices has never been more urgent. "Sustaining Soil and Water Quality in Changing Climate" emerges as a timely and essential guide, offering a comprehensive exploration of strategies to navigate the complexities of modern farming while prioritizing environmental stewardship. As stewards of the land, it is our responsibility to cultivate not only bountiful harvests but also a resilient and enduring relationship between humankind and the environment.

This book is born out of a collective commitment to address the pressing challenges faced by the agricultural sector, challenges that extend beyond the traditional realms of soil and water management. It reflects a holistic approach to sustainable agriculture, acknowledging the interconnectedness of various factors influencing farming practices. The preface sets the tone for this holistic perspective, emphasizing the importance of understanding the fundamental principles that govern soil health and water resources. The narrative unfolds against the backdrop of a changing climate, where the unpredictability of weather patterns and the intensification of environmental stressors demand innovative and adaptive approaches. This preface introduces the overarching theme of resilience — the capacity of agricultural systems to adapt and thrive in the face of adversity. It encourages readers to consider the book not merely as a manual of techniques but as a guide for cultivating a mindset of sustainability that permeates every aspect of agricultural decision-making.

The following chapters delve into innovative strategies for addressing environmental challenges. Chapter 3, "Adaptation Strategies for Climate-Resilient Wasteland Development and Management," focuses on revitalizing degraded lands in the face of climate change. Biological interventions take center stage in Chapter 4, "Biological Aspects of Soil and Water Conservation Techniques," highlighting the role of vegetation and microorganisms in preserving soil structure and preventing erosion.

Chapters 5 through 10 shift towards technological and holistic approaches. "Precision Agriculture and Technology Integration" explores the use of advanced technologies in restoring chemically degraded soils, while "Organic Farming Practices" advocates for eco-friendly methods. "Agroforestry and Sustainable Land Use" integrates trees and crops for enhanced soil fertility, and "Irrigation Management" emphasizes responsible water use. Chapter 9, "Soil Water Plant

Relationship in Climate Change," examines intricate interactions in the context of a changing climate. The book concludes with Chapter 10, "Rainwater Harvesting, Agricultural Drainage, and Water Conservation," offering innovative solutions for water harnessing and agricultural drainage.

It acknowledges the collective wisdom of agricultural communities, the expertise of researchers, and the responsibility of policymakers to create an inclusive and sustainable future for agriculture. The commitment to shared knowledge and experiences is woven into the fabric of the book, fostering a sense of community among those dedicated to advancing sustainable agricultural practices.

In lacuna, this book directs readers toward a deeper understanding of the intricate relationships between soil, water, and climate in the context of sustainable agriculture. It serves as an invitation to embark on a journey of discovery, reflection, and action — a journey that transcends the boundaries of conventional farming practices and embraces a vision of agriculture that is both resilient and regenerative.

Editors

Contents

Preface .... *vii*

1. Fundamentals of Soil Science **1**

Savita Jangde, B.N. Niranjan, Kh. Chandrakumar Singh Wajid Hasan and Sheetanshu Gupta

2. Water Resources in Agriculture **81**

Niranjan B N., Kh. Chandrakumar Singh, Savita Jangde Sheetanshu Gupta and Wajid Hasan

3. Adaptation Strategies for Climate-Resilient Wasteland Development and Management **111**

Savita Jangde, Wajid Hasan, Sheetanshu Gupta, Niranjan B.N. and Kh. Chandrakumar Singh

4. Biological Aspects of Soil and Water Conservation Techniques..143

B.N., Niranjan Kh. Chandrakumar Singh, Shubham Jeetendra Kumar, and Sheetanshu Gupta

5. Precision Agriculture and Technology Integration - Restoration of Chemically Degraded Soils **165**

Kh. Chandrakumar Singh, Savita Jangde, Niranjan B.N. Jeetendra Kumar and Wajid Hasan

6. Organic Farming Practices for the Management of Soil and Water **201**

Wajid Hasan, Sheetanshu Gupta, Deepa Rawat, Adarsh Pandey and Kh. Chandrakumar Singh

7. Agroforestry and Sustainable Land Use for Soil Management and Conservation **241**

Deepa Rawat, Kh. Chandrakumar Singh, Savita Jangde Wajid Hasan and Sheetanshu Gupta

8. Irrigation Management for Water Efficiency and Conservation 263

Niranjan B.N., Kh. Chandrakumar Singh, Savita Jangde Wajid Hasan and Sheetanshu Gupta

9. Soil Water Plant Relationship in Climate Change 285

Savita Jangde, Santosh Marahatta, Jeetendra Kumar, Sheetanshu Gupta and Adarsh Pandey

10. Rainwater Harvesting, Agricultural Drainage and Water Conservation 313

Niranjan B.N., Kh. Chandrakumar Singh, Savita Jangde Wajid Hasan and Er. Jeetendra Kumar

Index 357

1

Fundamentals of Soil Science

[1]*Savita Jangde*, [2]*B.N. Niranjan*, [3]*Kh. Chandrakumar Singh*
[4]*Wajid Hasan and* [5]*Sheetanshu Gupta*

[1]*Department of Plant Physiology, Institute of Agricultural Sciences B.H.U. Varanasi, U.P.*
[2]*Dakshina Kannada Milk Union (KMF), Mangalore, Karnataka*
[3]*VCSGU University of Horticulture and Forestry, Bharsar, Uttarakhand*
[4]*KVK Jehanabad, Bihar Agricultural University Sabour, Bihar*
[5]*Department of Biotechnology, Institute of Technology and Management BKT, Lucknow, U.P.*

1.1 Introduction to Soil Science

1.1.1 Definition and Scope of Soil Science

Definition: Soil science encapsulates the scientific exploration of one of Earth's most vital components—soil. It goes beyond the everyday understanding of soil as mere dirt, positioning it as a dynamic and complex system that supports life. At its core, soil science is an amalgamation of various scientific disciplines, each contributing to the comprehensive study of soil. This multidisciplinary approach ensures a holistic understanding that extends far beyond the surface.

Soil science is not confined to the physical substance beneath our feet; rather, it embraces a nuanced perspective that involves decoding the intricate language of the soil. It is the science of uncovering the secrets embedded in the Earth's outer layer—examining its history, composition, and the myriad interactions that shape its character.

Scope: The scope of soil science is vast, encompassing a rich tapestry of interconnected disciplines and phenomena. Each facet contributes to a deeper understanding of soil as a living, breathing entity:

1. Soil Formation (Pedogenesis): The journey of soil begins with the interplay of various factors—weathering, organic matter decomposition, and the influence of climate, topography, and parent material. Pedogenesis, the process of soil formation, is a dynamic saga that soil scientists unravel to interpret the unique characteristics of different soil profiles.

Soil formation is a continuous process influenced by geological forces, climate patterns, and the activities of living organisms. From the disintegration of rocks to the decomposition of plant and animal matter, pedogenesis shapes the diverse soil horizons that make each soil profile unique. Understanding these intricate processes is fundamental to deciphering the history encoded in the layers of soil.

2. Soil Composition and Structure: Delving into the physical attributes of soil—texture, structure, density, and porosity—lays the groundwork for understanding its functionality. These properties influence water retention, drainage, and the availability of nutrients, all of which are critical for sustaining life.

Soil texture, determined by the relative proportions of sand, silt, and clay, profoundly affects water retention and drainage. Sandy soils, with larger particles, allow water to pass through quickly, while clayey soils, with finer particles, retain water more effectively. The soil structure, referring to how individual particles aggregate, further influences root penetration, aeration, and nutrient distribution. This intricate dance of physical properties forms the basis of soil science's understanding of how soils support plant life and ecosystem functions.

3. Chemical Properties: Soil chemistry delves into the intricate composition of soil—a complex matrix of minerals, organic matter, and various chemical elements. Soil scientists analyze soil pH, nutrient content, and cation exchange capacity to unravel the intricacies of soil fertility.

Soil chemistry is akin to deciphering the chemical language of the soil. The pH level, a measure of soil acidity or alkalinity, profoundly influences nutrient availability. Soil nutrients, including essential elements like nitrogen, phosphorus, and potassium, are critical for plant growth. The cation exchange capacity, a measure of the soil's ability to retain and supply essential nutrients, further contributes to the understanding of soil fertility. Soil scientists employ advanced analytical techniques to unravel these chemical intricacies, providing a roadmap for sustainable agricultural practices.

4. Biological Components: Beneath the surface lies a bustling world of microorganisms, fauna, and flora, forming the biological tapestry of soil. Soil scientists explore the dynamic interactions between these biological entities and their impact on nutrient cycling, organic matter decomposition, and overall soil health.

The biological realm of soil is a captivating ecosystem teeming with life. Microorganisms, including bacteria and fungi, play crucial roles in decomposing organic matter, releasing nutrients, and suppressing harmful pathogens. Soil fauna, such as earthworms and insects, contribute to soil aeration and nutrient cycling.

Plant roots and their symbiotic relationships with mycorrhizal fungi further enhance nutrient uptake. Understanding these intricate interactions allows soil scientists to harness the power of soil biology for sustainable land management practices (fig 1).

Fig. 1: Soil Management Practices

5. Soil Classification and Taxonomy: Soil scientists are akin to the taxonomists of the Earth, classifying soils based on shared characteristics. This classification aids in understanding soil distribution globally. Systems like the USDA Soil Taxonomy provide a common language for soil scientists and land managers.

Soil classification is a systematic way of organizing the incredible diversity of soils found worldwide. The USDA Soil Taxonomy, for instance, classifies soils into orders, suborders, and soil families based on key characteristics such as soil texture, color, and horizon development. This classification not only provides a standardized nomenclature but also serves as a tool for predicting soil behavior and guiding land use decisions. Soil taxonomy is a dynamic field, evolving with advances in scientific understanding and technology.

6. Soil Management Practices: Armed with knowledge of soil properties, soil scientists guide the development of sustainable land management practices. Techniques such as conservation tillage, cover cropping, and crop rotation are rooted in soil science principles, aiming to preserve soil structure, fertility, and overall health.

Sustainable soil management is at the forefront of modern agriculture. Conservation tillage, which minimizes soil disturbance, helps prevent erosion and preserves soil structure. Cover cropping involves planting specific crops to protect and enrich the soil during periods when the main cash crop is not growing. Crop rotation, the practice of alternating crops in a specific sequence, helps break pest and disease cycles while maintaining soil fertility. These practices are not only grounded in scientific principles but also contribute to the broader goals of sustainable and regenerative agriculture.

7. Importance in Agriculture and Ecosystems: Recognizing the pivotal role of soil in supporting plant life and maintaining ecological balance is a cornerstone of soil science. It informs agricultural practices that sustainably harness the soil's potential for food production while acknowledging the broader environmental impacts.

The significance of soil in agriculture extends beyond being a mere substrate for plant growth. Soil provides the physical support and nutrient reservoir essential for plant development. The intricate relationships between soil, plants, and microorganisms create a dynamic ecosystem that influences crop health and productivity. Beyond agriculture, soil science acknowledges the broader ecosystem services provided by soils—water filtration, carbon sequestration, and habitat for diverse organisms. Understanding and respecting these services are crucial for achieving a balance between human activities and environmental sustainability.

1.1.2 Importance of Soil in Agriculture and Ecosystems

1.1.2.1 Soil as the Foundation of Agriculture

Agricultural Productivity: The importance of soil in agriculture cannot be overstated. Soil serves as the foundation for agricultural productivity, providing physical support, essential nutrients, and a hospitable environment for plant growth. The intricate interplay of soil properties—texture, structure, and nutrient content—directly influences crop health and yields.

The physical structure of soil affects root penetration and water movement, crucial for nutrient uptake and overall plant health. The soil's nutrient-holding capacity ensures a steady supply of essential elements for crops. Understanding these dynamics allows farmers to optimize soil conditions for various crops, enhancing agricultural sustainability and food security.

Soil Fertility: Soil fertility, a key aspect of soil health, determines its ability to support plant growth. Fertile soils are rich in essential nutrients, promoting robust plant development. Soil scientists employ various techniques, such as nutrient testing and soil amendments, to maintain and enhance soil fertility.

Soil fertility is a dynamic aspect influenced by the soil's chemical composition and biological activity. Sustainable agricultural practices, including cover cropping and organic matter incorporation, play a crucial role in maintaining soil fertility. By managing soil fertility effectively, farmers can mitigate the need for excessive fertilizers, reducing environmental impacts.

Water Regulation: Soil acts as a reservoir for water, playing a crucial role in regulating water availability for plants. The soil's capacity to retain and release water influences irrigation needs, drought resilience, and overall water-use efficiency in agriculture.

Different soil types exhibit varying water-holding capacities. Sandy soils drain quickly but may require more frequent irrigation, while clayey soils retain water for longer periods. Soil science guides farmers in optimizing irrigation practices, ensuring water conservation and minimizing the environmental impact of excessive water use.

1.1.2.2 Ecosystem Services Provided by Soil

Biodiversity Support: Soil is a habitat for a myriad of organisms, from microscopic bacteria to larger fauna. This diversity contributes to overall ecosystem health, supporting various trophic levels and ecological interactions (Fig 2).

Fig. 2: Biodiversity Support of Soil

Soil biodiversity is a key driver of ecosystem resilience. Microorganisms contribute to nutrient cycling and organic matter decomposition, enhancing soil fertility. Soil-dwelling organisms, such as earthworms, play a role in soil structure improvement through burrowing activities. Protecting soil biodiversity is essential for maintaining the delicate balance within ecosystems.

Carbon Sequestration: Soil serves as a significant reservoir for carbon, playing a crucial role in mitigating climate change. Through the decomposition of organic matter and the formation of stable organic compounds, soil acts as a carbon sink, sequestering carbon from the atmosphere.

Soil organic carbon is a vital component for soil structure and fertility. Sustainable land management practices, including cover cropping and reduced tillage, contribute to carbon sequestration in soils. Recognizing the role of soil in climate regulation highlights the importance of soil science in global efforts to address climate change.

Water Filtration and Purification: Soil plays a pivotal role in water filtration and purification. As water percolates through the soil profile, contaminants are filtered out, and essential nutrients are retained. This process contributes to the quality of groundwater and surface water.

Soil acts as a natural water filter, removing impurities and pollutants. Wetlands and riparian zones, characterized by specific soil types, play a critical role in purifying water before it enters larger aquatic systems. Understanding the soil's role in water quality is essential for maintaining healthy ecosystems and safeguarding human water supplies.

1.1.3 Historical Perspective of Soil Science

1.1.3.1 Early Observations and Practices

Ancient Agricultural Practices: The historical perspective of soil science traces back to ancient civilizations where agriculture played a central role in societal development. Early farmers observed and learned about the importance of soil fertility through practical experiences, such as crop rotations and the use of organic amendments.

Ancient agricultural practices often incorporated traditional wisdom passed down through generations. Farmers recognized the value of allowing fields to fallow to restore soil fertility and practiced rudimentary forms of crop rotation to manage nutrient levels. These observations laid the foundation for understanding the dynamic relationship between soil and agriculture.

Chinese and Roman Contributions: Ancient Chinese and Roman agricultural writings provide early insights into soil management. Chinese agricultural practices, documented in texts like the "Book of Agriculture," emphasized the importance of soil conservation and the use of organic materials. Similarly, Roman authors, such as Columella, discussed soil fertility and recommended specific crops for different soil types.

Chinese and Roman agricultural knowledge demonstrated an early awareness of soil properties and the need for sustainable land management. The recognition of soil conservation and fertility management in these ancient texts reflected an understanding of the long-term impacts of agricultural practices on soil health.

1.1.3.2 Emergence of Modern Soil Science

Agronomy and Soil Chemistry: The 19th century witnessed the emergence of modern soil science as a distinct scientific discipline. Agronomists and soil chemists, including Liebig and Boussingault, conducted pioneering research on the chemical composition of soils and the role of nutrients in plant growth.

Justus von Liebig's "Law of the Minimum" highlighted the importance of individual nutrients in limiting plant growth. This revolutionary concept laid the groundwork for modern fertilizer practices, emphasizing the role of specific nutrients in soil fertility. The integration of chemistry into soil science marked a significant shift towards a more systematic understanding of soil processes.

Soil Conservation Movement: The Dust Bowl era in the United States during the 1930s underscored the importance of soil conservation. Severe soil erosion and dust storms prompted the establishment of the Soil Conservation Service (now the Natural Resources Conservation Service), focusing on sustainable land management practices to prevent soil degradation.

The Dust Bowl was a turning point in soil science, highlighting the vulnerability of soils to improper land management. The soil conservation movement emphasized the need for practices such as contour ploughing, strip cropping, and afforestation to prevent soil erosion and maintain soil health. This period marked a shift towards proactive soil management strategies.

1.1.3.3 Advances in Soil Research and Technology

Soil Taxonomy and Classification: The mid-20th century saw the development of soil taxonomy and classification systems. The United States Department of Agriculture (USDA) Soil Taxonomy, introduced in the 1960s, provided a systematic framework for categorizing soils based on shared characteristics.

Soil taxonomy became a cornerstone of soil science, allowing scientists to communicate effectively about soil types and properties. The classification system facilitated research, land use planning, and conservation efforts by providing a common language for soil scientists and land managers worldwide.

Technological Advancements: Advances in technology, including remote sensing, GIS (Geographic Information System), and soil sensors, have revolutionized soil science. These tools enable scientists to collect detailed data on soil properties, analyze spatial patterns, and make informed decisions for precision agriculture.

Modern soil science benefits from the integration of cutting-edge technologies. Remote sensing allows researchers to monitor soil conditions from afar, while GIS helps map soil variability across landscapes. Soil sensors provide real-time data on moisture levels, temperature, and nutrient content, empowering farmers to make precise and informed decisions about their agricultural practices.

Environmental Concerns and Sustainable Practices: As environmental awareness grew in the late 20th century, soil science expanded its focus to address global challenges. Concerns about soil degradation, pollution, and the impact of agriculture on ecosystems led to a renewed emphasis on sustainable land management practices.

The latter part of the 20th century and the early 21st century witnessed a shift towards sustainable agriculture. Soil scientists actively engaged in research to develop practices that promote soil health, minimize environmental impact, and ensure long-term food security. The integration of ecological principles into soil science emphasized the importance of balancing human needs with environmental stewardship.

1.2 Soil Formation and Composition

1.2.1 Pedogenesis: How Soils Are Formed

Formation Processes: Pedogenesis, the intricate process of soil formation, is a dynamic and continuous journey that unfolds over geological timescales, shaping the Earth's surface into the complex and life-nurturing material we recognize as soil. This process involves an amalgamation of physical, chemical, and biological interactions, each contributing distinct threads to the rich tapestry of soil development.

Soil formation begins with **weathering processes**, where rocks transform the influence of various environmental forces. **Physical weathering** involves mechanical breakdown due to factors like temperature fluctuations, freeze-thaw cycles, wind abrasion, and the expansive force of plant roots. Simultaneously, **chemical weathering** facilitates mineral alteration through reactions with water, acids, and gases. These initial weathering processes set in motion the gradual release of minerals, providing the essential building blocks for soil.

Organic Matter Accumulation: A pivotal phase in pedogenesis is the accumulation of organic matter. The decomposition of plant and animal residues contributes organic material to the soil. This accumulation not only enriches the soil with vital nutrients but also plays a fundamental role in influencing soil structure, water retention, and microbial activity. The nature and quantity of organic matter vary based on a myriad of factors, including climate, vegetation types, and anthropogenic influences, contributing to the distinctive characteristics of each soil.

Climate and Topography Influence: The influence of climate is profound in determining the rate and nature of soil formation. In regions marked by distinct seasons, the freeze-thaw cycle accelerates physical weathering, fragmenting rocks into smaller particles. Tropical climates, characterized by high temperatures and consistent rainfall, foster rapid organic matter decomposition, influencing the nutrient content of soils. Additionally, **topography** introduces complexity, where factors such as slope impact drainage patterns, erosion rates, and the spatial distribution of soil types within a landscape.

1.2.2 Factors Influencing Soil Formation

Climate: Climate, a pervasive force in soil formation, encompasses temperature, precipitation, and seasonal variations. The role of temperature extends beyond its influence on physical weathering; it affects the rates of chemical reactions and biological activities. Likewise, precipitation patterns contribute to the leaching of minerals from the soil, influencing its composition. The intricate interplay between these climate factors results in diverse soil profiles across different climatic zones.

Parent Material: The geological foundation of soil, termed **parent material**, exerts a profound influence on soil composition. The breakdown of rocks or sediments serves as the primary source of mineral constituents for soils. Soils derived from granite parent material may contain minerals such as quartz, feldspar, and mica, imparting specific characteristics. In contrast, soils originating from limestone parent material might exhibit high calcium carbonate content, affecting pH and nutrient availability.

Biological Activity: Life beneath the surface actively participates in the transformative processes of pedogenesis. **Microorganisms**, ranging from bacteria to fungi, engage in crucial roles such as organic matter decomposition and nutrient cycling. Their activities contribute to the release of organic acids, facilitating mineral weathering. Plant roots, as agents of both physical and chemical weathering, play a fundamental role in shaping soil structure and composition.

Beyond microorganisms, the roots of plants contribute organic material through shedding and decay. The process of **root exudation** involves the release of organic compounds by plant roots, influencing soil pH, nutrient availability, and microbial activities. The intricate relationships between plants and soil microorganisms, such as mycorrhizal fungi, further enhance nutrient uptake by plants, contributing to the complexity of soil ecosystems.

Topography: The lay of the land, including factors like slope, aspect, and elevation, interacts with other influences to shape soils. **Sloped terrain** affects water drainage and erosion patterns, leading to the formation of distinctive soil profiles. Steeper slopes may experience more rapid erosion, resulting in the development of thinner topsoil layers. In contrast, flatter terrain may accumulate more organic matter, influencing nutrient content.

The aspect of a slope, representing its orientation relative to the sun, influences temperature and moisture conditions. South-facing slopes receive more sunlight, impacting soil temperature and vegetation patterns. This, in turn, affects organic matter decomposition rates and nutrient cycling. Additionally, high-elevation areas may exhibit unique soil characteristics influenced by reduced oxygen availability and temperature variations.

Time: Time is a silent but influential factor woven into the fabric of soil formation. Soils develop gradually over extended periods as a result of cumulative processes. The age of a soil profile influences its maturity, with older soils often exhibiting well-defined horizons and a more intricate composition.

The **chronological sequence of soil development** provides valuable insights into the history of landscapes and ecosystems. In the early stages, soils may lack well-defined horizons, displaying a relatively homogenous composition. Over time, distinct layers emerge, marking the progression of soil development. The O horizon (organic material), A horizon (topsoil), B horizon (subsoil), and C horizon (weathered parent material) collectively tell a story of the prevailing environmental conditions and the interplay of geological and biological processes.

1.2.3 Soil Horizons and Profiles

Beneath the surface of the Earth lies a hidden tapestry—an intricate arrangement of soil horizons that weaves a narrative of the landscape's history, composition, and the dynamic interplay of geological and biological processes. To explore the full richness of soil horizons and profiles is to embark on a journey through time and layers, uncovering the secrets held within the terrestrial skin of our planet (Table 1).

Table 1: Soil Horizons and Profiles

Soil Horizon	Description
O Horizon	Organic horizon at the surface, consisting of organic matter such as decomposed leaves and plant material. This horizon is rich in humus.
A Horizon	Topsoil, also known as the mineral horizon, where most plant roots are active. It contains a mix of organic material and mineral particles.
E Horizon	Eluviation horizon where minerals and nutrients are leached down, often resulting in a lighter color due to the removal of materials like iron and clay.
B Horizon	Subsoil horizon where minerals and nutrients accumulate, often referred to as the "zone of accumulation." It contains minerals leached from above horizons.
C Horizon	Parent material or unconsolidated material that is partially disintegrated or weathered but has not yet transformed into true soil.
R Horizon	Bedrock, the unweathered rock layer that lies beneath all the other horizons. It serves as the ultimate source of mineral materials for soil development.

O Horizon: Organic Material Layer

The O horizon, often referred to as the organic horizon, is more than a mere surface covering; it is a vibrant ecosystem in its own right. Composed of decomposed leaves, plant residues, and humus, this dark layer on the soil's surface is a testament to the ceaseless cycle of life and decay.

Role in Soil Fertility: Within the O horizon, a silent symphony of decomposition unfolds. Microorganisms, including bacteria and fungi, break down complex organic compounds, releasing nutrients into the soil. This organic material serves as a vital reservoir, ensuring the fertility of the soil and providing sustenance for the myriad life forms that call it home.

Biological Diversity: The O horizon, teeming with microbial life, is a hub of biological diversity. Insects, worms, and other microorganisms participate in the breakdown of organic matter, further contributing to nutrient cycling. The intricate interactions within this layer create a rich and fertile environment that supports the growth of plants and sustains the delicate balance of the soil ecosystem.

A Horizon: Topsoil

The A horizon, or topsoil, is the interface where the geological and biological realms converge. This layer, enriched with minerals, organic matter, and the activities of soil organisms, is the canvas upon which life unfolds.

Crucial for Plant Growth: Topsoil is the cradle of terrestrial life, providing the essential nutrients and support required for plant growth. It is within this layer that roots explore, establishing connections with soil microorganisms. The dark coloration, derived from organic matter, reflects the ongoing processes of decomposition and nutrient cycling that characterize the A horizon.

Microbial Abundance: The microcosm within the topsoil is a bustling city of microscopic life. Bacteria, fungi, and other soil organisms engage in complex relationships, breaking down organic matter, releasing nutrients, and contributing to the formation of soil structure. This layer is a testament to the interconnected web of life that sustains ecosystems above and below ground.

B Horizon: Subsoil

Beneath the vibrant topsoil lies the enigmatic B horizon, or subsoil—a realm where minerals accumulate, and geological legacies endure. This layer, shaped by processes such as leaching and translocation, unveils a story of transformation and persistence.

Mineral Accumulation: The subsoil serves as a repository for minerals leached from the overlying horizons. The distinctive coloration of this layer, often different from the topsoil, signifies the translocation of minerals over time. The mineral composition of the B horizon contributes to the overall structure and nutrient content of the soil, influencing the landscape's ecological potential.

Retaining Essential Elements: While lacking the organic richness of the topsoil, the subsoil plays a crucial role in retaining essential elements. Minerals that have undergone leaching from the O and A horizons accumulate here, influencing the overall nutrient availability in the soil. The B horizon, with its unique composition, serves as a reservoir that shapes the characteristics of the soil profile.

C Horizon: Weathered Parent Material

At the foundational level of the soil profile lies the C horizon, or weathered parent material. This layer provides a glimpse into the geological origins of the soil, offering insights into the bedrock or unconsolidated material from which soils develop. It is a testament to the relentless forces of weathering that initiate the journey of soil formation.

Geological Inheritance: The C horizon retains the essence of the parent material. It contains partially disintegrated or weathered rock fragments that showcase the geological history of the landscape. This layer, though less influenced by biological processes, sets the stage for the intricate dance of weathering, organic matter accumulation, and microbial activity that unfolds above.

Slow Contribution to Soil Formation: While the C horizon may not exhibit the biological vibrancy of upper layers, its composition gradually contributes to the overall soil formation process. Over extended periods, weathering processes break down rocks, releasing minerals that become part of the dynamic soil profile above. The C horizon anchors the soil to its geological roots, grounding the ecosystem in the enduring legacy of the Earth's crust.

Unveiling the Layers: Exploring Soil Profiles

Soil scientists and curious explorers alike unveil the layers of soil profiles through methods such as soil pits or auger sampling. These profiles provide a visual representation of the stratigraphy beneath the surface—a living canvas shaped by the factors influencing soil formation.

Observing Soil Profiles: A soil profile reveals the intricate arrangement of horizons, akin to turning the pages of a geological storybook. Soil scientists carefully study these profiles, noting the color, texture, and structure of each layer. The arrangement of horizons not only reflects the prevailing environmental conditions but also holds clues about the history of the landscape.

Diagnostic Properties: Each horizon possesses diagnostic properties that aid in soil classification and assessment. The color of the horizons, influenced by organic matter content and mineral composition, provides insights into drainage patterns and fertility. Texture, determined by the size of mineral particles, influences water retention and nutrient availability. Soil scientists decipher these properties to understand the unique identity of each soil and its suitability for various land uses.

Implications for Land Management: The knowledge of soil horizons and profiles guides land managers in making informed decisions about land use and conservation. Understanding the distribution of horizons across a landscape helps identify areas suitable for agriculture, forestry, or conservation. It also informs practices to mitigate soil erosion, enhance fertility, and promote sustainable land management.

1.3 Physical Properties of Soil

The physical properties of soil form the intricate groundwork upon which the dynamics of terrestrial ecosystems unfold. Among these properties, soil texture and structure stand out as pivotal components, influencing everything from agricultural productivity to environmental sustainability.

1.3.1 Soil Texture and Structure

Soil Texture: The Elemental Composition At the heart of soil texture lies the elemental composition, a mosaic intricately woven from particles of sand, silt, and clay. Each particle size contributes distinctive physical and chemical properties, collectively defining the textural class of the soil.

Sand: The Course Contributor Sand particles, characterized by their larger size (ranging from 0.05 to 2.0 millimeters), bring a coarse texture to the soil. The spaces between sand particles allow for excellent drainage, preventing water retention. However, this quality also means that sand soils have a lower capacity to retain nutrients, as water swiftly percolates through the coarse matrix. Sandy soils, often found in arid regions, exhibit characteristics such as rapid warming in spring, influencing plant growth and germination.

Silt: The Intermediate Intermediary Silt particles, smaller than sand and larger than clay (ranging from 0.002 to 0.05 millimeters), offer a compromise between drainage and nutrient retention. Silty soils exhibit a smoother texture than sand and moderate drainage capacities. While silts contribute to soil fertility by holding nutrients, their fine nature can lead to compaction issues and reduced drainage. Silty soils are commonly found in areas with balanced moisture levels.

Clay: The Finer Fraction Clay particles, the smallest among the three (<0.002 millimeters), create a soil matrix with high water and nutrient retention capabilities. However, the compact nature of clay can lead to poor drainage and aeration. Clay soils feel sticky and moldable when wet, forming hard clods as they dry. These soils are often found in humid climates, presenting challenges for plant root development and water movement.

Textural Classes: The Mosaic of Soil Types The proportions of sand, silt, and clay determine the textural class, leading to a wide array of soil types ranging from sandy to loamy to clayey soils. Loam, a balanced mixture of sand, silt, and clay, is often considered ideal for agriculture due to its optimal combination of drainage, water retention, and nutrient availability. Soil texture serves as a critical determinant for crop selection, irrigation practices, and land use planning.

Soil Structure: The Architectural Arrangement While soil texture deals with individual particles, soil structure explores their organization and aggregation, forming the architectural blueprint of the soil. Soil structure, manifested through the arrangement of soil particles into aggregates, profoundly influences water movement, root penetration, and overall soil health.

Aggregation: The Glue of the Soil Soil aggregation, akin to the glue holding particles together, is a key aspect of soil structure. Organic matter, microbial activity, and certain minerals act as binding agents, creating stable aggregates. These aggregates enhance soil structure, promoting improved porosity and soil stability. Well-aggregated soils have better water infiltration, root penetration, and resistance to erosion.

Types of Soil Structure: Crumbly, Granular, and Platy Different types of soil structure emerge based on various factors, including organic inputs, microbial activity, and land management practices. The crumb structure features small, loosely packed aggregates resembling breadcrumbs, offering excellent aeration and drainage. The granular structure consists of larger, rounded aggregates, fostering good soil porosity and root penetration. Platy structure, on the other hand, exhibits thin, flat layers that can impede water movement and root growth.

Factors Influencing Soil Structure: Organic Matter and Microbial Activity Organic matter plays a pivotal role in soil structure by acting as a binding agent. As organic materials decompose, they release substances that encourage the formation and stability of aggregates. Microorganisms, especially fungi, contribute to the creation of soil structure by producing sticky substances that bind particles together. Earthworms and other soil organisms further enhance soil structure through their burrowing activities, creating channels that improve aeration and water infiltration.

Implications for Agriculture: The Foundation of Productivity In agriculture, the physical properties of soil, particularly texture and structure, have profound implications. The choice of crops, irrigation practices, and soil management strategies depend on these properties. Sandy soils may require more frequent irrigation, while clayey soils may benefit from amendments to improve drainage. Understanding these dynamics is crucial for optimizing crop yields and ensuring sustainable agricultural practices.

Soil Conservation: Mitigating Erosion and Compaction Soil conservation strategies are intricately linked to soil texture and structure. Erosion control measures need to consider the stability of soil aggregates to prevent the loss of topsoil. Additionally, soil compaction, a common issue in agriculture, can be mitigated by promoting practices that enhance soil structure, such as incorporating organic matter and minimizing heavy machinery impact. Preserving soil structure is vital for maintaining the integrity of ecosystems and preventing soil degradation.

Urban Planning: Considering Soil Properties In urban planning, knowledge of soil texture and structure is vital for sustainable construction and landscaping. Compacted soils in urban areas may have limited drainage capacity, leading to increased runoff and flooding. Selecting appropriate plants and implementing green infrastructure practices can help improve soil health in urban environments, contributing to overall urban resilience.

Climate Change and Soil Physical Properties Climate change introduces dynamic challenges to soil physical properties. Changes in precipitation patterns may impact soil erosion, while shifts in temperature can influence microbial activity and organic matter decomposition. Understanding these interactions is essential for predicting the resilience of soils in the face of climate variability. Climate-smart soil management practices, informed by a deep understanding of soil physics, become imperative for mitigating the impacts of climate change on soil health.

Ongoing Research and Technological Advances: Unveiling New Frontiers Advancements in soil science, including sophisticated imaging technologies and molecular analyses, continue to deepen our understanding of soil texture and structure. High-resolution soil mapping techniques enable precise delineation of textural variations across landscapes, aiding in targeted land management. Molecular studies unravel the intricate interactions between soil microbes and organic matter, shedding light on the mechanisms underlying soil structure formation. These cutting-edge tools open new frontiers for sustainable land use planning, precision agriculture, and climate-resilient soil management.

In the symphony of soil, texture, and structure compose the foundational notes that reverberate through ecosystems, agriculture, and land management. The intricate dance of sand, silt, and clay particles, guided by the architectural arrangement of aggregates, shapes the character of the soil beneath our feet. As we continue to explore the physical properties of soil, we unravel a story of interconnectedness, where the elements of Earth converge to sustain life and foster resilience in the face of environmental challenges. Soil, in its diverse textures and structures, is not merely the substrate for life; it is a dynamic entity that holds the key to the sustenance and well-being of our planet. Ongoing research and technological advances propel us toward a future where soil management becomes a nuanced and responsive endeavor, ensuring the longevity of Earth's essential foundation.

1.3.2 Soil Density and Porosity

Soil Density

Soil density is a critical parameter that gauges the mass of soil per unit volume. It directly influences the arrangement of soil particles and, consequently, affects porosity, water movement, and root penetration. The density of soil is a delicate balance, influenced by factors such as particle size, organic matter content, and compaction due to external forces.

Factors Influencing Soil Density: A Multifaceted Interplay

- *Particle Size and Packing:* The size of soil particles significantly impacts soil density. Fine-textured soils, composed of smaller particles like clay, tend to have higher densities due to closer packing. Coarse-textured soils, dominated by larger particles like sand, generally exhibit lower densities.
- *Organic Matter Content:* Organic matter acts as a natural bulking agent, reducing soil density by creating pore spaces. Soils rich in organic matter tend to have lower densities and enhanced porosity, fostering a more favorable environment for plant growth.
- *Compaction:* External forces, often associated with human activities such as farming or construction, can lead to soil compaction. Compacted soils experience a reduction in porosity and an increase in density, restricting water infiltration and root development.

Soil density is not a static property; it can vary both vertically and horizontally in a soil profile. In the topsoil layer, where most plant roots proliferate, a balance between density and porosity is crucial for root penetration, nutrient uptake, and water absorption. As we delve deeper into the soil profile, density may increase due to the weight of overlying soil layers, impacting the movement of water and air.

Managing soil density involves adopting practices that mitigate compaction and promote aeration. Conservation tillage, cover cropping, and reduced machinery impact are strategies to maintain optimal soil density. Conservation tillage, in particular, minimizes soil disturbance and helps preserve soil structure, preventing excessive compaction. The use of cover crops, with their root systems penetrating and stabilizing the soil, contributes to maintaining an appropriate balance between soil density and porosity.

Porosity: The Voids Within

Porosity is the measure of the volume of open spaces, or voids, in the soil. It is a crucial indicator of a soil's capacity to hold and transmit water, air, and nutrients. The interconnected network of pores within soil determines its porosity, influencing its ability to support plant life and sustain various ecological processes.

Types of Porosity: Macro and Micro Pores

- *Macro Pores:* These are larger pores that facilitate the movement of air and water. Macro pores contribute to drainage, aeration, and the movement of roots through the soil. Their presence is vital for preventing waterlogging and ensuring aeration in the root zone.
- *Micro Pores:* Smaller pores create a capillary effect, holding water against the force of gravity. Micro pores contribute to water retention, crucial during dry periods when plants access stored soil moisture. The balance between macro and micro pores is essential for overall soil health.

Porosity is dynamic and influenced by soil texture, structure, and organic matter content. Different soil textures have varying porosities, with sandy soils generally having larger pores and higher drainage capacity compared to clayey soils. Soil structure, particularly the formation of aggregates, contributes to porosity by creating well-defined pore spaces. Organic matter acts as a binding agent, enhancing porosity by preventing soil particles from closely packing together.

Optimizing Soil Density and Porosity: Balancing Act for Agriculture

In agriculture, optimizing soil density and porosity is a delicate balancing act. While a certain degree of compaction is unavoidable due to farming practices, excessive compaction can hinder plant growth. Incorporating organic matter, practicing conservation tillage, and adopting precision agriculture techniques are strategies to maintain an optimal balance between soil density and porosity.

1.3.3 Soil Moisture and Water Retention

Soil moisture is a pivotal factor influencing plant growth, microbial activity, and overall soil health. It is the water present in the soil that is available for plants to absorb through their roots. The availability of soil moisture fluctuates with factors such as climate, soil type, and land use practices.

Factors Affecting Soil Moisture

- *Climate:* Precipitation patterns, temperature, and humidity directly influence soil moisture levels. Arid regions may experience low soil moisture, while humid areas may have higher and more consistent moisture content.
- *Soil Type:* Different soil textures impact moisture retention. Sandy soils, with larger particles, drain quickly and may have lower moisture retention, while clayey soils retain more moisture due to smaller particle sizes.
- *Vegetation Cover:* The presence of vegetation influences soil moisture through evapotranspiration. Plants extract water from the soil and release it into the atmosphere through transpiration, affecting overall moisture levels.

Understanding soil moisture dynamics is crucial for efficient water management in agriculture and natural ecosystems. Soil moisture content varies not only with depth but also over time. Monitoring changes in soil moisture throughout the growing season helps farmers make informed irrigation decisions, leading to improved water-use efficiency and crop yields.

Water Retention: The Soil's Reservoir Capacity

Water retention refers to the soil's ability to hold onto water against the force of gravity. It is a critical aspect of soil moisture management, influencing irrigation needs, drought resistance, and nutrient availability to plants.

Factors Influencing Water Retention: A Complex Matrix

- *Soil Texture:* The textural composition plays a crucial role in water retention. Clayey soils, with their smaller particles, have higher water retention capacities compared to sandy soils. The fine particles create a capillary effect that holds water against the gravitational pull.
- *Organic Matter Content:* Organic matter acts as a sponge, enhancing water retention. Soils rich in organic matter can hold more water, providing a buffer during dry periods and supporting microbial activity.
- *Soil Structure:* Well-aggregated soils with a good structure tend to have improved water retention. Aggregates create spaces that trap water, preventing rapid drainage and enhancing overall moisture availability.

Water retention is a dynamic process influenced by the interplay of soil properties and environmental conditions. Understanding the factors affecting water retention is essential for sustainable water management in agriculture and natural ecosystems. Soil texture, with its impact on capillary action, and organic matter, with its water-holding capacity, are key determinants of a soil's ability to retain water.

Managing Soil Moisture: Precision in Agriculture

In agriculture, managing soil moisture is a key aspect of optimizing crop yields. Precision irrigation techniques, soil moisture sensors, and cover cropping are tools employed to ensure efficient water use. Balancing the needs of crops with sustainable water management practices is crucial for long-term agricultural productivity.

Cover Cropping: Nature's Moisture Regulator

Cover cropping involves planting specific crops during the off-season to cover and protect the soil. Besides preventing erosion, cover crops contribute to soil moisture management. The root systems of cover crops help in creating channels for water infiltration, improving overall soil structure and water retention.

Soil Moisture Sensors: Real-time Monitoring

Advancements in technology have brought about the development of soil moisture sensors. These devices provide real-time information about soil moisture levels at different depths. Farmers can use this data to optimize irrigation schedules, ensuring that crops receive the right amount of water without wastage (fig 3).

Fig. 3: Soil Moisture Sensors

Precision Irrigation: Tailoring Water Delivery

Precision irrigation involves applying the right amount of water to specific areas of a field at the right time. This approach maximizes water-use efficiency, reduces water wastage, and minimizes the risk of over-irrigation, which can lead to waterlogging and nutrient leaching.

Balancing Act: Sustainable Water Management

Sustainable water management in agriculture requires a holistic approach that considers soil health, climate conditions, and crop water requirements. Integrating water-efficient practices, such as drip irrigation, rainwater harvesting, and soil conservation measures, contributes to long-term water sustainability and resilience in agriculture.

1.4 Chemical Properties of Soil

The chemical properties of soil are intricate and multifaceted, shaping the soil's capacity to support life and influencing vital ecological processes. Among these properties, soil pH stands out as a fundamental factor that intricately connects with nutrient availability, microbial activity, and overall soil health.

1.4.1 Soil pH and Nutrient Availability

Understanding Soil pH: The Acid-Base Balance

Soil pH serves as a key indicator of the acidity or alkalinity of the soil solution. It is a measure of the concentration of hydrogen ions in the soil water, determining whether a soil is acidic, neutral, or alkaline. The pH scale ranges from 0 to 14, with 7 being neutral. Values below 7 indicate acidity, while values above 7 indicate alkalinity. Soil pH is dynamic and can fluctuate due to various natural and anthropogenic factors.

Factors Influencing Soil pH: Dynamic Interactions

1. **Parent Material:** The geological origin of the soil, known as the parent material, significantly influences its initial pH. Soils derived from limestone rocks tend to be more alkaline, while those from granite or sandstone may lean towards acidity.
2. **Climate:** Weathering processes, influenced by climate conditions, can impact soil pH. In humid regions, leaching may contribute to soil acidity, while arid regions may experience alkaline soils due to less leaching.
3. **Organic Matter Decomposition:** The breakdown of organic matter by soil microbes can either increase or decrease soil pH. The decomposition of organic material tends to produce organic acids, contributing to soil acidity.
4. **Fertilizer Application:** Certain fertilizers, especially those containing ammonium nitrogen, can acidify the soil upon decomposition. Conversely, lime application is a common practice to raise soil pH.

Significance of Soil pH: Nutrient Availability and Microbial Activity

Soil pH profoundly influences the availability of essential nutrients to plants. The majority of plant nutrients are most accessible to plants within a specific pH range. For example, phosphorus tends to be more available in slightly acidic to neutral soils, while micronutrients like iron and manganese are more accessible in acidic soils.

Nutrient Availability at Different pH Levels

1. Acidic Soils (pH < 7):
   - *Increased Availability:* Aluminum, Iron, Manganese, Zinc.
   - *Decreased Availability:* Calcium, Magnesium, Phosphorus.
2. Neutral Soils (pH $= 7$):
   - Nutrient availability is generally balanced.
3. Alkaline Soils (pH > 7):
   - *Increased Availability:* Calcium, Magnesium, Phosphorus.
   - *Decreased Availability:* Iron, Manganese, Zinc.

Soil microbial activity is also influenced by pH. Microorganisms play a crucial role in nutrient cycling and organic matter decomposition. The activity of different microbial species varies with pH, impacting the overall health and fertility of the soil.

Managing Soil pH: Agricultural Practices

Agricultural practices often involve managing soil pH to create optimal conditions for plant growth. Lime application is a common method to raise pH in acidic soils. This involves adding calcium carbonate, which reacts with soil acidity, neutralizing it and releasing carbon dioxide.

Conversely, in soils with high alkalinity, acidifying agents such as sulfur may be applied. These amendments facilitate the release of hydrogen ions, reducing soil pH. Precision agriculture techniques, including variable rate applications based on soil pH maps, enable targeted management practices to optimize soil conditions.

Environmental Implications: Soil pH and Ecosystem Health

Beyond agriculture, soil pH plays a crucial role in shaping ecosystems. Acid rain, resulting from air pollution, can lead to soil acidification, negatively impacting plant and microbial communities. Understanding the environmental implications of soil pH is essential for conservation and sustainable land management.

Soil pH in Ecosystems: Biodiversity and Plant Adaptations

Different plant species exhibit varying tolerances to soil pH, contributing to the diversity of plant communities in natural ecosystems. Some plants thrive in acidic soils, while others prefer alkaline conditions. The pH of the soil plays a role in determining the composition of plant communities and influencing the interactions between plants and other organisms.

Acid Rain and Soil Acidification:

The phenomenon of acid rain, resulting from the emission of sulfur dioxide and nitrogen oxides into the atmosphere, has far-reaching consequences for soil pH. When acid rain falls on the ground, it can leach essential nutrients from the soil, contributing to soil acidification. This process has been identified as a global environmental concern, impacting both terrestrial and aquatic ecosystems (fig 4).

Fig 4: Acid Rain

Anthropogenic Influences: Industrial Activities and Soil pH

Human activities, especially industrial processes, can significantly alter soil pH. The release of pollutants and contaminants from industrial sources may contribute to soil acidification or alkalization, depending on the nature of the pollutants. Understanding the interplay between human activities and soil pH is crucial for mitigating the environmental impacts of industrialization.

Expanding the Horizon: Soil pH Research and Innovations

As we deepen our understanding of soil pH, ongoing research and technological innovations continue to unveil new dimensions of this critical soil property. High-resolution soil mapping technologies, coupled with advanced sensing techniques, enable researchers to explore variations in soil pH at unprecedented scales. Molecular studies delve into the intricate mechanisms by which microorganisms influence and respond to changes in soil pH, opening avenues for targeted interventions in soil management.

Tailoring pH Management

Precision agriculture, empowered by advancements in sensing technologies and data analytics, allows farmers to precisely manage soil pH on a site-specific basis. By identifying pH variations within a field, farmers can tailor lime or acidifying agent applications to specific areas, optimizing resource use and promoting sustainable farming practices.

Bioremediation: Microbial Solutions for Soil pH Management

Innovations in bioremediation focus on harnessing the capabilities of microbial communities to influence soil pH. Engineered microbes with the ability to produce organic acids or alkaline substances hold promise for localized interventions to modulate soil pH. These approaches aim to provide environmentally friendly alternatives to traditional soil amendments.

Climate-Smart Agriculture: Adapting to Changing pH Dynamics

As climate change introduces new challenges to soil health, the concept of climate-smart agriculture incorporates strategies to adapt to changing pH dynamics. Understanding the interactions between climate variables and soil pH is crucial for developing resilient agricultural practices that can withstand the impacts of shifting environmental conditions.

In the intricate tapestry of soil science, the chemical properties of soil, particularly pH, weave a story of interconnectedness between soil health, nutrient availability, and ecosystem dynamics. Recognizing the dynamic factors influencing soil pH and its cascading effects on nutrient availability is essential for sustainable land management and agricultural practices.

1.4.2 Cation Exchange Capacity (CEC)

Deciphering the Role of CEC in Soil Chemistry

Cation Exchange Capacity (CEC) is a fundamental property influencing soil fertility and nutrient availability. This intricate mechanism involves the exchange of positively charged ions (cations) between soil particles and the soil solution, primarily influencing plant nutrition and overall soil health.

Defining Cation Exchange Capacity: The Nutrient Exchange Hub

CEC quantifies the soil's ability to retain and exchange essential nutrient cations. Soil particles, especially clay and organic matter, carry negative charges on their surfaces. These negative charges attract and hold onto positively charged nutrient cations, preventing them from leaching away with water. CEC acts as a dynamic nutrient reservoir, releasing cations to plant roots as needed.

Factors Influencing CEC: Clay and Organic Matter Dynamics

1. **Clay Content:** The proportion of clay in soil significantly influences CEC. Clay particles, with their high surface area and negatively charged sites, contribute extensively to CEC. Soils with higher clay content generally exhibit higher CEC values, enhancing their nutrient retention capacity.

2. **Organic Matter:** Organic matter is another crucial factor shaping CEC. The organic material in soil, derived from plant and animal residues, contains negatively charged functional groups. Soils rich in organic matter have higher CEC, fostering nutrient availability and supporting microbial activity.

3. **Soil pH:** Soil pH plays a role in CEC dynamics. In acidic soils, aluminum and hydrogen ions may replace nutrient cations on soil particles, influencing CEC. Proper management of soil pH is essential for optimizing CEC and ensuring an environment conducive to nutrient exchange.

CEC in Action: Nutrient Availability to Plants

The interplay between CEC and nutrient availability is pivotal for plant growth. Nutrient cations held by soil particles can be released to plant roots as required. This controlled release mechanism contributes to sustained nutrient availability, playing a crucial role in influencing crop yields and overall ecosystem health.

Managing CEC: Implications for Agriculture

Agricultural practices often involve managing CEC to optimize nutrient availability for crops. Techniques include incorporating organic matter through cover cropping, green manuring, and adding compost, all of which enhance CEC. Balancing nutrient inputs based on CEC values ensures efficient use of fertilizers, minimizing environmental impact and promoting sustainable farming practices.

Innovations in CEC Research: Unraveling Soil Dynamics

Ongoing research in soil science explores innovative ways to understand and manipulate CEC. Advanced techniques, including molecular studies and imaging technologies, delve into the molecular interactions occurring at the soil particle level. These insights contribute to precision agriculture strategies that optimize CEC for sustainable and efficient nutrient management.

Precision Agriculture: Harnessing CEC for Sustainable Farming

Precision agriculture, facilitated by advancements in technology, allows farmers to precisely manage CEC on a site-specific basis. By understanding the spatial variability of CEC within a field, farmers can tailor their management practices, adjusting nutrient applications to match the soil's specific needs. This targeted approach optimizes resource use, promotes sustainable farming, and minimizes environmental impact.

Bioremediation: Microbial Solutions for CEC Enhancement

Innovations in bioremediation focus on harnessing the capabilities of microbial communities to influence and enhance CEC. Engineered microbes with the ability to produce organic acids or alter the surface charge of soil particles hold promise for localized interventions to modulate CEC. These approaches aim to provide environmentally friendly alternatives to traditional soil amendments.

Climate-Smart Agriculture: Adapting to Changing CEC Dynamics

As climate change introduces new challenges to soil health, the concept of climate-smart agriculture incorporates strategies to adapt to changing CEC dynamics.

Understanding the interactions between climate variables and CEC is crucial for developing resilient agricultural practices that can withstand the impacts of shifting environmental conditions.

1.4.3 Soil Fertility and Plant Nutrition

The Nexus of Soil Fertility and Plant Nutrition: A Holistic Perspective

Soil fertility and plant nutrition are intricately connected aspects that govern the productivity and health of ecosystems. The dynamic relationship between soil properties, nutrient availability, and plant uptake defines the complex interplay between soil fertility and plant nutrition.

Defining Soil Fertility: Soil fertility is the soil's capacity to provide essential nutrients to plants in optimal quantities. It involves a combination of factors, including nutrient content, soil structure, organic matter, and microbial activity. A fertile soil is capable of supporting vigorous plant growth without external nutrient supplementation.

The Role of Nutrients in Plant Nutrition: Plants require a range of nutrients for their growth and development. These nutrients can be broadly categorized into macronutrients and micronutrients. Achieving a balance of these nutrients is essential for optimal plant nutrition. Macronutrients, such as nitrogen, phosphorus, and potassium, are needed in larger quantities, while micronutrients, including iron, zinc, and copper, are required in smaller amounts (fig 5).

Fig. 5: Role of Soil Nutrients in Plant

Nutrient Cycling in Soil: Sustaining Fertility

The cycling of nutrients in the soil involves complex processes where organic matter is decomposed, releasing nutrients that are then taken up by plants. Microorganisms play a crucial role in nutrient cycling, breaking down organic matter and making nutrients available to plants. Sustainable agriculture practices focus on enhancing nutrient cycling to maintain soil fertility over the long term.

Soil Testing: A Diagnostic Tool for Fertility Management

Soil testing is a valuable tool in assessing soil fertility. By analyzing nutrient levels, pH, and other soil properties, farmers can make informed decisions about nutrient management. Soil tests guide the application of fertilizers, helping to tailor nutrient inputs based on the specific needs of the soil and crops.

Organic Farming: Nurturing Soil Fertility Naturally

Organic farming practices prioritize the use of natural inputs to enhance soil fertility. Incorporating organic matter through composting, cover cropping, and crop rotations enriches the soil with nutrients and promotes a healthy microbial community. These practices contribute to sustainable soil fertility management (fig 6).

Fig. 6: Organic Farming

Challenges to Soil Fertility: Erosion and Urbanization

Soil erosion and urbanization pose challenges to soil fertility. Erosion can lead to the loss of topsoil, which is rich in nutrients, while urban development can result in soil compaction and contamination. Implementing erosion control measures and adopting sustainable land-use practices are essential for preserving soil fertility.

The Future of Plant Nutrition: Precision Agriculture

The future of plant nutrition lies in precision agriculture, where technology is harnessed to optimize nutrient management. Remote sensing, GPS technology, and data analytics enable farmers to monitor crop health, assess nutrient needs, and apply fertilizers with precision. This targeted approach enhances nutrient use efficiency and minimizes environmental impact.

1.5 Biological Components of Soil

Soil, often seen as an inert substrate, is a vibrant and intricate ecosystem pulsating with life. This lively milieu is orchestrated by an ensemble of biological components, each playing a unique role in the complex symphony of soil ecology. At the heart of this biological tapestry are soil microorganisms, microscopic entities that wield immense influence over soil health and ecosystem dynamics (Table 2).

Table 2: Biological Components of Soil

S.No	Biological Component	Description and Function
1	Microorganisms	Bacteria, fungi, archaea, and protozoa are common microorganisms in soil. They play a crucial role in nutrient cycling, organic matter decomposition, and disease suppression.
2	Algae	Photosynthetic organisms that contribute to organic matter production and nutrient cycling. They are more common in the upper layers of soil.
3	Nematodes	Small, unsegmented worms that play a role in nutrient cycling and may act as both decomposers and predators of other soil organisms.
4	Arthropods	Insects and other arthropods are important for breaking down organic matter, controlling pest populations, and contributing to nutrient cycling.
5	Earthworms	Earthworms are ecosystem engineers that enhance soil structure, aeration, and nutrient availability through their burrowing and feeding activities.
6	Mycorrhizal Fungi	Form symbiotic relationships with plant roots, aiding in nutrient absorption, particularly phosphorus and some micronutrients. Enhance plant growth and resilience.
7	Actinomycetes	Filamentous bacteria with characteristics of both bacteria and fungi. They decompose complex organic compounds, contributing to the breakdown of plant residues.
8	Roots and Plant Debris	Living plant roots and decaying plant material contribute to soil organic matter, providing a food source for microorganisms and supporting the soil food web.

1.5.1 Soil Microorganisms

Unveiling the Microscopic World Beneath Our Feet

The realm of soil microorganisms is a hidden universe that defies the naked eye but holds profound significance in the intricate balance of soil ecosystems. Bacteria, fungi, archaea, protozoa, and viruses collectively form a microscopic orchestra, orchestrating processes that sustain soil fertility, nutrient cycling, and overall ecosystem health.

Bacteria

Bacteria, the unsung heroes of soil microbiology, are ubiquitous and abundant in every gram of soil. The sheer numbers are staggering, with estimates suggesting that a single gram of soil can harbor billions of bacterial cells. These microorganisms are versatile and resilient, adapting to a spectrum of environmental conditions. Within the soil, they act as pioneers, initiating processes crucial for nutrient cycling and organic matter decomposition. Some bacteria even exhibit the remarkable ability to fix atmospheric nitrogen, converting it into forms accessible to plants and thus enhancing soil fertility.

Fungi

Fungi, comprising molds and yeasts, stand as silent architects of the soil ecosystem. Through their intricate mycelial networks, these organisms facilitate nutrient transfer and water absorption. Mycorrhizal fungi establish symbiotic relationships with plant roots, enhancing nutrient uptake and creating a harmonious exchange. In the realm of decomposition, fungi play a vital role, breaking down complex organic compounds and contributing to the formation of humus, the organic component of soil that nurtures plant growth.

Archaea: Unveiling Ancient Microbial Players

While less studied compared to bacteria and fungi, archaea, with their ancient lineage, inhabit various soil ecosystems. These microorganisms are essential for nitrogen cycling, with some species contributing to methane production and oxidation. Archaea, with their unique metabolic activities, form a mosaic of microbial life that shapes the intricate balance of soil ecosystems.

Protozoa: Predators in the Microbial Food Web

Protozoa, single-celled eukaryotes, emerge as the predators in the microbial food web, feeding on bacteria, fungi, and other microorganisms. Their role is integral in regulating microbial populations, influencing nutrient cycling, and releasing nutrients in forms accessible to plants. Protozoa's impact reverberates through the soil ecosystem, contributing to its resilience and sustainability.

Viruses: Drivers of Microbial Diversity and Evolution

In the microscopic realm, viruses, although not conventional living organisms, play a vital role in driving microbial diversity and evolution. These entities infect bacteria and archaea, influencing the dynamics of microbial communities. Viruses contribute to nutrient cycling by releasing organic matter through the lysis of microbial cells, adding another layer of complexity to the intricate soil microbiome.

Roles of Soil Microorganisms: A Symphony of Functions

1. **Nutrient Cycling:** At the core of soil microbiology lies the essential function of nutrient cycling. Microorganisms, through processes like decomposition and nitrogen fixation, release vital nutrients in forms usable by plants, laying the foundation for soil fertility and sustained ecosystem productivity.

2. **Organic Matter Decomposition:** The breakdown of organic matter is a testament to the collaborative efforts of soil microorganisms. Bacteria and fungi, equipped with an arsenal of enzymes, collaborate in the decomposition of complex organic compounds, contributing to the formation of humus and enriching the soil with organic nutrients.

3. **Disease Suppression:** Certain soil microorganisms act as natural allies in the suppression of plant pathogens. This biological control mechanism helps maintain the health of crops and vegetation, reducing the reliance on synthetic pesticides and fostering a balanced ecosystem.

4. **Mycorrhizal Associations:** Mycorrhizal fungi engage in symbiotic relationships with plant roots, forming mycorrhizal associations. This mutualistic partnership enhances nutrient uptake by plants, particularly phosphorus and nitrogen, and provides resilience against environmental stresses.

5. **Biological Nitrogen Fixation:** Nitrogen, a crucial element for plant growth, undergoes a transformative journey facilitated by nitrogen-fixing bacteria. These bacteria convert atmospheric nitrogen into ammonia, contributing to soil fertility and reducing the need for synthetic nitrogen fertilizers.

Factors Influencing Soil Microbial Communities

1. **Soil pH:** The pH of soil plays a pivotal role in shaping the composition and diversity of microbial communities. Different microorganisms thrive within specific pH ranges, and variations in soil pH can impact their abundance and metabolic activities.

2. **Soil Moisture:** The availability of water in the soil environment is a critical factor influencing microbial activity. Microorganisms require water for metabolic processes, and variations in soil moisture content can influence their abundance and functions.

3. **Organic Matter Content:** The quantity and quality of organic matter in the soil significantly influence microbial communities. Organic matter serves as a nutrient source for microorganisms, sustaining their growth, and contributing to the overall health of the soil.

4. **Temperature:** Soil temperature serves as a thermostat for microbial activity. Microbial processes generally increase with rising temperatures, up to a certain threshold. Fluctuations in temperature can impact the timing and efficiency of microbial activities.

5. **Land Use and Management Practices:** Anthropogenic activities, such as agricultural practices and land-use changes, can significantly alter soil microbial communities. Intensive agriculture, in particular, may lead to shifts in microbial diversity and functions, highlighting the importance of sustainable land management.

Interactions in the Soil Microbial Food Web

The soil microbial food web is a complex network of interactions among microorganisms, plants, and other soil-dwelling organisms. This web encompasses multiple trophic levels, including primary producers (plants), primary consumers (bacteria, fungi), secondary consumers (protozoa, nematodes), and tertiary consumers (microarthropods). The interactions within this food web influence nutrient cycling, energy flow, and overall ecosystem dynamics.

Challenges and Opportunities in Soil Microbiology Research

1. **Microbial Diversity Studies:** Advances in molecular biology techniques, particularly DNA sequencing, have revolutionized our understanding of microbial diversity in soil. Ongoing research endeavors aim to unravel the vast diversity of microorganisms, including previously unculturable species.

2. **Functional Metagenomics:** The exploration of functional metagenomics involves studying the genetic potential of entire microbial communities. This innovative approach provides insights into the functional capabilities of microbial assemblages, offering a deeper understanding of their roles in nutrient cycling and ecosystem processes.

3. **Microbial Biotechnology:** The application of microbial biotechnology in agriculture holds significant promise. Researchers explore the development of biofertilizers, biopesticides, and microbial inoculants that harness the beneficial capabilities of microorganisms to promote plant growth and health.

4. **Climate Change Impacts:** The responses of soil microorganisms to climate change represent a critical area of research. Changes in temperature, precipitation patterns, and atmospheric carbon dioxide levels can influence microbial activity, potentially altering ecosystem functions.

1.5.3 Soil Biota's Role in Nutrient Cycling

Soil biota, comprising microorganisms, fauna, and flora, engage in a dynamic dance of nutrient cycling—a fundamental process that sustains soil fertility and ecosystem health.

Microbial Contributions to Nutrient Cycling

1. Decomposition Dynamics

In the complex orchestra of nutrient cycling, soil microorganisms, including bacteria and fungi, emerge as the principal conductors of decomposition. Their enzymatic prowess orchestrates the breakdown of complex organic matter, releasing elemental nutrients back into the soil. This intricate ballet ensures the recycling of organic materials, fostering soil fertility and supporting the vitality of terrestrial ecosystems.

The microbial-driven decomposition process is a captivating display of biochemical transformations. As microorganisms break down organic matter, they secrete enzymes that cleave complex molecules into simpler forms. Carbohydrates, proteins, and lignin undergo a metamorphic journey, releasing carbon, nitrogen, phosphorus, and other essential elements. This orchestrated decay not only contributes to soil fertility but also shapes the physical structure of the soil, influencing water retention and aeration.

2. Nitrogen Fixation: Enriching Soils with Essential Nutrients

Within the nutrient cycling symphony, certain bacteria take center stage in the performance of nitrogen fixation. This process, often associated with leguminous plants, involves the conversion of atmospheric nitrogen into ammonia—a key nutrient for plant growth. The nitrogen-fixing bacteria form harmonious partnerships with plants, contributing to soil fertility and reducing the dependence on synthetic nitrogen fertilizers.

The nitrogen-fixing prowess of bacteria such as Rhizobia and Frankia is a remarkable ecological collaboration. These microorganisms form nodules on the roots of legumes, creating a specialized microenvironment. Within these nodules, atmospheric nitrogen is transformed into ammonia, becoming a readily available source of nitrogen for the host plant. This symbiotic dance not only benefits the legume but also enriches the surrounding soil, creating a virtuous cycle of nutrient availability.

3. Mycorrhizal Partnerships

Mycorrhizal fungi, forming symbiotic partnerships with plant roots, engage in a dance of nutrient exchange. In this intricate pas de deux, the fungi extend their hyphal threads, creating a network that expands the reach of plant roots. In return for sugars provided by the plant, the mycorrhizal fungi enhance the plant's nutrient uptake—particularly phosphorus and nitrogen. This mycorrhizal association is a harmonious exchange that elevates nutrient cycling efficiency in the soil.

Mycorrhizal associations are an exemplification of mutualistic synergy in soil ecosystems. Arbuscular mycorrhizal fungi, commonly found in the roots of the majority of plant species, form intimate connections with plant cells. This symbiosis not only enhances nutrient acquisition but also plays a crucial role in the plant's resilience to environmental stresses. As the mycorrhizal hyphae explore the soil matrix, they extract nutrients from distant regions, contributing to the overall nutrient balance in the rhizosphere.

Faunal Influence on Nutrient Dynamics

1. Earthworm Castings

In the grandeur of the soil nutrient symphony, earthworms, often regarded as ecosystem engineers, play a pivotal role. Their burrowing activities create channels in the soil, improving aeration and water infiltration. However, it is their castings—nutrient-rich excreta—that steal the spotlight. Earthworm castings are a treasure trove of plant-available nutrients, enhancing soil fertility and supporting overall soil health.

The actions of earthworms extend beyond soil structure enhancement. The intricate interactions between earthworms and microorganisms in their digestive tract contribute to the transformation of organic matter. As plant residues pass through the earthworm's digestive system, they undergo a process of microbial digestion. The resulting castings not only contain higher concentrations of essential nutrients but also harbor beneficial microbial communities, further influencing nutrient cycling dynamics.

2. Macroarthropod Decomposition

Macroarthropods, such as beetles and ants, step into the limelight as performers in the organic matter decomposition spectacle. Their activities contribute to the breakdown of plant residues, accelerating the release of nutrients back into the soil. These decomposers play a vital role in nutrient cycling, ensuring the continuous recycling of organic materials in the soil.

The decompositional efforts of macroarthropods extend the reach of nutrient cycling into various soil horizons. Their activities fragment organic matter, increasing its surface area and making it more accessible to microbial decomposers. This fragmentation not only expedites the decomposition process but also facilitates the incorporation of organic matter into the soil, enriching the nutrient pool.

Plant-Soil Feedbacks: The Harmonious Dialogue

1. Rhizosphere Interactions

The rhizosphere, where plant roots and soil biota engage in a dynamic dialogue, is a focal point of nutrient cycling. Root exudates, comprising sugars, organic acids, and other compounds, attract a diverse microbial community. This vibrant exchange influences nutrient cycling dynamics, creating a microenvironment where nutrient availability is finely tuned to the plant's needs.

The rhizosphere is a bustling arena of biochemical interactions. Root exudates not only nourish soil microorganisms but also shape microbial communities based on the plant's nutritional requirements. The release of organic acids by plant roots plays a role in nutrient solubilization, making previously unavailable nutrients accessible to both plants and microorganisms. This dynamic interplay in the rhizosphere sets the stage for a harmonious exchange that supports plant growth and overall soil health.

2. Plant-Microbe Symbiosis

The collaboration between plants and soil microbes, particularly mycorrhizal fungi and nitrogen-fixing bacteria, is a mutualistic symphony that echoes through the soil ecosystem. These symbiotic partnerships enhance nutrient acquisition, improve plant resilience, and contribute to the overall nutrient cycling ballet in the soil. The intricate dance between plants and microbes ensures a harmonious and sustainable nutrient cycle.

Mycorrhizal associations extend the nutrient-absorbing capacity of plant roots, enabling them to explore larger soil volumes. This enhanced nutrient uptake not only benefits the individual plant but also contributes to the overall nutrient cycling efficiency in the ecosystem. Similarly, nitrogen-fixing bacteria forge alliances with leguminous plants, providing a supplementary source of nitrogen that nourishes both the plants and the surrounding soil.

Challenges and Resilience in Nutrient Cycling

1. Anthropogenic Influences

In the ongoing performance of nutrient cycling, human activities cast both challenges and opportunities. Anthropogenic influences, including agricultural practices and land-use changes, can disrupt the natural rhythm of nutrient cycling. Understanding these impacts is crucial for mitigating nutrient imbalances and preserving soil health.

Modern agricultural practices often involve the application of synthetic fertilizers and pesticides. While these inputs aim to enhance crop productivity, they can disrupt the delicate balance of nutrient cycling in the soil. Nutrient runoff, soil erosion, and changes in microbial communities are among the consequences of intensive agriculture. Sustainable land management practices, such as agroecology and organic farming, emerge as pathways to minimize anthropogenic impacts and foster resilient nutrient cycling.

2. Climate Change Impacts

The unfolding impacts of climate change introduce an additional layer of complexity to the nutrient-cycling symphony. Shifts in temperature, precipitation patterns, and extreme weather events can influence microbial activities, nutrient availability, and overall ecosystem dynamics.

Climate change alters the timing and intensity of precipitation, leading to changes in soil moisture levels. This, in turn, affects microbial activities and nutrient cycling rates. Additionally, rising temperatures can influence the metabolism of soil organisms, potentially accelerating or decelerating nutrient cycling processes. Understanding the nuances of climate change impacts on soil nutrient cycling is crucial for adapting land management strategies and ensuring the resilience of soil ecosystems.

Future Perspectives

1. Sustainable Agriculture Practices

The future of nutrient cycling lies in the adoption of sustainable agricultural practices that prioritize soil health. Cover cropping, crop rotation, reduced tillage, and agroforestry are among the practices that contribute to nutrient cycling efficiency.

Cover cropping involves planting cover crops during fallow periods to protect and enrich the soil. The roots of cover crops prevent soil erosion, while their residues contribute organic matter upon decomposition. Crop rotation, on the other hand, disrupts pest and disease cycles, promoting a diverse soil microbiome. Reduced tillage minimizes soil disturbance, preserving soil structure and microbial habitats. Integrating these practices into agricultural systems fosters a harmonious relationship between crop production and nutrient cycling.

2. Biotechnological Innovations

Advancements in biotechnology offer promising avenues for optimizing nutrient cycling in agricultural landscapes. The development of microbial inoculants, precision agriculture technologies, and biofertilizers allows for targeted interventions that enhance nutrient availability.

Microbial inoculants, consisting of beneficial microorganisms, can be applied to soils to enhance nutrient cycling processes. These microorganisms may include nitrogen-fixing bacteria, mycorrhizal fungi, and plant growth-promoting bacteria.

Precision agriculture technologies, such as sensor-based nutrient management, enable farmers to tailor fertilizer applications based on real-time soil conditions. Biofertilizers, derived from organic sources, provide a sustainable alternative to synthetic fertilizers. These innovations align with the principles of precision agriculture and environmental sustainability.

3. Ecosystem-Based Approaches

Adopting ecosystem-based approaches to soil management involves recognizing the interconnectedness of soil biota and nutrient cycling. Preserving biodiversity, restoring degraded ecosystems, and minimizing disturbances contribute to resilient nutrient dynamics.

Biodiversity conservation is essential for maintaining a diverse soil biota, each playing a unique role in nutrient cycling. Restoring degraded ecosystems involves implementing practices that enhance soil structure, such as reforestation and sustainable land use. Minimizing disturbances, such as excessive land clearing and habitat destruction, protects the delicate balance of soil life. Ecosystem-based approaches acknowledge the intricate web of interactions within soil ecosystems and seek to nurture these relationships for sustainable nutrient cycling.

1.6 Soil Classification and Taxonomy

Soil classification and taxonomy are fundamental frameworks that facilitate our understanding of the vast diversity of soils worldwide. This chapter is a comprehensive exploration of the hierarchical levels of soil classification, providing insights into soil orders, suborders, soil families, and the overarching soil taxonomy systems, with a primary focus on the USDA Soil Taxonomy.

1.6.1 Soil Orders, Suborders, and Soil Families

1. Soil Orders: Defining Global Soil Characteristics

- Soil orders represent the highest level of soil classification, providing a broad categorization based on global soil characteristics influenced by climate, vegetation, parent material, and other factors.
- The twelve recognized soil orders, including Alfisols, Andisols, Aridisols, Entisols, Gelisols, Histosols, Inceptisols, Mollisols, Oxisols, Spodosols, Ultisols, and Vertisols, showcase the diverse array of soils present on Earth.
- Alfisols, characterized by a horizon with accumulations of clay, iron, or aluminum, are often conducive to agriculture. Gelisols, found in cold environments, exhibit permafrost within two meters of the surface, impacting soil development and structure.

2. Soil Suborders: Refining Soil Characteristics

- Within each soil order, suborders offer a more nuanced classification, refining the characterization based on additional properties such as temperature regimes, moisture levels, and specific soil features.
- Suborders within Andisols, such as Cryands (cold) and Vitrands (non-cold), highlight variations in temperature regimes. This finer classification captures the intricacies of soil development and properties within a broader order.

3. Soil Families

- Soil families represent a more detailed level of classification, zooming into distinctive features and properties that define specific groups of soils within a suborder.
- Taking Mollisols as an example, soil families like Argiudolls and Hapludolls highlight variations in soil properties. Argiudolls, characterized by a significant horizon with clay illuviation, are conducive to agriculture, while Hapludolls may exhibit less clay illuviation, impacting their suitability for certain crops.

1.6.2 Soil Taxonomy Systems (e.g., USDA Soil Taxonomy)

USDA Soil Taxonomy: A Comprehensive Soil Classification System

1. Introduction to USDA Soil Taxonomy

- The USDA Soil Taxonomy, developed by the United States Department of Agriculture, provides a comprehensive framework for categorizing soils based on key properties. This system serves as a universal language for scientists, land managers, and policymakers.
- The system classifies soils hierarchically, aligning with global soil orders but further refining them to suit the needs of soil classification in the United States.

2. Key Components of USDA Soil Taxonomy

- The USDA Soil Taxonomy incorporates several key components in its classification, including soil properties related to soil moisture, temperature, and morphology. These properties help define each taxonomic level, offering a comprehensive description of a soil's characteristics.
- Soil moisture regimes, such as aridic and ustic, reflect the availability of water and its impact on soil development. Temperature regimes, encompassing thermic and frigid, provide insights into the role of temperature in shaping soil properties. Morphological characteristics, including horizon

development and soil texture, contribute to the detailed classification of soils within the taxonomy.

3. Soil Series: Identifying Specific Soil Entities

- At the finest level of classification in the USDA Soil Taxonomy, soil series represent specific entities that share similar characteristics. Soil series are named based on the location where they were first identified, providing a standardized nomenclature for precise communication.
- An example of a soil series is "Tunica clay," representing a specific soil in Tunica County, Mississippi. This naming convention allows for the clear identification and communication of specific soil types.

1.7 Soil Management Practices

In the intricate dance of agriculture, the way we manage our soil holds the key to both bountiful harvests and ecological resilience. This chapter embarks on a journey through various soil management practices, beginning with a profound exploration of the art and science behind conservation tillage.

1.7.1 Conservation Tillage

Introduction

In the vast canvas of agricultural practices, conservation tillage emerges as a sustainable brushstroke, delicately minimizing soil disturbance while nurturing a plethora of benefits. It stands in stark contrast to traditional tillage methods, aiming not only for a rich harvest but also for the preservation of soil health.

Key Principles of Conservation Tillage

1. Minimum Soil Disturbance

- Conservation tillage is an ode to minimalism in soil management. It delicately treads the earth, disrupting its structure just enough to foster growth without sacrificing the intricate balance within.

2. Retention of Crop Residues

- Picture a field after the harvest – instead of stripping it bare, conservation tillage embraces the wisdom of leaving behind the remnants of the previous crop. Stalks and leaves become a protective shield, shielding the soil from erosive forces.

3. Increased Water Retention

- In the ballet of water and soil, conservation tillage plays a pivotal role in choreographing water retention. By minimizing disturbance, it helps the soil embrace water, mitigating the impact of drought and fostering a more resilient ecosystem.

Benefits of Conservation Tillage

1. Erosion Control

- The artistry of conservation tillage is most evident in its ability to control erosion. Keeping crop residues in place, it shields the soil from the erosive whispers of wind and water, ensuring that the intricate layers of earth remain undisturbed.

2. Improved Soil Structure

- Soil is a symphony of aggregates, and conservation tillage seeks to preserve this harmony. By minimizing disruption, it allows these aggregates to stand tall, paving the way for improved structure, enhanced water infiltration, and robust fertility.

3. Water Conservation

- The thirst of the soil is quenched through the practice of conservation tillage. By cradling water in its undisturbed embrace, it reduces the reliance on irrigation, contributing to the conservation of this precious resource.

4. Enhanced Organic Matter

- Conservation tillage is akin to an artist adding strokes of organic matter to the canvas of soil. The retention of crop residues becomes a gift, fostering microbial life and enriching the soil with the nutrients necessary for life to thrive.

Types of Conservation Tillage

1. No-Till Farming

- Imagine a field where the plow never sings its disruptive tune. No-till farming is the embodiment of this vision, where seeds find their home in untilled soil, cradled by the protective cover of crop residues.

2. Reduced Tillage

- This is the nuanced dance of reduced tillage. While it acknowledges the need for some disturbance, it does so with grace, disturbing only a portion of the soil surface and allowing the rest to continue its undisturbed symphony.

Challenges and Considerations

1. Weed Management

- Every masterpiece has its challenges. Conservation tillage, with its reduced disturbance, may foster the persistence of certain weed species. Managing this delicate balance becomes an art in itself.

2. Transition Period

- As farmers transition to the realm of conservation tillage, there may be an initial period of adjustment. The soil, accustomed to the rhythms of traditional tillage, needs time to adapt to this new, less disruptive cadence.

Adoption and Global Impact

1. Increasing Adoption

- Across the world, farmers are increasingly embracing the principles of conservation tillage. It is a testament to the realization that this practice is not just a farming technique but a pledge to be stewards of the land.

2. Global Impact

- The collective adoption of conservation tillage resonates with global efforts toward sustainable agriculture. It becomes a cornerstone in the mosaic of initiatives aimed at soil conservation, climate resilience, and a more harmonious coexistence with the Earth.

1.7.2 Cover Cropping

Introduction

In the rhythm of agricultural cycles, cover cropping emerges as a transformative act, a gentle interlude between harvests that plays a crucial role in maintaining soil health and vitality. The essence of cover cropping lies in sowing crops not for immediate harvest but to cover and protect the soil during times of rest.

Key Principles of Cover Cropping: Greening the Fields

1. Soil Cover and Erosion Control

- Envision a field cloaked in a vibrant quilt of greenery. This living cover, crafted by crops like clover, rye, or legumes, acts as a shield against erosive forces, safeguarding the earth from the relentless whispers of wind and water.

2. Nutrient Cycling and Soil Enrichment

- Beyond their role as mere cover, these crops become custodians of nutrients. As they grow, cover crops draw up essential elements from the soil, preventing leaching and making them available for subsequent crops. When these crops decompose, they return the favor, enriching the soil with organic matter and nutrients.

3. Weed Suppression

- In the intricate theater of agriculture, where weeds seek to steal the spotlight, cover crops act as green sentinels. Their dense foliage shades the soil, suppressing the growth of invasive weeds and providing a natural defense mechanism.

4. Improving Soil Structure

- The roots of cover crops delve into the soil, breaking up compacted layers and creating channels for improved aeration and water infiltration. They are the architects of soil structure, fostering an environment conducive to healthy plant growth.

Benefits of Cover Cropping: A Tapestry of Rewards

1. Biodiversity and Habitat Creation

- The green tapestry of cover crops extends its benefits beyond soil health. It becomes a haven for biodiversity, offering habitat and sustenance for beneficial insects, microorganisms, and even small mammals.

2. Disease Suppression

- Some cover crops release compounds that act as natural pesticides, contributing to disease suppression. This not only protects the current crop but sets the stage for healthier future plantings.

3. Temperature Regulation

- Like a natural thermostat, cover crops with moderate soil temperature. They shield the earth from the extremes of heat and cold, fostering a more stable and conducive environment for plant growth.

4. Drought Mitigation

- In the face of changing climates, cover crops become allies against drought. Their extensive root systems improve water retention, offering a lifeline during dry spells and contributing to the resilience of the agricultural ecosystem.

Types of Cover Crops

1. Legumes

- Clover, peas, and vetch are leguminous cover crops that form symbiotic relationships with nitrogen-fixing bacteria. This enriches the soil with nitrogen, a vital nutrient for plant growth.

2. Grasses: Fibrous Roots for Soil Structure

- Rye, barley, and oats are common grass-cover crops. Their fibrous root systems contribute to soil structure, erosion control, and the overall health of the soil.

3. Brassicas: Breaking Soil Compaction

- Radishes, mustard, and turnips belong to the Brassica family. These cover crops have deep taproots that help break up compacted soil, improving aeration and water infiltration.

Challenges and Considerations

1. Timing and Integration

- The effectiveness of cover cropping lies in its timing and seamless integration into the crop rotation cycle. Careful planning ensures that cover crops maximize their benefits without impeding the growth of cash crops.

2. Management of Cover Crop Residues

- Once cover crops have played their role, managing their residues becomes a crucial aspect. Balancing decomposition rates and nutrient release is a delicate act that requires consideration for the needs of subsequent crops.

Adoption and Global Impact: A Green Revolution

1. Increasing Adoption: From Farms to Fields

- Farmers worldwide are increasingly embracing cover cropping as an integral part of sustainable agriculture. The realization of its multifaceted benefits has sparked a green revolution, turning fields into canvases of sustainable practices.

2. Global Impact: Beyond Individual Fields

- The global impact of cover cropping extends beyond individual fields. It contributes to larger sustainability goals, promoting practices that enhance soil health, conserve water, and mitigate the environmental impacts of agriculture on a broader scale.

1.8. Understanding Soil Erosion in Changing Climate

Definition of Soil Erosion

Soil erosion is a complex geological process wherein the top layer of soil is detached, transported, and deposited elsewhere. This phenomenon results from the interplay of natural forces such as wind and water, as well as human activities that disturb the delicate balance of the Earth's surface. It involves the gradual or sometimes rapid removal of soil particles from their original location, altering the landscape and potentially leading to significant environmental consequences (Table 3).

Table 3: Types of Soil Erosion

S.No	Type of Soil Erosion	Description and Causes
1	Sheet Erosion	Thin layers of soil are removed uniformly across a large area, often due to rainfall and surface runoff. It is common on sloping lands with little or no vegetation cover.
2	Rill Erosion	Small channels (rills) are formed on the soil surface by the concentrated flow of water. It occurs when runoff water accumulates and concentrates in certain areas, creating small channels.
3	Gully Erosion	Larger and deeper channels (gullies) are formed, often caused by the merging of multiple rills. Gully erosion can result in significant soil loss and is more severe than rill erosion.
4	Bank Erosion	Erosion of riverbanks or shorelines due to water flow. It can be caused by factors such as water currents, wave action, or human activities. Bank erosion can contribute sediment to nearby water bodies.
5	Wind Erosion	Soil particles are lifted and transported by the wind. Wind erosion is common in arid and semi-arid regions where vegetation cover is sparse. It can lead to the formation of sand dunes and the loss of topsoil.
6	Tillage Erosion	Erosion caused by human activities such as ploughing, cultivation, and other forms of soil manipulation. Improper tillage practices can lead to the exposure of soil to erosive forces, resulting in soil loss.
7	Landslide Erosion	Downward movement of rock and soil on a slope. Landslides can occur due to factors such as heavy rainfall, earthquakes, or human activities, leading to the rapid displacement of large volumes of soil.
8	Freeze-Thaw Erosion	Soil is loosened and moved due to the expansion of water as it freezes and contracts as it thaws. This type of erosion is common in cold climates where freezing and thawing cycles occur.

Natural erosion processes are exacerbated by human-induced factors, including deforestation, agriculture, construction, and poor land management practices. Soil erosion poses a threat to soil fertility, water quality, and the overall health of ecosystems. Recognizing the dynamics of soil erosion is fundamental to developing effective conservation strategies and sustainable land use practices.

Classification of Soil Erosion

1. Based on Agents

A. Water Erosion

Water erosion, an elemental force sculpting the Earth's surface, intricately weaves its narrative across landscapes. As we unravel the complexities of this geological process, we delve into two distinct yet interrelated forms: sheet erosion and rill erosion. These manifestations of water erosion are shaped by a myriad of factors and hold profound implications for soil health, land productivity, and ecosystem sustainability (Fig 7).

Fig 7: Soil Erosion by water

1. Sheet Erosion

Definition

Sheet erosion, often imperceptible but far-reaching, is characterized by the uniform removal of thin layers of soil across a large area. Unlike the dramatic carving of channels seen in rill or gully erosion, sheet erosion operates on a broader scale, gradually wearing away the topsoil without the formation of distinct channels or rivulets.

Processes

The genesis of sheet erosion lies in the intricate interplay of rainfall and the soil surface. Raindrops, impacting the exposed soil, dislodge individual particles. As rainfall intensity surpasses the soil's infiltration capacity, water begins to flow over the surface, carrying with it the detached soil particles. This imperceptible movement of soil, akin to a subtle dance, constitutes the essence of sheet erosion.

Factors Influencing Sheet Erosion

1. **Slope Gradient:** The slope gradient plays a pivotal role in sheet erosion. Steeper slopes increase the velocity of water runoff, intensifying the erosive forces acting on the soil.
2. **Soil Texture:** The texture of the soil significantly influences its susceptibility to sheet erosion. Fine-textured soils, such as clay, tend to be more prone due to their reduced water infiltration capacity.
3. **Vegetation Cover:** Vegetation stands as a natural shield against sheet erosion. The protective canopy of plants intercepts raindrops, reducing their impact on the soil, while root systems stabilize the soil structure, minimizing erosion.
4. **Rainfall Intensity:** The intensity and duration of rainfall events directly influence sheet erosion. Heavy rainfall, particularly in areas with insufficient vegetation cover, can trigger the detachment and transport of soil particles.

Environmental Implications

The consequences of sheet erosion extend beyond the apparent loss of topsoil. Reduced soil fertility, compromised water retention capacity, and altered nutrient cycling are among the environmental implications. Moreover, the transported sediments may find their way into water bodies, contributing to sedimentation and impacting water quality. The seemingly inconspicuous nature of sheet erosion belies its potential for far-reaching ecological consequences.

Mitigation Strategies

1. **Contour Plowing:** Plowing along the contour of the land is a fundamental strategy to minimize sheet erosion. This technique helps to break the flow of water downhill, reducing the formation of water channels.
2. **Cover Cropping:** Introducing cover crops provides a protective layer on the soil surface. The plant canopy mitigates the impact of raindrops, while the root systems enhance soil structure, reducing the risk of sheet erosion.
3. **Terracing:** Particularly effective in hilly terrain, terracing breaks the slope, reducing water velocity and minimizing the potential for sheet erosion. Terraces act as physical barriers, slowing down water movement and allowing for better infiltration.

2. Rill Erosion

Definition

Rill erosion marks a more advanced stage in the erosive journey, characterized by the formation of small channels or rills on the soil surface. Unlike the diffuse erosion of sheet erosion, rill erosion introduces a visible dimension, where water carves out distinct pathways across the landscape.

Processes

Rill erosion emerges when the volume and intensity of water exceed the soil's absorption capacity, leading to the concentration of water into rivulets. These rivulets, initially shallow, gradually deepen and widen as they flow across the landscape. Rill erosion often evolves from sheet erosion when the force of water becomes more concentrated.

Factors Influencing Rill Erosion

1. **Slope Gradient:** While steeper slopes are conducive to the formation of rills, moderate slopes are more likely to facilitate the development and enlargement of these distinct channels.
2. **Soil Texture:** The texture of the soil continues to play a role in rill erosion. Coarser soils may experience rill erosion when exposed, as the larger particles are more resistant to detachment.
3. **Vegetation Cover:** The presence or absence of vegetation is a critical factor in rill erosion. While vegetation helps prevent both sheet and rill erosion, the absence of adequate cover can contribute to the development of rills.
4. **Rainfall Intensity:** Intense or prolonged rainfall events can lead to the concentration of water flow, triggering the formation and enlargement of rills. The erosive force of water becomes more focused and pronounced.

Environmental Implications

Rill erosion exacerbates the environmental consequences initiated by sheet erosion. As channels deepen and widen, the potential for soil loss increases, impacting soil fertility, water retention, and overall ecosystem health. Additionally, the transported sediments may contribute to sedimentation in nearby water bodies, affecting aquatic habitats and water quality. The visible nature of rill erosion underscores its potential for altering landscapes and ecosystems.

Mitigation Strategies

1. **Vegetative Cover:** Maintaining or restoring vegetation cover is a fundamental strategy for preventing and mitigating rill erosion. The roots of plants provide stability to the soil, reducing the likelihood of channel formation.
2. **Check Dams:** Strategically placed check dams can help slow down water flow, reducing the energy needed to form and enlarge rills. Check dams act

as physical barriers, allowing sedimentation to occur and preventing further downstream erosion.

3. **Terracing:** Terracing, effective in addressing both sheet and rill erosion, breaks the slope and minimizes the velocity of water flow. Terraces act as structural defenses, interrupting the formation and progression of rills.

3. Gully Erosion

Definition

Gully erosion stands as a formidable manifestation of water erosion, characterized by the development of larger and deeper channels known as gullies. Unlike the subtle rills seen in earlier stages, gullies are more pronounced and represent a significant escalation in erosional processes. Gully erosion often arises as a consequence of the continuous and intensified erosion of pre-existing rill channels.

Processes

The genesis of gully erosion is intricately tied to the cumulative effects of water flow. Initially starting as smaller channels or rills, the erosive forces, if left unchecked, gradually transform these pathways into larger and more prominent gullies. The concentration of water, coupled with the progressive detachment of soil particles, leads to the deepening and widening of the channels, eventually forming distinct gullies that scar the landscape.

Factors Influencing Gully Erosion

1. **Slope Gradient:** Gully erosion is more likely to occur on steep slopes, where the force of water runoff is intensified. The gradient provides the necessary energy for the development and enlargement of gullies.

2. **Soil Composition:** The composition of the soil plays a crucial role in gully erosion. Soils that are less resistant to erosion, such as loose or sandy soils, are more susceptible to gully formation.

3. **Vegetation Cover:** The absence or removal of vegetation significantly contributes to gully erosion. Plant roots stabilize the soil, and the canopy mitigates the impact of raindrops, preventing the initiation of gullies.

4. **Water Flow:** The volume and velocity of water flow are critical factors in gully erosion. High-intensity rainfall events or continuous water flow can exacerbate the erosion of existing channels, leading to the development of gullies.

Environmental Implications

Gully erosion has profound environmental consequences, transcending the impacts of sheet and rill erosion. The formation of gullies results in extensive soil loss, altered topography, and compromised land productivity. Sediments carried by gully erosion can contribute to downstream sedimentation, affecting aquatic ecosystems and water quality. The scars left by gullies on the landscape are enduring symbols of the transformative power of water erosion.

Mitigation Strategies

1. **Vegetative Stabilization:** Reintroducing or maintaining vegetation cover is a crucial strategy to prevent and mitigate gully erosion. The roots of plants provide stability to the soil, reducing the likelihood of gully formation.
2. **Contour Plowing and Terracing:** Implementing contour ploughing and terracing helps break the slope, reducing the velocity of water runoff and minimizing the potential for gully erosion. Terraces act as physical barriers, preventing the concentrated flow of water.
3. **Check Dams:** Strategically placed check dams can be effective in slowing down water flow, reducing the energy needed to form and enlarge gullies. These dams allow sedimentation to occur, preventing further downstream erosion.

4. Streambank Erosion

Definition

Streambank erosion is a distinctive form of water erosion that occurs along the banks of rivers or streams. It is characterized by the gradual wearing away of the soil and rock material that constitutes the riverbank. This erosion is a natural process driven by the flow of water and is influenced by various environmental factors.

Processes

The dynamics of streambank erosion involve the continuous interaction between flowing water and the soil structure along riverbanks. The force of the water, particularly during periods of high flow or flooding, leads to the detachment of soil particles. As these particles are transported downstream, the shape and composition of the riverbank transform, contributing to the reshaping of the riverine landscape.

Factors Influencing Streambank Erosion

1. **Water Flow:** The intensity and volume of water flow play a central role in streambank erosion. High-flow events, such as floods, exert a greater force on the riverbanks, accelerating erosion.

2. **Bank Composition:** The composition of the riverbank materials influences the susceptibility to erosion. Banks with loose or poorly cohesive soils are more prone to erosion than those with stable, cohesive materials.
3. **Vegetation Cover:** Vegetation along riverbanks provides stability by anchoring the soil with its root systems. The removal of riparian vegetation increases the vulnerability of riverbanks to erosion.
4. **Human Activities:** Alterations to natural river flow patterns, such as dam construction or channelization, can significantly impact streambank erosion. Human infrastructure and activities in the riparian zone can also exacerbate erosion.

Environmental Implications

Streambank erosion contributes to the natural evolution of river channels, but excessive erosion can have adverse effects. Sedimentation downstream can alter aquatic habitats and water quality. Changes in riverbank morphology may impact adjacent ecosystems, affecting plant and animal communities along the river corridor. Mitigating streambank erosion requires a balanced approach that considers both natural processes and human interventions.

Mitigation Strategies

1. **Riparian Vegetation Management:** Maintaining and restoring riparian vegetation is crucial for stabilizing riverbanks. Well-established vegetation with deep root systems helps bind the soil and reduce erosion.
2. **Bioengineering Techniques:** Implementing bioengineering techniques involves using natural materials and vegetation to create erosion-resistant structures along riverbanks. These techniques enhance bank stability while promoting ecological benefits.
3. **Bank Stabilization Structures:** Installing engineered structures, such as retaining walls or riprap, can provide immediate stabilization to eroding riverbanks. However, these interventions should be carefully designed to minimize environmental impacts.
4. **Land Use Planning:** Responsible land use planning considers the dynamics of river systems, avoiding activities that exacerbate erosion. Establishing buffer zones and protecting riparian areas are integral components of effective land use practices.

Gully erosion and streambank erosion, while distinct in their manifestations, share the common thread of water's transformative power. Gullies carve deep scars across landscapes, mirroring the relentless force of water, while streambank erosion reshapes the delicate interface between land and river. Understanding these processes, their influencing factors, and the associated mitigation strategies is paramount for holistic land management, environmental conservation, and sustainable development. As we navigate the delicate dance between water and land, a comprehensive approach that embraces the intricacies of each erosion type becomes imperative for fostering resilient ecosystems and preserving the dynamic equilibrium of our natural world.

B. Wind Erosion

Wind erosion, a force of nature most prominently witnessed in arid and semi-arid landscapes, unfolds its intricate processes through deflation and abrasion (Fig 8).

Fig 8: Soil Erosion by Wind

Deflation

Definition

Deflation, a process emblematic of wind erosion, involves the selective removal of fine, dry soil particles from the Earth's surface by the force of the wind. It is a subtle yet transformative mechanism that can significantly influence the topography of arid regions.

Processes

Deflation initiates with the lifting of fine soil particles from the ground into the air. Wind, acting as the sculptor, carries these airborne particles across the landscape. The erosive force selectively removes the lighter, smaller particles, leaving behind coarser materials that are less susceptible to wind transport. Over time, the ground surface undergoes lowering, creating depressions and contributing to the formation of distinctive landforms, such as sand dunes.

Factors Influencing Deflation

1. **Surface Material:** The susceptibility to deflation is heightened in loose and fine-textured soils. Deserts with extensive sandy or silty surfaces are particularly vulnerable to this erosional process.

2. **Vegetation Cover:** The presence of vegetation acts as a natural shield against deflation. Plants, with their root systems and canopies, stabilize the soil, reducing wind speed at the ground level and impeding the detachment of fine particles.

3. **Wind Speed:** Wind velocity is a critical factor influencing the efficacy of deflation. In arid climates, where wind speeds can be substantial, the erosive potential of deflation is heightened, especially in areas with sparse or no vegetation.

Environmental Implications

Deflation contributes to the redistribution of soil particles across landscapes, influencing soil fertility, nutrient cycling, and land productivity. The creation of sand dunes is a visible outcome of deflation, impacting local ecosystems and creating distinctive features in arid regions.

Mitigation Strategies

1. **Vegetative Cover:** Establishing and maintaining vegetation cover is a fundamental strategy to combat deflation. The presence of plants acts as a natural barrier, reducing wind speed at ground level and stabilizing the soil against erosive forces.

2. **Windbreaks:** Artificial windbreaks, such as planting rows of trees or erecting barriers, help reduce wind velocity, minimizing the potential for deflation. These structures serve as protective shields against the erosive impact of wind.

3. **Cover Crops:** In agricultural settings, planting cover crops can protect the soil surface, reducing the impact of wind and preventing the detachment of fine particles. The establishment of ground cover enhances soil stability and resilience.

Abrasion

Definition

Abrasion, a complementary process to deflation, involves the impact of larger soil particles against surfaces, wearing them down over time. This mechanism, prevalent in arid and semi-arid regions, contributes to the sculpting of surfaces and the creation of distinctive landforms.

Processes

The abrasive action of wind-borne particles, including sand and pebbles, occurs when these particles collide with exposed surfaces. Rocks, landforms, and even human-made structures are subject to the erosive impact of abrasion. Over time, this continuous impact leads to the smoothing, polishing, or carving of surfaces, shaping the landscape.

Factors Influencing Abrasion

1. **Particle Size:** The size of wind-borne particles influences the extent of abrasion. Larger particles, such as sand and gravel, have greater potential for abrasion, especially when driven by strong winds.
2. **Wind Velocity:** The erosive potential of abrasion is heightened in regions with consistent and strong winds. Higher wind speeds enhance the impact of larger particles on exposed surfaces.
3. **Surface Characteristics:** The composition and hardness of exposed surfaces play a crucial role in abrasion. Softer materials are more susceptible to wear, contributing to the creation of distinctive landforms.

Environmental Implications

Abrasion, over time, contributes to the shaping of landforms in arid regions. Rocks may exhibit polished or faceted surfaces, and landscapes bear the signature of this erosional process. The visual impact of abrasion on exposed surfaces is a testament to the dynamic forces at play.

Mitigation Strategies

1. **Windbreaks and Shelterbelts:** Planting windbreaks or shelterbelts helps reduce wind speed, minimizing the impact of wind-borne particles and mitigating abrasion. These structures serve as protective barriers, shielding vulnerable surfaces from the erosive forces of the wind.
2. **Surface Cover:** Implementing ground cover, such as gravel or other protective materials, can shield surfaces from direct impact and reduce the erosive effects of abrasion. This is particularly relevant in areas where the exposure of surfaces to wind-borne particles is a concern.
3. **Structural Protection:** Applying protective coatings or materials to vulnerable structures helps shield them from abrasion. This preventive measure is crucial, especially in regions where human-made structures are exposed to the continuous erosive action of wind-borne particles.

Wind erosion, through the tandem processes of deflation and abrasion, crafts the intricate landscapes of arid and semi-arid regions. The subtlety of deflation, as fine particles dance in the wind, and the artistry of abrasion, as larger particles sculpt surfaces over time, reveal the dynamic interplay between wind and land. Understanding these processes and their environmental implications is pivotal for implementing effective mitigation strategies that balance the preservation of fragile ecosystems with human activities in arid regions. As the winds continue their timeless choreography, our responsibility lies in harmonizing our actions with the natural rhythms of these windswept terrains, ensuring a sustainable coexistence with the forces of wind erosion.

2. Based on Intensity

Erosion Intensity

Erosion, a natural geomorphic process, manifests in varying intensities, shaping landscapes and influencing ecosystems in diverse ways. The spectrum of erosion intensity, from light to severe, provides a nuanced understanding of its ecological consequences and underscores the need for tailored mitigation strategies.

1. Light Erosion

Definition

Light erosion is characterized by the subtle and gradual detachment of minimal soil particles, often escaping immediate notice. Though its impacts may seem inconspicuous, the cumulative effects of light erosion can influence soil health and the dynamics of ecosystems.

Key Characteristics

Light erosion is typically associated with gentle slopes, occasional surface runoff, and light rain. The erosive processes are subtle, involving the gradual movement of small soil particles. The changes in the landscape may be minor and go unnoticed in the short term.

Environmental Impact

While light erosion results in negligible soil loss, its long-term consequences can contribute to the depletion of topsoil and impact nutrient cycling. Ecosystems, given time and favorable conditions, can often recover naturally from the mild effects of light erosion.

Mitigation Strategies

Mitigating light erosion involves implementing practices such as maintaining ground cover, employing contour ploughing to minimize runoff, and encouraging the use of cover crops to stabilize the soil structure.

2. Moderate Erosion

Definition

Moderate erosion causes noticeable soil loss, drawing attention due to its more apparent impact on the landscape. This level of erosion can compromise soil fertility, alter structure, and potentially affect neighboring ecosystems.

Key Characteristics

Moderate erosion involves a more significant detachment and transport of soil particles compared to light erosion. It is often associated with moderate rainfall, steeper slopes, and periodic intense weather events. The landscape may exhibit visible signs of erosion, such as gullies or increased sedimentation in water bodies.

Environmental Impact

The consequences of moderate erosion are more pronounced, including reduced soil fertility, altered water retention capacity, and potential sedimentation in nearby water bodies. Ecosystems may experience stress, and agricultural productivity can be compromised, necessitating more active mitigation measures.

Mitigation Strategies

To mitigate moderate erosion, a combination of strategies is required. These may include contour ploughing to control water flow, cover cropping to protect the soil surface, and implementing structural measures like check dams to effectively manage runoff.

3. Severe Erosion

Definition

Severe erosion is characterized by extensive soil loss, leading to significant damage to the landscape. This level of erosion poses a critical threat to agricultural productivity, ecosystem health, and overall environmental integrity.

Key Characteristics

Severe erosion involves the rapid detachment and transport of large volumes of soil. Contributing factors include heavy rainfall, steep slopes, poor land management practices, and the absence of protective vegetation. Distinctive features such as deep gullies, loss of topsoil, and altered topography are evident.

Environmental Impact

The environmental impact of severe erosion is profound, with compromised soil structure, loss of essential nutrients, and altered drainage patterns. Agricultural lands may experience drastically reduced yields, and ecosystems can face long-term degradation. Sedimentation in water bodies can lead to water quality issues, affecting aquatic habitats.

Mitigation Strategies

Mitigating severe erosion demands urgent and comprehensive measures. Strategies include the implementation of terracing to reduce slope gradient, reforestation to restore vegetation cover, construction of check dams to slow down water flow, and the adoption of sustainable land management practices to rehabilitate and protect the landscape.

Recognizing the intensity of erosion, from light to severe, provides a crucial foundation for devising effective mitigation strategies. As we navigate the complexities of erosion, a nuanced approach is essential, considering the severity of soil loss and the unique characteristics of each landscape. Implementing sustainable land management practices tailored to the specific intensity of erosion ensures the preservation of ecosystems and fosters resilient landscapes in the face of this dynamic natural process.

3. Based on Geographical/Topographical Context

Erosion in Geographical and Topographical Context

Erosion, a relentless force shaping the Earth's surface, takes on diverse characteristics as it interacts with the geographical and topographical features of different landscapes.

1. Hill Slope Erosion

Definition

Hill slope erosion is a geological phenomenon primarily occurring on sloped surfaces, where the influence of gravity becomes a significant player in shaping the landscape. Often associated with rainfall or overland flow, this form of erosion is common in regions characterized by varied topography (fig 9).

Fig 9: Hill Slope Erosion

Key Characteristics

Hill slope erosion involves the detachment and transport of soil particles downslope. Intense rainfall events contribute to the erosive forces, leading to the formation of rills and gullies. Gravity amplifies the movement of water down the slope, accentuating the erosional processes and shaping the topography over time.

Environmental Impact

The environmental impact of hill slope erosion is multifaceted. Fertile topsoil is lost, altering soil structure and potentially causing downstream sedimentation. Local ecosystems may be affected as habitats undergo changes due to soil displacement, affecting plant and animal species.

Mitigation Strategies

Mitigating hill slope erosion necessitates a combination of strategic interventions. Contour ploughing helps break the slope continuity, reducing the velocity of water runoff. Terracing is effective in stabilizing slopes and preventing the formation of deep gullies. Afforestation plays a vital role in stabilizing the soil structure and minimizing erosion risks.

2. Valley Erosion

Definition

Valley erosion encompasses erosional processes within valleys or low-lying areas, influencing riverbanks and valley floors. It is a dynamic force that shapes the topography of these areas, influencing river channels and the surrounding landscape.

Key Characteristics

Valley erosion involves the removal and transportation of sediments within the confines of a valley. The lateral movement of water contributes to riverbank erosion, altering the course of rivers and reshaping the valley floor. Over time, valleys undergo continuous transformation driven by the erosive forces of flowing water.

Environmental Impact

The environmental impact of valley erosion extends beyond altered river courses. Changes in sediment distribution can disrupt aquatic and riparian habitats. Sedimentation downstream affects water quality and poses challenges to the health of ecosystems relying on river dynamics.

Mitigation Strategies

Mitigating valley erosion demands a holistic approach. Riverbank stabilization through the introduction of vegetation helps reduce erosive forces. Check dams strategically placed along river courses control water flow, minimizing erosion risks. Land use planning that considers the dynamic nature of river systems is essential for long-term erosion management.

3. Coastal Erosion

Definition

Coastal erosion results from the erosive forces of waves, tides, and storm surges along coastlines. It is a dynamic process that shapes the coastal environment, influencing landforms, and impacting both natural and human-made structures.

Key Characteristics

Coastal erosion is driven by the relentless action of waves, tides, and storm surges. It involves the removal and transportation of sediments along the shoreline, reshaping beaches and altering coastal morphology. Vulnerable coastal areas may experience land loss and changes in the dynamics of coastal ecosystems.

Environmental Impact

The environmental impact of coastal erosion is extensive. Land loss along coastlines can lead to habitat disruption and changes in beach dynamics. The threat extends to human infrastructure, with buildings and developments near coastlines at risk. Coastal ecosystems, including dunes and wetlands, may face significant challenges.

Mitigation Strategies

Mitigating coastal erosion requires a multi-faceted approach. Beach nourishment, involving the introduction of sediments to restore beach profiles, helps counteract land loss. The construction of protective structures such as seawalls and groins provides defense against erosive forces. Sustainable coastal management practices, considering the dynamic nature of coastal processes, are essential for long-term resilience.

Integrating Strategies for Holistic Erosion Management

Common Themes and Cross-Cutting Solutions

1. Vegetative Cover

- Implementing and maintaining vegetative cover emerges as a universal strategy across hill slope, valley, and coastal erosion contexts. Plants act as natural stabilizers, reducing the impact of erosive forces and promoting soil retention (fig 10).

Fig. 10: Prevention of erosion with vegetative Cover

2. Land Use Planning

- Thoughtful land use planning is essential for minimizing erosion risks. Identifying vulnerable areas, establishing buffer zones, and implementing measures that consider the natural dynamics of the landscape contribute to sustainable development.

3. Sustainable Practices

- Adopting sustainable land management practices ensures a balance between human activities and environmental preservation. This includes responsible agriculture, forestry, and urban planning that mitigate erosion risks and promote long-term ecological health.

4. Structural Interventions

- The construction of check dams, terraces, and protective structures such as seawalls or groins provides immediate solutions to erosion challenges. These interventions should be designed with a thorough understanding of local ecosystems to minimize unintended consequences.

5. Community Engagement

- Involving local communities in erosion management is crucial. Community engagement fosters awareness, encourages sustainable practices, and ensures that mitigation strategies are culturally and socially acceptable.

Erosion, when viewed through the prism of geographical and topographical contexts, reveals itself as a dynamic force, continuously shaping the Earth's surface. The processes associated with hill slope, valley, and coastal erosion underscore the complexity of interactions between natural forces and human activities.

A holistic and adaptive approach to erosion management is imperative. Integrating strategies that consider the specific characteristics of each erosion type, while recognizing their interconnected nature, ensures the preservation of landscapes, ecosystems, and the sustainability of human activities in diverse geographical settings. As we navigate this dynamic landscape, the quest for effective erosion management becomes a journey of balance, resilience, and coexistence with the ever-changing forces shaping our planet.

4. Based on Human Activities

Erosion, a natural geomorphic process, takes on new dimensions when influenced by human activities. The interaction between human actions and the Earth's surface can lead to specific forms of erosion, each with distinct characteristics and consequences.

1. Agricultural Erosion

Definition

Agricultural erosion arises from farming practices that include ploughing, overgrazing, and improper irrigation. These activities, while essential for food production, can lead to the degradation of arable land, impacting soil structure and fertility.

Key Characteristics

The mechanical processes associated with agriculture, such as ploughing, can disrupt the natural arrangement of soil particles. Overgrazing by livestock can expose soil to erosive forces, and improper irrigation practices may contribute to water runoff, carrying away essential topsoil.

Environmental Impact

Agricultural erosion poses risks to soil health, depleting nutrients and reducing fertility. Sedimentation in water bodies can lead to water quality issues, affecting aquatic ecosystems. The loss of arable land due to erosion may also contribute to land degradation.

Mitigation Strategies

Implementing conservation practices such as contour ploughing, cover cropping, and adopting sustainable irrigation methods can mitigate agricultural erosion. Terracing and agroforestry techniques contribute to soil stabilization, ensuring a balance between food production and environmental preservation.

A deeper understanding of agricultural erosion reveals the intricacies of its environmental impact and the challenges associated with sustainable land management. Contour ploughing, a technique where furrows follow the natural contours of the land, helps slow down water runoff, minimizing soil erosion. This method also aids in retaining moisture and nutrients in the soil, fostering a more conducive environment for plant growth.

Cover cropping, another essential practice, involves planting crops specifically for the purpose of protecting and improving the soil. Cover crops act as a protective blanket, shielding the soil from erosive forces and promoting biodiversity. Their root systems help bind the soil together, preventing excessive runoff and reducing the risk of erosion.

In addition to these techniques, adopting sustainable irrigation practices plays a pivotal role in mitigating agricultural erosion. Over-irrigation can lead to waterlogging and increased runoff, carrying away fertile topsoil. Employing precision irrigation methods, such as drip or sprinkler systems, ensures targeted water delivery, reducing the impact of erosive forces.

Moreover, the promotion of agroforestry—a practice that integrates trees and shrubs into agricultural landscapes—offers a holistic approach to combat erosion. The presence of trees provides additional canopy cover, reducing the impact of raindrops on the soil surface. The root systems of trees contribute to soil stabilization, preventing erosion and enhancing overall ecosystem health.

Educational initiatives and outreach programs are crucial components of sustainable agricultural practices. Farmers need access to information and resources that empower them to adopt erosion control measures effectively. Collaborative efforts between governmental agencies, non-profit organizations, and local communities can contribute to the widespread implementation of sustainable agricultural practices.

2. Urban Erosion

Definition

Urban erosion is a consequence of construction activities, deforestation, and altered land use patterns in urban areas. This form of erosion affects soil in both residential and industrial settings, often intensifying due to the impervious surfaces created by urbanization.

Key Characteristics

Construction activities disturb natural soil cover, exposing surfaces to erosive forces. Deforestation in urban settings can remove vegetation that stabilizes soil. Altered land use, such as the creation of impermeable surfaces, contributes to increased runoff and erosion.

Environmental Impact

Urban erosion can lead to increased sedimentation in water bodies, affecting water quality. Erosion-induced sedimentation in stormwater systems and rivers poses challenges to aquatic ecosystems. Soil loss in urban areas may also impact the stability of structures.

Mitigation Strategies

Integrating green infrastructure, such as permeable pavements and green roofs, helps reduce runoff and erosion in urban areas. Vegetation restoration projects and afforestation initiatives contribute to stabilizing soil in urban landscapes. Effective stormwater management practices are crucial for mitigating erosion risks.

The challenges of urban erosion extend beyond the immediate impacts on soil and water bodies. The altered hydrology in urban areas, characterized by increased impervious surfaces, amplifies the intensity and frequency of runoff. This shift in water flow dynamics not only contributes to erosion but also poses a threat to urban infrastructure and public safety.

Permeable pavements, designed to allow water to pass through the surface into the underlying soil, emerge as a sustainable solution to mitigate urban erosion. These surfaces reduce runoff by promoting infiltration, decreasing the erosive forces acting on soil. Additionally, green roofs, which involve planting vegetation on building rooftops, contribute to soil stabilization while providing other environmental benefits, such as temperature regulation and biodiversity support.

Vegetation restoration projects in urban environments involve the strategic planting of trees, shrubs, and green spaces. These initiatives not only enhance the aesthetic appeal of urban areas but also play a crucial role in preventing soil erosion. Trees, with their extensive root systems, help bind soil particles together, reducing the risk of erosion caused by rainfall or runoff.

Afforestation programs, aimed at establishing forests in urban regions, can be instrumental in combating erosion. Urban forests provide habitat for diverse plant and animal species, contribute to air quality improvement, and act as a natural buffer against soil erosion. Proper land use planning that incorporates green spaces and preserves natural features is essential for fostering urban resilience against erosion.

The integration of sustainable stormwater management practices is paramount in urban areas. Techniques such as the construction of retention basins, vegetated swales, and infiltration trenches help capture and slow down stormwater runoff, minimizing erosive impacts. Municipalities should invest in modernizing stormwater infrastructure to align with sustainable practices and ensure the long-term health of urban ecosystems.

Educating urban residents about the importance of responsible land use practices and water management is vital. Community engagement programs can empower citizens to actively participate in erosion control initiatives, promoting a sense of environmental stewardship and fostering a resilient urban landscape.

3. Mining Erosion

Definition: Mining erosion results from operations that expose soil and rock surfaces to erosion. The extraction of minerals and resources can lead to the degradation of landscapes, altering the topography and contributing to soil instability.

Key Characteristics: Mining operations often involve the removal of vegetation and topsoil, exposing underlying surfaces to erosive forces. Excavation activities, transportation of mined materials, and the creation of waste piles can contribute to soil erosion in mining areas.

Environmental Impact: Mining erosion can result in the loss of fertile topsoil and alteration of natural drainage patterns. Sedimentation in nearby water bodies poses risks to aquatic ecosystems. The creation of barren landscapes in mined areas may lead to long-term ecological disruption.

Mitigation Strategies: Implementing effective reclamation and restoration plans after mining activities are essential for mitigating erosion. Soil stabilization measures, including the use of cover crops and erosion control blankets, contribute to the rehabilitation of mined areas. Monitoring and regulating mining practices help minimize environmental impacts.

Mining, a crucial industry for resource extraction, often comes with significant environmental consequences, especially regarding erosion. Understanding the complex interplay between mining activities and erosion unveils opportunities for sustainable practices and rehabilitation.

The reclamation of mined areas stands out as a critical strategy for mitigating erosion and promoting ecological recovery. Reclamation involves reshaping the terrain, replacing topsoil, and revegetating the area with native plant species. This process aims to restore the landscape's functionality and biodiversity, reducing the risk of soil erosion.

Cover crops play a vital role in the reclamation process, acting as living shields against erosion. These crops provide immediate ground cover, preventing the exposure of bare soil to erosive forces. Additionally, cover crops contribute organic matter to the soil, enhancing its structure and fertility during the early stages of reclamation.

Erosion control blankets, made from natural or synthetic materials, offer an effective physical barrier against soil erosion in mined areas. These blankets stabilize the soil surface, prevent the detachment of particles, and facilitate the establishment of vegetation. Their use is particularly valuable on steep slopes prone to erosion.

Technological advancements in mining practices can significantly reduce the environmental footprint of extraction activities. Innovations such as precision mining and real-time monitoring systems help minimize the disturbance of large areas, decreasing the extent of exposed soil surfaces vulnerable to erosion. Employing best practices, including proper waste disposal and minimizing soil disturbance during excavation, contributes to sustainable mining.

Community engagement and collaboration between mining companies, regulatory bodies, and local communities are essential for responsible resource extraction. Transparent communication about mining plans, potential environmental impacts, and mitigation measures fosters trust and enables communities to actively participate in decision-making processes. Sustainable mining practices that prioritize environmental stewardship and social responsibility contribute to long-term landscape health.

4. Forestry Erosion

Definition: Forestry erosion is caused by logging and deforestation activities in forested regions. These activities disrupt natural ecosystems, exposing soil to erosive forces and affecting soil stability in forested landscapes.

Key Characteristics: Logging involves the removal of trees, altering the forest canopy and vegetation cover. Without the protective canopy, soil becomes vulnerable to erosive forces, especially during rainfall events. The construction of logging roads can further contribute to soil disturbance.

Environmental Impact: Forestry erosion can lead to the loss of topsoil, impacting soil fertility and nutrient cycling. Sediment runoff can affect nearby water bodies, posing risks to aquatic habitats. The disruption of natural ecosystems may also contribute to habitat degradation.

Mitigation Strategies: Implementing sustainable logging practices, such as reduced-impact logging and selective cutting, helps minimize soil disturbance. Reforestation efforts contribute to restoring vegetation cover and stabilizing soil. Developing erosion control plans and utilizing erosion control measures during logging activities are essential for sustainable forestry management.

Balancing the extraction of valuable timber resources with the preservation of forest ecosystems requires nuanced approaches and careful consideration of erosion risks associated with forestry activities.

Reduced-impact logging (RIL) emerges as a key strategy for minimizing the environmental impact of timber harvesting. RIL focuses on selectively harvesting trees while minimizing damage to the surrounding vegetation and soil. By avoiding clear-cutting and large-scale disturbance, RIL helps maintain soil structure and reduces the risk of erosion.

The implementation of best management practices during logging operations is crucial for erosion control in forestry. These practices include constructing water bars and sediment basins along logging roads to divert water and prevent sedimentation in nearby water bodies. Buffer zones around water bodies and sensitive ecosystems act as natural barriers, preserving water quality and minimizing the impact of erosive forces.

Reforestation efforts play a pivotal role in mitigating forestry erosion and promoting the sustainable management of forest resources. Strategic replanting of native tree species helps restore the protective canopy cover and stabilizes soil, preventing erosion in areas previously impacted by logging activities. The establishment of diverse and resilient forest ecosystems contributes to long-term landscape health.

Incorporating technology into forestry management enhances the precision and sustainability of logging practices. Geographic Information System (GIS) mapping facilitates the identification of high-risk erosion areas, enabling targeted intervention strategies. Real-time monitoring systems track logging activities, ensuring compliance with environmental regulations and minimizing unintended consequences.

Community involvement in forestry management is integral to achieving sustainable outcomes. Collaborative efforts between forestry companies, local communities, and conservation organizations can lead to the development of responsible logging practices. Educational programs and outreach initiatives raise awareness about the importance of sustainable forestry, fostering a shared commitment to preserving forest ecosystems.

Integrating Solutions for Sustainable Coexistence

As human activities continue to shape the landscape, understanding the intricate relationship between our actions and erosion processes becomes crucial for fostering sustainable coexistence. Common themes and cross-cutting solutions emerge, highlighting the need for a comprehensive and integrated approach to erosion management.

Cross-Cutting Solutions

1. Soil Conservation Practices

- Implementing soil conservation practices such as contour ploughing, cover cropping, and agroforestry across agricultural, urban, mining, and forestry contexts helps stabilize soil, reduce erosion risks, and promote sustainable land use.

2. Revegetation and Afforestation

- Introducing vegetation through reforestation and afforestation initiatives aids in stabilizing soil in various landscapes. The protective cover provided by plants mitigates erosive forces and contributes to ecosystem restoration.

3. Erosion Control Structures

- The construction of erosion control structures, including check dams, sediment basins, and retaining walls, plays a vital role in minimizing the impacts of erosion. These structures help manage water flow and prevent soil displacement.

4. Sustainable Land Management

- Adopting sustainable land management practices in agriculture, urban planning, mining, and forestry involves balancing human activities with environmental conservation. Thoughtful planning and regulation contribute to long-term soil and ecosystem health.

5. Community Engagement

- Engaging local communities in erosion management fosters awareness, encourages responsible practices, and ensures that mitigation strategies align with social and cultural values. Community involvement enhances the success and sustainability of erosion control initiatives.

5. Based on Temporal Scale

A. Long-Term Erosion

Definition: Long-term erosion is a natural geomorphic process that occurs over extended periods, shaping landscapes as part of the Earth's gradual evolution. This temporal scale spans geological epochs, encompassing the patient forces that mold terrains over millions of years.

Key Characteristics: Long-term erosion is characterized by a gradual and persistent transformation of landscapes. Factors such as weathering, tectonic movements, and the erosive forces of wind and water contribute to the slow but profound changes that define the geological features of our planet.

Environmental Impact: The impact of long-term erosion is observed in the grand geological features that shape Earth's surface. Mountains are worn down, valleys are carved, and coastlines evolve over epochs. While the immediate effects may seem subtle, the cumulative influence of long-term erosion is a testament to the dynamic and ever-changing nature of our planet.

Mitigation Strategies: Mitigating long-term erosion is inherently challenging due to its natural and ongoing nature. Conservation and preservation efforts play a crucial role in maintaining the delicate balance of ecosystems shaped by long-term erosion. Sustainable land use practices, reforestation, and measures to prevent excessive exploitation of natural resources contribute to the preservation of landscapes shaped over geological time scales.

Accelerated Erosion

The Anthropocene Influence

The era of human dominance, known as the Anthropocene, has left an indelible mark on the Earth's surface. Accelerated erosion, a manifestation of this influence, is driven by activities that alter landscapes at a pace unparalleled in natural geological processes. Deforestation, a prime contributor, strips away protective vegetation cover, leaving soil vulnerable to erosive forces.

Urbanization, another catalyst for accelerated erosion, transforms natural landscapes into impervious surfaces. Pavements and buildings disrupt the natural flow of water, leading to increased runoff and soil displacement. In agricultural landscapes, intensive farming practices, such as excessive ploughing and improper irrigation, exacerbate erosion rates, impacting soil fertility and agricultural sustainability.

Immediate Environmental Impacts

The consequences of accelerated erosion are felt acutely in affected regions. Loss of topsoil, rich in nutrients essential for plant growth, compromises agricultural productivity. Sedimentation in water bodies, a byproduct of accelerated erosion, poses threats to aquatic ecosystems, disrupting the delicate balance of aquatic life and water quality.

Increased runoff, a characteristic feature of accelerated erosion, contributes to flash floods, which can have devastating effects on both natural and human-made environments. Erosion-induced landslides further compound the risks, endangering communities and infrastructure. The immediate and tangible impacts of accelerated erosion demand urgent attention and mitigation efforts.

Mitigation Strategies for Accelerated Erosion

Addressing accelerated erosion requires a combination of regulatory measures, sustainable land management practices, and community engagement initiatives.

1. Afforestation and Reforestation

- Planting trees helps restore vegetation cover, reducing soil exposure to erosive forces. Afforestation involves establishing forests in areas devoid of tree cover, while reforestation focuses on replanting trees in deforested or degraded areas.

2. Contour Plowing

- Implementing contour ploughing, where furrows follow the natural contours of the land, helps slow down water runoff. This technique reduces the speed and erosive potential of flowing water, minimizing soil loss.

3. Erosion Control Structures

- Constructing erosion control structures, such as check dams, sediment basins, and retaining walls, helps manage water flow and prevent soil displacement. These structures act as physical barriers, reducing the impact of erosive forces.

4. Regulatory Measures

- Enforcing regulations that limit deforestation, promote sustainable land use practices, and prevent overgrazing is essential for curbing activities contributing to accelerated erosion. Zoning and land use planning play key roles in mitigating human-induced changes to landscapes.

5. Community Education and Outreach

- Educating local communities about the consequences of accelerated erosion fosters awareness and encourages responsible land management practices. Engaging communities in erosion control initiatives ensures the sustainability and success of mitigation efforts.

Geological Forces at Play: Long-term erosion unfolds over geological time scales, driven by the patient forces of weathering, tectonic movements, and the erosive power of wind and water. This process shapes the Earth's surface in ways that reflect the cumulative impact of natural forces acting over millions of years.

Landforms Shaped by Time: The grandeur of landscapes shaped by long-term erosion is evident in diverse landforms. Mountains, once towering peaks, are gradually worn down by weathering and erosive forces. Valleys, carved by the persistent flow of rivers and streams, showcase the transformative power of water over time. Coastlines evolve as waves sculpt the land, showcasing the intricate dance between land and sea.

The Role of Weathering: Weathering, a fundamental component of long-term erosion, involves the breakdown of rocks into smaller particles over time. Physical weathering, driven by factors like temperature changes and frost action, contributes to the disintegration of rocks. Chemical weathering, involving processes like dissolution and oxidation, further transforms rocks into sediments.

Tectonic Forces: Shaping Continents Slowly

Tectonic movements, occurring over vast time scales, play a pivotal role in long-term erosion. Continental drift, the slow movement of Earth's tectonic plates, results in the gradual rearrangement of continents. Mountain-building processes, driven by tectonic collisions, create majestic ranges that bear witness to the ongoing geological drama.

Mitigation Strategies for Long-Term Erosion

Mitigating long-term erosion is inherently challenging due to its natural and ongoing nature. Conservation and preservation efforts, guided by an understanding of geological processes, contribute to maintaining the delicate balance of ecosystems shaped by long-term erosion.

1. Preservation of Natural Habitats

- Protecting natural habitats, including forests, wetlands, and ecosystems, is crucial for maintaining the resilience of landscapes shaped by long-term erosion. Conservation efforts should focus on minimizing human disturbances and preserving biodiversity.

2. Sustainable Land Use Planning

- Implementing sustainable land use planning practices helps prevent excessive exploitation of natural resources. Zoning regulations that safeguard environmentally sensitive areas and restrict intensive development contribute to the preservation of landscapes shaped by geological forces.

3. Educational Initiatives

- Educating the public about the geological processes shaping landscapes over geological time scales fosters an appreciation for the Earth's dynamic history. Understanding the interconnectedness of geological forces and ecosystems encourages responsible stewardship.

4. Geotourism and Geological Conservation

- Promoting tourism, which emphasizes responsible travel to geological sites, can raise awareness about the beauty and significance of landscapes shaped by long-term erosion. Establishing geological conservation areas ensures the protection of key sites for future generations.

Bridging Past and Present: Understanding the intricate interplay between accelerated and long-term erosion demands a comprehensive approach that acknowledges both the immediate impacts of human activities and the enduring forces that shape Earth's geological history.

Conservation Ethics: Conservation ethics serve as a unifying principle in addressing erosion across temporal scales. Whether mitigating the accelerated erosion induced by contemporary human activities or preserving landscapes shaped by eons of geological forces, a shared commitment to conservation underpins sustainable practices.

Community Engagement: Fostering Stewardship: Engaging communities in erosion mitigation efforts is essential for long-term success. Local communities, equipped with knowledge about the consequences of erosion, become active participants in preserving their landscapes. Collaborative initiatives, driven by a sense of stewardship, contribute to the shared responsibility of protecting Earth's natural heritage.

Technological Innovations: Tools for Mitigation: Harnessing technological innovations is instrumental in addressing erosion challenges. Satellite imagery, remote sensing, and Geographic Information System (GIS) mapping provide valuable tools for monitoring erosion dynamics. Real-time data facilitates informed decision-making and allows for the timely implementation of mitigation strategies.

Policy Advocacy: Shaping Sustainable Practices: Advocacy for policies that promote sustainable land management, enforce regulations against deforestation and over-exploitation, and incentivize conservation efforts is crucial. Governments, in collaboration with environmental organizations, play a pivotal role in shaping policies that balance human needs with ecological preservation.

Scientific Research: Guiding Mitigation Strategies: Continued scientific research enhances our understanding of erosion processes, enabling the development of targeted mitigation strategies. Interdisciplinary approaches that integrate geological studies, ecological assessments, and socio-economic analyses provide holistic insights into erosion dynamics.

Education for Future Stewardship: Educational initiatives, spanning formal curricula and public outreach programs, are vital for cultivating a sense of environmental stewardship. By instilling an understanding of erosion processes, their consequences, and the interconnectedness of human actions with geological forces, education becomes a catalyst for responsible behavior and decision-making.

As stewards of the Earth, our responsibility lies in fostering a harmonious coexistence with these erosional forces. Mitigating accelerated erosion demands immediate action to curb detrimental human activities, while embracing the enduring nature of long-term erosion calls for conservation and sustainable practices. By navigating this complex tapestry, we embark on a journey of preservation, ensuring that the landscapes we inherit remain resilient and awe-inspiring for generations to come.

Management of Soil and Nutrient Losses: Shifting Cultivation - Principles, Extent, and Impact

Management of Soil and Nutrient Losses

1. Conservation Tillage

Conservation tillage is a set of practices designed to minimize soil disturbance during planting. This includes techniques like no-till farming, reduced tillage, and the use of conservation tillage machinery. The principle is to reduce erosion by preserving the soil structure and organic matter. Conservation tillage has gained widespread adoption in modern agriculture due to its effectiveness in retaining soil health.

Conservation tillage techniques, such as no-till farming, involve minimal soil disturbance during planting. This approach helps to maintain the soil structure, reduce erosion, and enhance water retention. In traditional tillage, the soil is plowed or cultivated before planting, which can lead to soil erosion, loss of organic matter, and disruption of soil structure. On the other hand, conservation tillage practices leave the previous crop residue on the field, acting as a protective cover that shields the soil from the impact of raindrops, minimizes water runoff, and promotes water infiltration.

The extent of impact of conservation tillage is notable in regions where sustainable farming practices are prioritized. Farmers adopting these methods observe improvements in soil health, reduced erosion, and enhanced water retention. No-till farming, in particular, has become a key component of sustainable agriculture, contributing to long-term soil conservation and promoting overall environmental resilience.

2. Cover Cropping

Cover cropping involves planting specific crops, such as legumes or grasses, during periods when the main crop is not growing. This practice protects the soil surface, prevents erosion, and enhances nutrient retention. Commonly used in sustainable agriculture and organic farming, cover cropping is an effective method for improving soil health and reducing nutrient losses.

Cover crops play a crucial role in soil conservation by providing ground cover during periods when the main cash crop is not actively growing. Leguminous cover crops, such as clover or vetch, have the additional benefit of fixing nitrogen from the atmosphere into the soil, enriching it with essential nutrients. Grass-cover crops, like rye or oats, create a protective layer that shields the soil from erosion caused by wind and water.

The extent of the impact of cover cropping is evident in the improved soil structure, increased organic matter content, and enhanced water infiltration observed in fields where this practice is employed. Beyond erosion control, cover crops contribute to weed suppression, pest management, and overall biodiversity in agricultural landscapes.

3. Contour Plowing and Terracing

Contour ploughing is a technique where farmers plow along the contour lines of the land to slow water runoff and minimize soil erosion. Terracing involves creating steps on hilly terrain to reduce water flow and soil displacement. Both practices are crucial for preventing water-induced erosion, particularly in regions with sloping topography.

Contour ploughing is a soil conservation method that follows the natural contour lines of the land, creating furrows that run perpendicular to the slope. By ploughing along these contour lines, water runoff is slowed down, reducing its erosive force and allowing it to infiltrate into the soil. This method helps to prevent the formation of gullies and rills, common indicators of water erosion in sloping landscapes.

Terracing, on the other hand, involves creating steps on hilly terrain to reduce water flow and soil displacement. Terraces act as physical barriers that slow down water runoff, providing a series of level surfaces that help prevent soil erosion. This practice is particularly important in regions with steep slopes where conventional farming practices could lead to significant soil loss.

The extent of impact of contour ploughing and terracing is substantial in regions where these practices are widely adopted. Not only do they help in preventing soil erosion, but they also contribute to water conservation, as the slowed down runoff allows for better infiltration into the soil. These practices are essential components of sustainable land management, particularly in areas prone to water-induced erosion.

4. Agroforestry

Agroforestry integrates trees into agricultural systems to provide ground cover, reduce wind and water erosion, and enhance nutrient cycling. This practice not only contributes to soil conservation but also promotes biodiversity and sustainable land use. Agroforestry is implemented in various agroecological zones globally to improve soil health and enhance overall environmental resilience.

Agroforestry is a holistic approach that combines the cultivation of trees or shrubs with agricultural crops or livestock within the same land area. The integration of trees into farming systems offers multiple benefits for soil conservation. The canopy of trees provides shade and protection, reducing the impact of rainfall on the soil surface and preventing water-induced erosion.

The extent of impact of agroforestry is widespread in regions where this practice is embraced. Beyond erosion control, agroforestry contributes to improved soil fertility through nutrient cycling, increased organic matter content, and enhanced water retention. The presence of trees in agricultural landscapes also promotes biodiversity, providing habitats for diverse plant and animal species.

Agroforestry systems vary based on the specific environmental conditions of a region. Alley cropping, where rows of trees are integrated into crop fields, and silvopastoral systems, combining trees with livestock grazing, are examples of how agroforestry can be adapted to different contexts. The adoption of agroforestry is a testament to its effectiveness in achieving sustainable land management and conservation goals.

5. Soil Conservation Structures

Building structures like check dams, retention ponds, and contour bunds helps manage water flow and prevent soil erosion. These structures are commonly used in watershed management projects to control erosion and sedimentation. By regulating water movement, soil conservation structures contribute to maintaining the stability of agricultural landscapes.

Soil conservation structures are engineered features designed to control water runoff, minimize soil erosion, and manage sedimentation. Check dams are barriers built across water channels to slow down the flow of water and reduce its erosive force. Retention ponds capture excess water, preventing it from causing downstream erosion and allowing sediment to settle. Contour bunds are embankments constructed along the contour lines of the land to slow down water runoff and promote infiltration.

The extent of impact of soil conservation structures is particularly evident in regions prone to water-induced erosion, such as those with steep slopes or intense rainfall. Watershed management projects often incorporate these structures to address soil conservation at a broader scale. By strategically placing check dams, retention ponds, and contour bunds, the erosive power of water is effectively mitigated, safeguarding agricultural lands and preventing sedimentation in water bodies.

6. Organic Farming Practices

Organic farming emphasizes the use of organic inputs, crop rotations, and green manure to maintain soil fertility and structure. By avoiding synthetic chemicals and promoting natural nutrient cycling, organic farming helps reduce nutrient losses and promotes sustainable agricultural practices. The adoption of organic farming methods is growing globally, driven by a focus on environmental sustainability.

Organic farming is an agricultural approach that prioritizes ecological balance, biodiversity, and the use of natural inputs to enhance soil fertility. The principles of organic farming include avoiding synthetic pesticides and fertilizers, practicing crop rotation, using compost and green manure, and fostering a holistic and sustainable farming system.

The extent of impact of organic farming practices is evident in the improved soil health, increased microbial activity, and enhanced nutrient availability observed in organic fields. By avoiding the use of synthetic chemicals, organic farming minimizes the risk of soil and water contamination, contributing to overall environmental conservation. Organic practices also promote the development of resilient and biodiverse agroecosystems, fostering long-term sustainability.

Organic farming is a viable alternative for farmers seeking to adopt environmentally friendly and sustainable agricultural practices. The growing consumer demand for organic products has further incentivized the adoption of organic farming methods. As the global agricultural landscape evolves, organic farming continues to play a crucial role in promoting soil conservation, sustainable land management, and overall environmental well-being.

7. Soil Erosion Modeling

Soil erosion modeling involves using computer models to simulate and predict soil erosion. This approach aids in designing effective conservation strategies by assessing erosion risks and understanding the factors influencing soil loss. Soil erosion modeling is applied in research and land management planning to make informed decisions about soil conservation measures.

Soil erosion modeling is a sophisticated tool used to simulate and predict the dynamic processes of soil erosion. By utilizing computer-based models, researchers and land managers can assess the potential risk of soil erosion in specific areas, identify contributing factors, and develop targeted conservation strategies. These models consider various parameters, including topography, soil type, land cover, rainfall patterns, and land management practices.

The extent of the impact of soil erosion modeling is seen in its applications across different scales. At the local level, it helps farmers and land managers make informed decisions about soil conservation practices. On a larger scale, it contributes to watershed management plans and regional strategies to mitigate soil erosion and protect valuable agricultural lands.

Soil erosion modeling incorporates the principles of geospatial technology, hydrology, and soil science. Geographic Information System (GIS) mapping, remote sensing, and real-time data are integrated into these models to enhance accuracy and reliability. The adoption of soil erosion modeling represents a forward-thinking approach to soil conservation, leveraging technology to address the complex and dynamic nature of erosion processes.

Shifting Cultivation: Principles, Extent, and Impact

Principles

Shifting cultivation, also known as slash-and-burn agriculture, involves a cyclical process of clearing a piece of land, planting crops for a few years until fertility declines, and then abandoning it for the land to regenerate. The rotational cultivation and fallow periods are key principles, allowing for the sustainable use of land.

Shifting cultivation is rooted in the principle of rotational cultivation and fallow periods. Farmers clear a section of land, usually through slash-and-burn methods, plant crops for a limited duration, and then leave the land fallow for a period of natural regeneration. This cyclical process allows the soil to recover its fertility, and the land is reused after a few cycles.

Extent

Shifting cultivation is practiced in tropical regions across the world, with significant prevalence in parts of Africa, Southeast Asia, and South America. It holds cultural significance and is often associated with indigenous communities relying on traditional agricultural practices.

The extent of shifting cultivation is considerable, especially in regions where it aligns with the cultural practices of indigenous communities. In Southeast Asia, shifting cultivation is prevalent in upland areas, while in parts of Africa, it is practiced by various ethnic groups. The extent of shifting cultivation is influenced by factors such as population density, land availability, and cultural traditions.

Impact

Positive Impacts

1. **Biodiversity:** The fallow periods in shifting cultivation support the regeneration of diverse plant species, enhancing overall biodiversity in the ecosystem.
2. **Cultural Heritage:** Shifting cultivation is deeply embedded in the traditions and cultures of many indigenous communities, forming an integral part of their way of life.
3. **Local Food Security:** In certain regions, shifting cultivation provides subsistence livelihoods, supporting local food security and the well-being of communities.

Negative Impacts

1. **Deforestation:** The initial clearing of land for shifting cultivation can lead to deforestation and the loss of natural habitats, impacting overall ecosystem health.

2. **Soil Erosion:** Intensive cultivation followed by abandonment during fallow periods can result in soil erosion, particularly in regions with inadequate land management practices.

3. **Land Degradation:** Continuous shifting cultivation without proper land management can lead to soil fertility decline and overall land degradation over time.

4. **Limited Crop Yields:** Crop yields in shifting cultivation systems may be lower compared to more intensive agricultural systems, posing challenges for food production.

Sustainable Practices in Shifting Cultivation

1. **Agroforestry Integration:** Planting trees within shifting cultivation systems can help stabilize soils, provide additional income through timber, and enhance overall sustainability.

2. **Crop Diversification:** Introducing a variety of crops in the cultivation cycle can improve soil fertility and reduce the risk of nutrient depletion.

3. **Fallow Management:** Implementing controlled fallow periods and avoiding excessive burning during the fallow phase can mitigate negative environmental impacts.

4. **Community-Based Management:** Involving local communities in decision-making and sustainable land management practices is crucial for the success of shifting cultivation while ensuring that it remains environmentally sustainable.

Agroforestry Integration in Shifting Cultivation

Agroforestry, the integration of trees into agricultural systems, offers a sustainable solution to mitigate the negative impacts associated with shifting cultivation. By incorporating tree species that complement the cropping cycle, agroforestry enhances soil fertility, reduces erosion, and provides additional socio-economic benefits.

1. **Stabilizing Soils:** The root systems of trees play a crucial role in stabilizing soils, preventing erosion during both cultivation and fallow periods. Trees contribute to the overall structural integrity of the soil, reducing the risk of landslides and soil displacement.

2. **Nutrient Cycling:** Agroforestry promotes nutrient cycling by facilitating the transfer of essential nutrients between trees, crops, and soil. Nitrogen-fixing tree species contribute to soil fertility, reducing the reliance on external inputs and enhancing the long-term sustainability of shifting cultivation systems.

3. **Biodiversity Enhancement:** The integration of diverse tree species in agroforestry systems fosters biodiversity. This diversity benefits not only the ecosystem but also contributes to improved pest management and resilience to environmental changes.

4. **Timber and Non-Timber Products:** Agroforestry provides additional sources of income for farmers through the sustainable harvesting of timber and non-timber forest products. This economic diversification can reduce dependency on shifting cultivation as the sole source of livelihood.

Methods of Soil Erosion Control: Vegetative and Mechanical Measures

Soil erosion, a complex and widespread environmental issue, demands multifaceted solutions for effective control. Beyond the conventional methods explored earlier, further insights into the intricacies of erosion control measures, along with their ecological and socio-economic implications, provide a more comprehensive understanding. This extended exploration aims to delve into additional aspects of both vegetative and mechanical measures, emphasizing their application, challenges, and the imperative need for integrated approaches.

Vegetative Measures

6. Agroforestry Integration

Agroforestry, a practice that integrates trees and shrubs with agricultural crops, goes beyond afforestation. This method aims not only to prevent soil erosion but also to enhance overall agricultural sustainability.

Principle: Agroforestry combines the principles of agriculture and forestry, allowing for the simultaneous cultivation of crops and the growth of trees. The diverse ecosystem created provides numerous benefits, including soil stabilization and erosion control.

Effectiveness: The integration of trees in agroforestry systems contributes significantly to soil conservation. The roots of trees help bind soil particles, preventing erosion. Furthermore, the presence of trees enhances biodiversity, creating a resilient and sustainable agroecosystem.

Practical Application: Farmers can adopt agroforestry by strategically planting trees alongside crops. Alley cropping, silvopasture, and windbreaks are common agroforestry practices. These systems offer economic and ecological benefits while effectively addressing soil erosion concerns.

7. Crop Diversification

Diversifying crops within a field or farm is an approach that not only contributes to soil erosion control but also enhances overall agricultural resilience.

Principle: Crop diversification involves growing a variety of crops on the same piece of land. This practice disrupts pest and disease cycles, improves soil health, and mitigates the risk of erosion associated with monoculture.

Effectiveness: The varied root structures, nutrient needs, and growth patterns of different crops enhance soil stability. Crop diversification also reduces the susceptibility of the entire crop system to specific pests or diseases, minimizing the need for chemical interventions that can contribute to soil degradation.

Practical Application: Farmers can implement crop diversification by rotating crops based on their seasonal and rotational characteristics. This practice helps maintain soil fertility, reduces erosion risks, and promotes a more sustainable and resilient farming system.

Mechanical Measures

6. Check Dams and Retention Ponds

Check dams and retention ponds are engineered structures designed to control water flow, reduce erosion, and capture sediment.

Principle: Check dams and retention ponds operate on the principle of slowing down water runoff. Check dams are typically small structures built across gullies or channels to impede water flow, while retention ponds trap and store water, allowing sediment to settle.

Effectiveness: Check dams and retention ponds are highly effective in preventing soil erosion by controlling water movement. They reduce the speed of runoff, preventing gully formation, and provide an opportunity for sediment to settle, protecting downstream areas.

Practical Application: The construction of check dams involves placing materials such as stones or concrete across water channels. Retention ponds are excavated or constructed to capture and store water. Both structures are strategically placed in areas prone to erosion.

7. Soil Conservation Structures

Beyond the individual measures explored earlier, the construction of soil conservation structures encompasses a broader set of engineered interventions aimed at preventing erosion and maintaining soil health.

Types of Structures: These structures include contour bunds, terraces, sediment basins, and gabions. Contour bunds are embankments built along contour lines, terraces transform slopes into steps, sediment basins capture sediment-laden runoff, and gabions are wire mesh containers filled with rocks.

Principle: The overarching principle of soil conservation structures is to modify the landscape to intercept and manage water flow, preventing soil erosion. Each type of structure addresses specific erosion challenges based on topography and land use.

Effectiveness: Soil conservation structures are effective in minimizing erosion by controlling the movement of water and sediment. They provide physical barriers that protect against gully formation, reduce surface runoff, and enhance overall landscape stability.

Practical Application: The selection and design of soil conservation structures depend on the characteristics of the landscape and erosion risk. Engineers and land managers collaborate to implement these structures strategically, considering the specific needs of each location.

Integrated Approaches

3. Community-Based Management

Community-based management involves actively engaging local communities in decision-making processes and implementing erosion control measures.

Principle: The principle is rooted in empowering local communities to take ownership of their landscapes. By involving residents in planning and executing erosion control strategies, the success and sustainability of these measures are significantly increased.

Effectiveness: Community-based management fosters a sense of responsibility and stewardship among local communities. When people feel a direct connection to the land, they are more likely to adopt and sustain erosion control practices, leading to long-term benefits.

Practical Application: Community-based management requires collaborative efforts between community members, local authorities, and relevant organizations. It involves training, awareness campaigns, and the establishment of communal agreements to ensure the effective implementation of erosion control measures.

4. Precision Agriculture

Precision agriculture leverages technology and data-driven approaches to optimize farming practices, contributing to soil health and erosion control.

Principle: The principle of precision agriculture is to use technology, including GPS, sensors, and data analytics, to precisely manage inputs such as water, fertilizers, and pesticides. This targeted approach minimizes environmental impact and reduces the risk of soil erosion.

Effectiveness: Precision agriculture allows farmers to tailor their practices based on real-time data and site-specific conditions. By optimizing resource use, precision agriculture minimizes soil disturbance and erosion associated with excessive or improper application of inputs.

Practical Application: Farmers can adopt precision agriculture by investing in technologies like GPS-guided tractors, soil sensors, and data analytics tools. These tools enable more precise and efficient farming practices, contributing to both productivity and sustainability.

Challenges and Considerations

1. Socio-Economic Considerations

The successful implementation of erosion control measures requires a nuanced understanding of socio-economic factors that influence land management decisions.

Challenges: Issues such as land tenure, economic constraints, and cultural practices can pose challenges to the adoption of erosion control measures. For example, farmers may be hesitant to implement certain practices if they perceive short-term economic losses.

Considerations: Tailoring erosion control strategies to align with local customs, economic realities, and community needs is crucial. Incorporating traditional knowledge and involving communities in decision-making can enhance the acceptability and effectiveness of erosion control initiatives.

2. Monitoring and Adaptation

Erosion control measures should be continually monitored, and adjustments made based on evolving environmental conditions.

Challenges: Environmental factors, climate change, and land use dynamics can impact the effectiveness of erosion control measures over time. Failure to adapt strategies accordingly may result in diminished effectiveness.

Considerations: Regular monitoring and assessment of erosion control initiatives are essential. Integrating adaptive management practices allows for the modification of strategies in response to changing conditions, ensuring sustained efficacy.

Conclusion

Expanding our exploration of soil erosion control measures reveals the dynamic and interconnected nature of these strategies. From agroforestry integration to precision agriculture, the array of approaches showcases the diversity required to combat soil erosion effectively. The challenges and considerations underscore the importance of context-specific solutions that are adaptable, socially inclusive, and ecologically sustainable.

As global concerns regarding soil degradation intensify, the implementation of these varied erosion control measures becomes paramount. The integration of technological advancements, traditional knowledge, and community engagement forms a holistic approach that addresses not only the symptoms but also the root causes of soil erosion. In the pursuit of sustainable land use, acknowledging the multifaceted nature of soil erosion and adopting integrated, context-specific solutions is imperative for the well-being of our planet and future generations.

2

Water Resources in Agriculture

[1]Niranjan B N., [2]Kh. Chandrakumar Singh, [3]Savita Jangde
[4]Sheetanshu Gupta and [5]Wajid Hasan

[1]Dakshina Kannada Milk Union (KMF), Mangalore, Karnataka
[2]VCSGU University of Horticulture and Forestry, Bharsar, Uttarakhand
[3]Department of Plant Physiology, Institute of Agricultural Sciences B.H.U. Varanasi, U.P
[4]Department of Biotechnology, Institute of Technology and Management BKT, Lucknow, U.P
[5]KVK Jehanabad, Bihar Agricultural University Sabour, Bihar

1. Introduction to Water Resources in Agriculture

Water, the quintessential life force for agriculture, intricately intertwines its significance throughout the fabric of global farming practices.

1.1 Significance of Water in Agricultural Practices

1.1.1 Sustenance of Plant Life

- Beyond its role as a life sustainer, water emerges as the very essence that propels plant development. It acts as the vital catalyst for photosynthesis, the engine of plant growth. This intricate process is the heart of the plant's ability to convert sunlight into energy, driving robust development from seed germination to harvest.

Water's significance in sustaining plant life goes beyond the visible; it is the unseen force that enables plants to thrive in their dynamic journey through various growth stages.

1.1.2 Nutrient Transport

- The journey of nutrients from the soil to the plant involves water as a meticulous carrier. This intricate transport system is akin to nature's bloodstream, facilitating the delivery of essential elements crucial for the synthesis of proteins, enzymes, and other vital compounds fundamental to crop health.

In this complex choreography, water acts as a conduit for life-enriching nutrients, ensuring that plants receive a balanced diet for optimal health and productivity.

1.1.3 Temperature Regulation

- Water, in its transformative journey through plants, becomes a natural thermostat. The process of transpiration, where plants release water vapor, not only cools the plant but also regulates the temperature of the surrounding environment. This delicate balance ensures an optimal atmosphere for growth.

The dance of temperature regulation orchestrated by water is a symphony that maintains an equilibrium essential for the intricate biochemical processes within plant cells.

1.1.4 Disease Prevention

- Water's contribution extends to the realm of disease prevention. Hydrated plants exhibit resilience against stress, creating an environment less conducive to the proliferation of pathogens. This aspect of water's role underscores its protective function in maintaining crop health.

Water, acting as a shield, fortifies plants against the onslaught of diseases, establishing a robust defense mechanism that contributes to the overall health of agricultural ecosystems.

1.1.5 Soil Structure and Microbial Activity

- Water's influence on soil structure is akin to a master sculptor refining the canvas. The formation of soil aggregates and improved aeration are the outcomes, creating an environment conducive to microbial activity. Water fosters a dynamic rhizosphere where beneficial microorganisms thrive, contributing to nutrient cycling and soil health.

In the subterranean realm, water collaborates with soil particles to craft an environment teeming with microbial life. This unseen partnership is essential for the sustenance of a thriving agricultural ecosystem.

1.2 Role of Water in Crop Growth and Yield (Table 1)

Table 1: Role of Water in Crop Growth and Yield

S.No	**Aspect**	**Description**
1	Hydration	Water is a vital component of plant cells, facilitating nutrient uptake and biochemical processes essential for plant growth.
2	Photosynthesis	Water is a key reactant in photosynthesis, the process by which plants convert light energy into chemical energy, producing glucose and oxygen. Adequate water availability enhances photosynthetic efficiency.
3	Nutrient Transport	Water serves as a medium for the transport of nutrients from the soil to the plant roots. It helps in the movement of essential minerals, facilitating their absorption by the plant.
4	Cell Expansion	Water uptake by plant cells creates turgor pressure, leading to cell expansion. This contributes to plant growth, turgidity, and overall structural support.
5	Temperature Regulation	Water has a moderating effect on temperature, preventing extreme fluctuations. Adequate soil moisture helps regulate the temperature around plant roots, optimizing enzyme activity and metabolic processes.
6	Dissolving Minerals	Water solubilizes minerals in the soil, making them available for plant uptake. This is crucial for the supply of essential nutrients required for various physiological processes in plants.
7	Stress Resistance	Sufficient water helps plants cope with environmental stress, such as drought or high temperatures. Well-hydrated plants are more resilient and can better withstand adverse growing conditions.
8	Transpiration	The process by which water is drawn up through the plant and released into the atmosphere through small pores (stomata) in the leaves. Transpiration helps regulate water movement and cool the plant.
9	Fruit and Seed Formation	Water availability during the reproductive stages is critical for the development of fruits and seeds. Insufficient water during this period can lead to poor fruit set and reduced crop yield.
10	Yield and Crop Quality	Adequate water supply is directly linked to crop yield and quality. Water stress during critical growth stages can result in yield reduction, smaller fruit size, and lower nutritional quality of crops.

1.2.1 Germination and Establishment

- The role of water during germination extends beyond initiation; it orchestrates a symphony of biochemical reactions. Enzymatic activation, triggered by water, sets the stage for the conversion of stored energy in seeds, ensuring the successful emergence of seedlings and their establishment in the soil.

Water's role in germination is akin to a conductor initiating a grand performance. It triggers a cascade of events that mark the beginning of a plant's journey toward maturity.

1.2.2 Vegetative Growth

- In the expansive canvas of vegetative growth, water acts as both artist and medium. It fuels the expansion of leaves, stems, and roots, creating a sturdy foundation for plants. The development of a robust root system, facilitated by ample water, empowers plants to explore the soil for nutrients and anchor themselves securely against environmental challenges.

Water, as the brushstroke of growth, paints a vibrant picture of vegetative development. Its presence ensures the creation of a resilient framework for plants to flourish.

1.2.3 Reproductive Stage

- Water assumes a leading role during the reproductive phase, directing the ballet of flowering, pollination, and fruit development. Adequate water availability at this critical juncture is paramount, as it ensures the seamless progression of these intricate stages. Insufficient water during this phase can lead to poor fruit set and diminished crop yields.

The reproductive stage, choreographed by water, is a delicate dance where every drop contributes to the formation of fruits—a testament to the synergy between water and the continuity of plant life.

1.2.4 Yield Formation and Maturation

- The symphony of water's influence crescendos during yield formation and maturation. Its presence or absence during these critical stages determines the abundance or scarcity of the harvest. Water stress during these pivotal moments can result in reduced yields and compromised crop quality, highlighting its paramount importance.

As crops transition towards maturity, water emerges as the conductor guiding the final movements. Its role in yield formation shapes the grand finale of the agricultural lifecycle.

1.2.5 Water Use Efficiency

- Maximizing water use efficiency emerges as a modern-day agricultural imperative. Water, when used judiciously through optimized application techniques such as precision irrigation, becomes a valuable resource. Enhancing water use efficiency ensures that crops receive the optimal amount of water required for growth, minimizing wastage and contributing to sustainable agricultural practices.

In the quest for sustainable agriculture, water becomes the maestro orchestrating a harmonious balance between resource use and crop productivity. Precision irrigation emerges as a key player in this endeavor.

2. Types of Water Sources for Agriculture

In the vast realm of agricultural water resources, the diverse origins of water play a pivotal role in shaping the sustainability and productivity of farming practices (Table 2).

Table 2: Types of Water Sources for Agriculture

S. No	Water Source	Description
1	Surface Water	- Water from rivers, lakes, ponds, and reservoirs that is collected on the Earth's surface. Surface water is often used for irrigation through channels, canals, and other conveyance systems.
2	Groundwater	- Water stored beneath the Earth's surface in aquifers. Groundwater is accessed through wells and is a significant source of irrigation water, particularly in areas where surface water is limited.
3	Rainwater Harvesting	- Collection and storage of rainwater for agricultural use. This can be done through the use of rainwater harvesting systems, such as catchment surfaces, gutters, and storage tanks.
4	Reservoirs and Dams	- Man-made structures designed to store and regulate water flow. Reservoirs and dams are used to capture and store water during periods of high flow for later release during dry seasons for irrigation.
5	Wells	- Vertical or inclined holes drilled into the ground to access groundwater. Different types of wells, such as tube wells and dug wells, are used to extract water for agricultural purposes.
6	Canals and Channels	- Artificial waterways designed to transport water from rivers, reservoirs, or other sources to agricultural fields. Canals and channels are used to distribute water for irrigation over large areas.
7	Ponds and Tanks	- Small reservoirs or impoundments created by constructing earthen embankments. Ponds and tanks store water for agricultural use, providing a local and decentralized water source for irrigation.
8	Desalination	- Process of removing salt and other impurities from seawater to make it suitable for irrigation. Desalination is employed in arid coastal regions where conventional freshwater sources are limited.
9	Recycled or Treated Water	- Treated wastewater from urban or industrial sources that is reused for agricultural irrigation. Recycling and treating water contribute to sustainable water management practices in agriculture.
10	Snowmelt Water	- Water released from melting snow during the spring. In mountainous regions, snowmelt water contributes to streamflow and can be harnessed for irrigation in downstream agricultural areas.

2.1 Surface Water: Rivers, Lakes, and Reservoirs

Rivers in Agriculture

Rivers, with their meandering flow and life-sustaining waters, form the backbone of agricultural civilizations. Their significance in providing a continuous source of water for crops dates back to the earliest days of agriculture.

Rivers are nature's conduits, channeling freshwater from highland areas to the plains, creating fertile valleys along their banks. In the agricultural narrative, rivers become the arteries that nourish vast expanses of cropland. The cycle begins in the upstream reaches, where precipitation and melting snow contribute to the river's flow. As the river winds its way through diverse landscapes, it becomes a lifeline for agricultural communities.

Lakes: Reservoirs for Agricultural Prosperity

Lakes, with their tranquil expanses, are not only picturesque but also vital reservoirs that sustain agricultural communities. The significance of lakes in agriculture extends beyond their aesthetic appeal; they are strategic sources of water storage and supply.

Lakes serve as natural reservoirs, collecting and holding water over time. In regions where rainfall is sporadic, lakes become invaluable assets for farmers. Their ability to store water during wet periods ensures a continuous supply during dry spells, mitigating the impact of water scarcity on crops. Lakes, like pearls on the agricultural necklace, add a layer of resilience to farming practices.

Agricultural Reservoirs as Water Banks

Reservoirs, human-engineered water bodies, stand as testaments to our ability to manipulate and harness water resources for agricultural prosperity. These artificial lakes play a crucial role in water management strategies, serving as reservoirs that store water for controlled release.

Constructed with the dual purpose of flood control and water storage, reservoirs are key components of agricultural infrastructure. Their construction involves the creation of a dam, which not only prevents downstream flooding but also creates a large storage area. This stored water can then be released strategically to irrigate crops during dry periods, turning arid landscapes into flourishing farmlands.

Navigating the Agricultural Waterways: The Role of Surface Water in Farming

The utilization of surface water in agriculture is a dynamic interplay between natural sources and human interventions. Understanding the intricate dance of rivers, lakes, and reservoirs is essential for sustainable water management practices.

2.1.1 Harnessing River Dynamics: From Tributaries to Farmlands

Rivers, with their vast networks of tributaries, weave intricate patterns that irrigate vast agricultural landscapes. The utilization of river water in agriculture involves a nuanced understanding of river dynamics and strategic interventions to optimize its benefits.

Tributaries, the smaller streams that contribute to river flow, are integral to agricultural water supply. Their management involves a delicate balance between upstream and downstream needs. Upstream regions, where precipitation and snowmelt occur, impact the volume and timing of water flow downstream. Managing this variability is essential for ensuring a consistent and reliable water supply for agriculture.

2.1.2 Lakes as Agricultural Reservoirs

Lakes, with their natural beauty, serve as more than visual delights—they are strategic reservoirs that contribute to the sustainable management of agricultural water resources.

The utilization of lakes in agriculture involves a dual strategy: maintaining the ecological balance of the lake while extracting water for irrigation. Sustainable practices include regulating water withdrawal to prevent ecological harm and promoting recharging mechanisms, such as rainwater and runoff, to maintain the lake's water levels. This delicate equilibrium ensures that lakes continue to play their role as dependable water sources for agriculture.

2.1.3 Reservoir Management: Engineering Water Security

Reservoirs, engineered to control water flow and availability, are pivotal in ensuring water security for agriculture. The management of reservoirs involves a careful orchestration of storage, release, and conservation to meet the demands of both crops and ecosystems.

Striking a balance between storing enough water for agricultural needs and maintaining sufficient reserves for ecological systems is a challenge in reservoir management. Techniques such as controlled releases, which mimic natural flow patterns, aim to minimize the impact on downstream ecosystems while meeting the water demands of farmlands. Additionally, incorporating sedimentation management strategies ensures the long-term viability of reservoirs.

2.1.4 Challenges and Opportunities: Navigating the Surface Water Landscape

The utilization of surface water in agriculture comes with a set of challenges, ranging from water scarcity to environmental impacts. However, these challenges also present opportunities for innovative solutions and sustainable practices.

Challenges

- Water Scarcity: In regions with erratic rainfall patterns and increasing water demand, ensuring a consistent supply of surface water for agriculture becomes a significant challenge. Climate change exacerbates this issue, leading to unpredictable precipitation and more frequent droughts.
- Ecological Impact: Unregulated water withdrawal from rivers, lakes, and reservoirs can have detrimental effects on aquatic ecosystems. Altering natural flow patterns, reducing water levels, and disrupting sediment transport can harm fish habitats and degrade water quality.
- Infrastructure Maintenance: Reservoirs and dams, essential for managing surface water, require regular maintenance. The accumulation of sediment, structural wear and tear, and changes in sedimentation patterns necessitate ongoing investments in infrastructure.

Opportunities

- Water Conservation Practices: Implementing water conservation practices, such as precision irrigation and soil moisture management, can optimize water use efficiency in agriculture. These practices not only enhance crop productivity but also contribute to sustainable water management.
- Technological Innovation: Harnessing technology, such as remote sensing and data analytics, allows for real-time monitoring of water availability and demand. Smart irrigation systems and sensor-based technologies enable farmers to make informed decisions, minimizing water wastage.
- Ecosystem-Friendly Approaches: Adopting ecosystem-friendly approaches, such as environmental flow management, helps mitigate the ecological impact of water extraction. Balancing the needs of agriculture with the preservation of natural habitats ensures long-term sustainability.

2.2 Groundwater: Wells and Aquifers

Unlocking the Depths: Groundwater in Agricultural Resilience

Groundwater, hidden beneath the Earth's surface, stands as a silent but powerful ally in ensuring water security for agriculture. This section unravels the dynamics of groundwater, focusing on wells and aquifers as essential components of agricultural water management.

2.2.1 Wells: Tapping into Subsurface Wealth

Harvesting from Below: The Role of Wells in Agriculture

Wells, timeless structures that pierce the veil of the Earth, are critical conduits for accessing groundwater. Their significance in agriculture lies in their ability to tap into subsurface reservoirs, providing a reliable and consistent source of water for crops.

Wells have been integral to agricultural practices throughout history. The process of drilling or digging wells to access groundwater is a testament to human ingenuity in harnessing the Earth's hidden resources. Shallow wells, penetrating the upper layers of aquifers, are suitable for smaller-scale agricultural operations. These wells provide access to water close to the surface, supporting crops with more immediate hydration.

On the other hand, deep wells reach further into the depths of aquifers, extracting water from more significant depths. These wells are instrumental in meeting the water demands of larger agricultural areas, ensuring a more abundant and stable supply. The drilling of deep wells involves advanced techniques and equipment to reach below the Earth's surface and access groundwater reservoirs.

2.2.2 Aquifers: Earth's Underground Reservoirs

The Subterranean Treasure: Understanding Aquifers

Aquifers, vast underground geological formations, serve as natural reservoirs for groundwater. This subsection explores the characteristics of aquifers and their role in sustaining agriculture.

Aquifers are geological formations capable of storing and transmitting groundwater. They consist of permeable materials, such as gravel, sand, or rock, allowing water to flow through and be stored in significant quantities. Understanding the types of aquifers is crucial for effective groundwater management in agriculture.

Confined aquifers are bounded by impermeable layers, creating a distinct geological structure. The impermeable layers act as a barrier, confining the water within the aquifer. When wells tap into confined aquifers, the pressure within the aquifer may force water to rise to the surface spontaneously, a phenomenon known as artesian flow. This natural flow is advantageous for agriculture, providing a self-sustaining water supply.

Unconfined aquifers, in contrast, lack impermeable layers above them, allowing water to interact more directly with surface conditions. Wells accessing unconfined aquifers may not exhibit artesian flow, but they still provide valuable water resources. Understanding the dynamics of unconfined aquifers is crucial for sustainable groundwater extraction, ensuring that water levels are maintained to support agricultural needs.

2.3 Rainwater Harvesting and Management

Nature's Bounty: Harnessing Rainwater for Agriculture

Rainwater, a gift from the skies, holds untapped potential for sustainable agriculture. This section explores the art and science of rainwater harvesting and its integral role in agricultural water management.

2.3.1 Principles of Rainwater Harvesting

Capturing Liquid Gold: The Essence of Rainwater Harvesting

Rainwater harvesting involves collecting and storing rainwater for later use. This subsection delves into the fundamental principles that guide effective rainwater harvesting practices.

Key Components

- **Catchment Area:** Surfaces such as rooftops or impermeable ground where rainwater is collected.

  The catchment area serves as the starting point for rainwater harvesting. Rooftops of buildings, greenhouses, or any impermeable surface can act as effective catchment areas. Rainwater falling on these surfaces is directed towards the collection system, initiating the harvesting process.

- **Conveyance System:** Gutters and downspouts that channel rainwater from the catchment area.

  The conveyance system plays a crucial role in efficiently channeling rainwater from the catchment area to storage. Gutters, installed along the edges of rooftops, collect rainwater and direct it to downspouts. These downspouts transport the water downwards, guiding it to storage tanks or other collection points.

- **Storage Tanks:** Containers for holding harvested rainwater, ensuring a steady supply during dry periods.

  Storage tanks are the reservoirs that hold the collected rainwater. These tanks come in various sizes and materials, ranging from small barrels to large underground cisterns. Properly designed storage systems ensure that rainwater is available when needed, even during periods of limited or no rainfall.

2.3.2 Techniques for Rainwater Management

From Rooftops to Fields: Applying Rainwater in Agriculture

Effectively managing harvested rainwater involves strategic techniques for its distribution and utilization in agricultural settings.

Distribution Techniques

- **Drip Irrigation:** Precisely delivering small amounts of water to individual plants, maximizing efficiency and minimizing wastage.

  Drip irrigation is a highly efficient technique that delivers water directly to the root zone of plants. A network of tubing and emitters ensures that water is applied precisely where needed, reducing evaporation and minimizing water wastage. This method is particularly beneficial for crops that require precise and controlled irrigation.

- **Micro-Sprinklers:** Emitting fine droplets, micro-sprinklers provide uniform coverage, suitable for a variety of crops.

  Micro-sprinklers are designed to distribute water in fine droplets, creating a mist-like spray. This method ensures uniform coverage over a larger area, making it suitable for a variety of crops. The even distribution of water helps in maintaining consistent soil moisture levels, promoting healthy plant growth.

On-Farm Rainwater Management

- **Contour Plowing:** Shaping fields to follow the natural contours of the land, preventing rainwater runoff and promoting absorption.

  Contour ploughing is a land management technique that involves shaping fields along the natural contours of the landscape. This practice helps in preventing rainwater runoff, allowing water to be absorbed into the soil. By minimizing soil erosion and enhancing water retention, contour ploughing contributes to sustainable agricultural practices.

- **Rain Gardens:** Landscaped areas designed to capture and utilize rainwater, promoting biodiversity and reducing soil erosion.

  Rain gardens are intentional landscaping features designed to capture and manage rainwater runoff. These gardens are strategically positioned to collect rainwater, preventing it from flowing into storm drains. The collected water supports the growth of vegetation in the garden, contributing to biodiversity and reducing soil erosion.

3: Water Management Practices in Agriculture

3.1 Irrigation Systems and Techniques

Nurturing the Fields: Precision in Water Delivery

Irrigation, the deliberate application of water to soil, stands as a cornerstone in the arsenal of agricultural water management. This section explores various irrigation systems and techniques, each a testament to human ingenuity in optimizing water use for crop growth.

3.1.1 Traditional Furrow Irrigation: Channeling the Flow

A Time-Honored Approach: Furrows as Water Conduits

Traditional furrow irrigation traces its roots back through centuries, embodying simplicity in water distribution. In this method, water flows through shallow channels or furrows between crop rows, providing hydration to the soil. While a historic practice, furrow irrigation faces challenges related to water efficiency and soil erosion.

Challenges

- **Uneven Water Distribution:** Traditional furrow irrigation often leads to uneven water distribution, with areas closer to the water source receiving more water than those farther away. This inconsistency can impact crop growth and yield.
- **Soil Erosion:** The flow of water in furrows has the potential to erode soil, especially on sloping terrain. Soil erosion poses threats to both the structural integrity of the field and water quality.

Optimization Techniques

- **Leveling Fields:** Ensuring uniform field topography through leveling helps mitigate the issue of uneven water distribution. Level furrow systems enhance water efficiency and promote uniform crop growth.
- **Cover Crops:** Introducing cover crops in furrow-irrigated fields helps prevent soil erosion. The cover crops act as a protective layer, reducing the impact of water flow on the soil surface.

Evolution of Furrow Irrigation

Traditional furrow irrigation, a practice rooted in ancient agricultural traditions, has undergone evolution with the integration of modern technologies. Today, precision leveling equipment ensures more uniform furrow depths, addressing the challenge of uneven water distribution. Additionally, the use of cover crops is being optimized through research into crop selection and planting patterns, enhancing soil conservation efforts.

3.1.2 Sprinkler Irrigation: Precise Aqueous Showers

The Aerial Symphony: Sprinklers as Water Maestros

Sprinkler irrigation transforms the act of watering into a mesmerizing display, with water droplets dancing through the air. This method involves spraying water over the crops in a manner that mimics natural rainfall. Sprinkler systems offer advantages in water efficiency, adaptability, and reduced soil erosion.

Advantages

- **Uniform Water Distribution:** Sprinklers excel in providing uniform water coverage across the entire field. The controlled spray ensures that each plant receives a consistent amount of water, promoting even growth.
- **Adaptability to Various Crops:** Sprinkler systems are versatile and can be adapted to different crop types and field sizes. Their flexibility makes them suitable for a wide range of agricultural settings.
- **Reduced Soil Erosion:** Unlike furrow irrigation, sprinkler systems minimize soil erosion since the water is applied directly to the crop canopy. This protects the soil structure and prevents runoff.

Types of Sprinkler Systems

- **Rotary Sprinklers:** These systems have rotating heads that release water in a circular pattern. They are suitable for large fields and provide even coverage.
- **Fixed Sprinklers:** Stationary sprinklers are positioned in specific locations and cover a fixed area. They are ideal for smaller fields or areas with irregular shapes.

Enhancements in Sprinkler Technology

Sprinkler technology has evolved to incorporate innovations aimed at maximizing water efficiency. Advanced nozzle designs and pressure regulation systems contribute to more precise water distribution. Additionally, the use of weather-based sensors and automation enhances the adaptability of sprinkler systems to changing environmental conditions, optimizing water use.

3.1.3 Drip Irrigation: Targeted Hydration Precision

Pinpoint Hydration: Drip Systems as Water Artisans

Drip irrigation epitomizes precision in water delivery, providing targeted hydration directly to the root zone of plants. This method minimizes water wastage and optimizes resource use, making it an increasingly popular choice for sustainable agriculture.

Advantages

- **Water Efficiency:** Drip irrigation significantly reduces water wastage by delivering water directly to the base of plants. This targeted approach minimizes evaporation and runoff.
- **Nutrient Management:** Drip systems can be integrated with nutrient delivery, allowing for precise control over fertilizers. This enhances nutrient uptake by plants, promoting healthy growth.

- **Weed Control:** Since water is delivered only to the intended crop area, weed growth is minimized. Drip irrigation contributes to weed control, reducing competition for water resources.

Components of Drip Systems

- **Drip Emitters:** Devices that release water droplets at a controlled rate, ensuring uniform distribution.
- **Tubing and Pipes:** Conduits that transport water from the source to the emitters. These components are designed for durability and resistance to clogging.
- **Filters and Pressure Regulators:** Essential components that prevent clogging and regulate water pressure, ensuring consistent performance.

Drip System Design Considerations

- **Spacing of Emitters:** The distance between emitters determines the coverage area. Proper spacing ensures that each plant receives sufficient water.
- **Water Quality:** Since drip systems are susceptible to clogging, maintaining water quality is crucial. Filtration systems help prevent debris and sediment from clogging emitters.

Technological Advancements in Drip Irrigation

Drip irrigation has witnessed significant advancements driven by technology. Smart irrigation systems equipped with sensors and controllers allow for real-time monitoring and adjustment of water delivery. This integration of precision agriculture principles enhances the efficiency and sustainability of drip irrigation, making it a vital tool in modern water management.

3.2 Efficient Water Use and Conservation

Harmonizing Agriculture with Nature: The Quest for Water Efficiency

Efficient water use in agriculture is not just a goal but a necessity in the face of growing water scarcity and the imperative of sustainable practices. This section explores strategies and technologies aimed at maximizing water utilization while minimizing waste.

3.2.1 Precision Agriculture: Targeting Water Application

Fine-Tuning Water Delivery: Precision Agriculture as a Paradigm Shift

Precision agriculture leverages technology to tailor farming practices to the specific needs of each plant. In terms of water management, it involves the precise application of water where and when it is needed. The integration of sensors, GPS technology, and data analytics enables farmers to make informed decisions, optimizing water use and minimizing environmental impact.

Components of Precision Agriculture

- **Sensor Technologies:** Soil moisture sensors, weather stations, and aerial imagery contribute real-time data, allowing farmers to assess the moisture content of the soil and plan irrigation accordingly.
- **Variable Rate Technology (VRT):** This technology enables the adjustment of water application rates based on field variability, ensuring that each section receives the optimal amount of water.
- **Decision Support Systems:** Analytical tools that process data and provide recommendations, guiding farmers in making informed decisions about water application and crop management.

Benefits of Precision Agriculture

- **Water Conservation:** By precisely tailoring water application, precision agriculture reduces water wastage and promotes optimal plant growth.
- **Resource Efficiency:** The targeted use of water, fertilizers, and other inputs enhances overall resource efficiency, contributing to sustainable farming practices.
- **Yield Optimization:** Precision agriculture can lead to improved crop yields as it maximizes the effectiveness of water and other inputs.

Challenges and Innovations

Despite the benefits, the adoption of precision agriculture faces challenges such as initial investment costs, technological literacy among farmers, and data security concerns. Ongoing research aims to address these challenges and further enhance the accessibility and effectiveness of precision agriculture tools.

3.2.2 Water Harvesting and Storage

Preserving Raindrops: Water Harvesting as a Sustainable Practice

Water harvesting involves capturing and storing rainwater for agricultural use. This ancient practice has seen a resurgence with modern techniques and technologies, offering a sustainable solution to augment water resources for farming.

Methods of Water Harvesting

- **Contour Trenches:** Dug along the contour lines of the land, these trenches capture and slow the flow of rainwater, allowing it to percolate into the soil.

- **Check Dams:** Small dams constructed across ephemeral streams or gullies help impound water during rainfall, allowing it to infiltrate the soil and recharge groundwater.
- **Roof Water Harvesting:** Collecting rainwater from rooftops and storing it in tanks for later agricultural use.

Benefits of Water Harvesting

- **Groundwater Recharge:** Water harvesting contributes to the replenishment of groundwater, ensuring a sustainable supply for wells and boreholes.
- **Reduced Soil Erosion:** By slowing down the movement of rainwater, water harvesting minimizes soil erosion and preserves the topsoil.
- **Community Resilience:** In regions with erratic rainfall patterns, water harvesting provides a reliable water source, enhancing the resilience of farming communities.

Innovations in Water Harvesting

Ongoing innovations in water harvesting technologies include the development of smart harvesting systems that optimize the collection and distribution of rainwater. These systems may incorporate sensors to monitor rainfall patterns, automated valves for precise water release, and data analytics for efficient water management.

3.3 Challenges and Solutions in Water Management

Navigating the Waters: Addressing Challenges in Agricultural Water Use

As agriculture grapples with evolving climatic conditions and growing demands, addressing challenges in water management becomes paramount. This section explores the hurdles faced by farmers and presents innovative solutions to safeguard water resources.

3.3.1 Climate Change Impacts on Water Availability

Rising Temperatures, Shifting Rainfall: Adapting to a Changing Climate

Climate change poses significant challenges to water availability for agriculture. Increasing temperatures, altered precipitation patterns, and more frequent extreme weather events necessitate adaptive measures to secure water resources.

Adaptive Strategies

- **Drought-Resistant Crops:** Research and development efforts focus on breeding crop varieties that can withstand water scarcity and thrive in challenging climatic conditions.

- **Climate-Responsive Irrigation Scheduling:** Utilizing advanced forecasting models and real-time weather data, farmers can optimize irrigation schedules to align with changing climate patterns.
- **Water Storage Infrastructure:** Constructing resilient water storage facilities, such as ponds and reservoirs, helps buffer against variability in precipitation and ensures a stable water supply.

Innovations in Climate Adaptation

Research institutions and agricultural organizations are actively engaged in developing climate-resilient farming practices. This includes the introduction of climate-smart crops, precision weather forecasting tools, and community-based adaptation strategies that empower farmers to mitigate the impacts of climate change on water resources.

3.3.2 Balancing Water Demand and Supply

Striking the Equilibrium: Meeting Water Needs Sustainably

The growing demand for water in agriculture necessitates a delicate balance between consumption and conservation. This challenge requires innovative solutions that promote sustainable water use while meeting the food demands of a burgeoning global population.

Solutions for Balance

- **Water-Efficient Technologies:** The adoption of precision irrigation, drip systems, and other water-efficient technologies minimizes water wastage and enhances overall agricultural productivity.
- **Crop Diversification:** Exploring and promoting crops that are well-suited to the local climate and require less water can contribute to balancing water demand.
- **Water Pricing and Policies:** Implementing fair and effective water pricing mechanisms, coupled with supportive policies, encourages judicious water use and incentivizes conservation practices.

Community Engagement and Education

Empowering farming communities with knowledge about water conservation practices is vital. Educational programs, extension services, and farmer cooperatives play a crucial role in disseminating information about sustainable water management practices. Additionally, fostering a sense of community responsibility encourages collective efforts in water conservation.

4: Impact of Climate Change on Water Resources

Climate change stands as a formidable force reshaping the dynamics of water resources, challenging traditional patterns, and presenting novel hurdles for agriculture.

4.1 Changing Precipitation Patterns

Dancing Raindrops: The Flux of Precipitation in a Warming World

The symphony of changing precipitation patterns orchestrated by climate change significantly impacts water availability for agriculture. Historically reliable rainfall patterns are becoming increasingly erratic, posing challenges for farmers worldwide.

Impact on Agriculture

Changing precipitation patterns disrupt the timing and distribution of rainfall, influencing planting seasons, crop growth, and harvest cycles. Prolonged droughts, intensified storms, and altered monsoon patterns are manifestations of this climate-induced shift, directly affecting agricultural productivity.

Mitigation Strategies

1. **Water Harvesting Systems:** Investing in robust water harvesting infrastructure becomes imperative to capture and store rainfall during periods of abundance. This stored water serves as a crucial buffer during dry spells, ensuring a continuous water supply for crops.

2. **Diversified Crop Selection:** Adapting to changing precipitation patterns involves reevaluating crop choices. Opting for drought-resistant varieties and exploring alternative crops better suited to evolving climate conditions can mitigate the risks associated with unpredictable rainfall.

3. **Smart Irrigation Practices:** Implementing precision irrigation techniques, such as drip irrigation and soil moisture sensors, helps optimize water use. This ensures that crops receive adequate hydration, even when traditional rainfall patterns become unreliable.

4. **Community-Based Rainwater Management:** Encouraging communities to adopt rainwater management practices fosters self-reliance. This can include communal rainwater harvesting structures and educational programs that empower communities to harness and manage rainwater effectively.

4.2 Temperature Effects on Water Availability

Rising Mercury, Shifting Tides: Temperature's Impact on Water Resources

As global temperatures soar, the intricate balance of water availability is further tipped. The interplay between rising temperatures and water resources poses multifaceted challenges to agriculture, demanding innovative solutions to safeguard crop health.

Impact on Agriculture

1. **Increased Evaporation:** Elevated temperatures accelerate water evaporation from soil and plant surfaces, intensifying the demand for irrigation. This puts additional stress on water sources, particularly in regions already grappling with water scarcity.

2. **Altered Snowmelt Patterns:** In colder regions, rising temperatures affect the timing and magnitude of snowmelt. This has downstream implications for water availability, affecting both agricultural irrigation and the replenishment of rivers and aquifers.

Adaptation Strategies

1. **Climate-Resilient Crop Varieties:** Developing and promoting crop varieties that can withstand higher temperatures is crucial. Research efforts focus on breeding heat-tolerant crops that maintain productivity even in the face of rising mercury.

2. **Efficient Irrigation Technologies:** Precision irrigation technologies, such as sprinkler and drip systems, become paramount. These technologies reduce water wastage through targeted water delivery, mitigating the impact of increased evaporation.

3. **Water Conservation Practices:** Implementing on-farm water conservation practices, such as mulching and cover cropping, helps retain soil moisture. These practices enhance the water-holding capacity of the soil, offering a buffer against the effects of increased temperatures.

4. **Urban Planning for Temperature Moderation:** Integrating green spaces and water bodies in urban planning helps mitigate the urban heat island effect. This, in turn, contributes to maintaining more moderate temperatures in surrounding agricultural areas.

4.3 Adaptation Strategies for Agriculture

Navigating Uncertain Waters: Adaptive Measures for Sustainable Agriculture

In the face of a changing climate, adaptation becomes the linchpin for the resilience of agriculture. This section explores holistic strategies that farmers can employ to adapt to the evolving landscape and secure water resources for sustainable food production.

Integrated Water Management

1. **Integrated Water Harvesting Systems:** Building on traditional water harvesting methods, modern integrated systems combine rainwater harvesting, runoff management, and groundwater recharge. These multifaceted systems enhance overall water availability on farms.

2. **Diversified Crop Management:** Embracing crop diversification strategies ensures that farms remain resilient to changing environmental conditions. Farmers can explore cultivating a mix of crops with varying water requirements, allowing for flexibility in response to climatic uncertainties.

3. **Smart Irrigation Solutions:** Leveraging technology, smart irrigation systems equipped with sensors, weather data, and automation optimize water use. These systems dynamically adjust irrigation schedules based on real-time conditions, promoting efficient water management.

4. **Climate-Adaptive Crop Practices:** Implementing climate-smart agricultural practices involves aligning crop management with climate predictions. This may include adjusting planting times, choosing heat-tolerant varieties, and adopting agroecological approaches tailored to changing environmental conditions.

5. **Community Engagement and Knowledge Transfer Initiatives:** Empowering farmers with knowledge about climate-resilient practices is pivotal. Extension services, farmer education programs, and knowledge-sharing platforms foster a collaborative approach, enabling communities to collectively adapt to changing climate patterns.

6. **Policy Support for Adaptation:** Governments play a crucial role in facilitating adaptation through supportive policies. Incentivizing sustainable water management practices, providing financial support for technological adoption, and integrating climate resilience into agricultural policies contribute to effective adaptation.

5. Water Quality in Agricultural Systems

5.1 Importance of Water Quality for Crop Health

Nourishing Crops, Ensuring Health: The Vital Role of Water Quality

In the intricate dance of agriculture, water quality emerges as a silent conductor, orchestrating the well-being of crops. Its significance lies not only in quenching the thirst of plants but in the profound influence it exerts on various facets of crop health.

Impact on Crop Growth

1. **Nutrient Absorption:** The journey of nutrients from soil to plant is intricately linked to water quality. High-quality water is the carrier, ensuring the seamless absorption of essential nutrients by plant roots. Conversely, water contaminated with excess salts or pollutants disrupts this nutrient highway, leading to deficiencies that manifest in stunted growth and diminished yields.
2. **Soil Structure:** Water quality extends its influence below the ground, shaping the very structure of the soil. Poor-quality water laden with salts or pollutants can degrade soil quality, jeopardizing its ability to provide a conducive environment for root development. The consequences are profound, affecting not only crop growth but also the long-term sustainability of agricultural lands.
3. **Disease Prevention:** Beyond nourishment, clean water is a powerful guardian against waterborne diseases that can afflict crops. Pathogens, microbes, and contaminants in water can serve as vectors for diseases, posing a dual threat to crop yield and quality. Recognizing the intimate link between water quality and disease prevention is pivotal for sustainable agriculture.

Technological Monitoring

1. **Sensor Technologies:** The marriage of agriculture and technology heralds a new era of precision. In the realm of water quality, sensors play a pivotal role. These technological sentinels monitor crucial parameters such as pH, nutrient levels, and microbial content. Real-time data empowers farmers with actionable insights, enabling timely interventions to maintain the optimal health of crops.
2. **Precision Irrigation:** The marriage of precision agriculture with water quality management is transformative. Precision irrigation, facilitated by sensor technologies, ensures that crops receive water precisely tailored to their needs. This not only conserves water but also minimizes the risk of exposing crops to contaminants present in excessive irrigation water.

5.2 Contaminants and Pollutants in Agricultural Water

Hidden Threats: Identifying and Understanding Water Contaminants

Agricultural water sources, once pristine, are vulnerable to an array of contaminants and pollutants. These hidden threats lurk in the very lifeline of agriculture, posing risks to both crop health and the safety of the food produced.

Common Contaminants

1. **Heavy Metals:** The echoes of industrial activities resonate in the water, carrying heavy metals like lead, cadmium, and mercury into agricultural landscapes. These silent intruders, when present in excessive amounts, insidiously accumulate in crops, posing not only a threat to plant health but also a potential hazard to consumers.

2. **Pesticides and Herbicides:** The tools of agriculture—pesticides and herbicides—often come at a cost. Runoff from treated fields can carry these chemical warriors into water sources, creating a cocktail of contaminants. While these chemicals serve the purpose of protecting crops, their unintended presence in water poses risks to aquatic ecosystems and the broader environment.

3. **Nitrogen and Phosphorus:** The very nutrients that sustain plant life can turn into agents of disruption. Excessive fertilizer use contributes to elevated levels of nitrogen and phosphorus in water. This nutrient runoff, while fueling algal blooms, sets in motion a cascade of events leading to eutrophication—a phenomenon that disrupts aquatic ecosystems and imperils water quality.

Microbial Contaminants

1. **Bacteria and Pathogens:** In the microscopic realms of water, unseen threats abound. Harmful bacteria and pathogens find their way into water sources, creating a perilous journey for both crops and humans. E. coli, a notorious waterborne bacterium, can stealthily contaminate fruits and vegetables through irrigation water, posing risks to consumers.

2. **Fungi and Mold:** Standing water and waterlogged soils create a haven for unseen adversaries—fungi and mold. These microbial menaces, thriving in damp conditions, can infiltrate crops, leading to post-harvest losses, compromised quality, and economic repercussions for farmers.

5.3 Water Treatment and Remediation Measures

Purifying the Lifeline: Techniques for Ensuring Clean Agricultural Water

In response to the relentless assault on water quality, farmers deploy an arsenal of techniques for water treatment and remediation. These measures stand as guardians, protecting both crops and the delicate balance of ecosystems.

Water Treatment Technologies

1. **Filtration Systems:** The journey to purify water often begins with filtration. Physical filtration through sand or mesh filters becomes the first line of defense. By sieving out particulate matter and sediment, these systems enhance water clarity, reducing the risk of clogging irrigation systems and safeguarding the integrity of water.
2. **Reverse Osmosis:** In the intricate dance of water purification, reverse osmosis emerges as a maestro. This advanced technique employs semi-permeable membranes to selectively remove ions, contaminants, and pathogens from water. The result is pristine water, purified and ready for its crucial role in sustaining crops.
3. **Activated Carbon Filtration:** The sorcery of activated carbon lies in its ability to attract and capture contaminants. In agricultural water treatment, activated carbon filtration proves invaluable. It effectively absorbs organic contaminants, pesticides, and certain chemicals, delivering water of enhanced quality by reducing their concentrations.

Bioremediation Techniques

1. **Constructed Wetlands:** Nature's wisdom becomes a blueprint for bioremediation. Constructed wetlands, inspired by the intricate ecosystems of natural wetlands, harness biological processes to break down and absorb contaminants in water. This harmonious interplay between flora, fauna, and microorganisms contributes to the restoration of water quality.
2. **Phytoremediation:** The plant kingdom, often underestimated, reveals its remedial prowess. Phytoremediation capitalizes on the ability of certain plants to absorb and accumulate contaminants from water. Strategically planting these green sentinels in agricultural landscapes becomes a sustainable approach to purifying water.

Integrated Water Management Practices

1. **Crop Rotation and Cover Cropping:** The fields become a canvas for strategic maneuvers. Crop rotation and cover cropping emerge as time-tested strategies for integrated water management. Beyond their role in soil health, these practices curtail soil erosion and runoff, reducing the transport of contaminants into water sources.

2. **Precision Irrigation:** The symphony of precision extends its melody to the realm of water management. Precision irrigation systems, finely tuned to the needs of crops, not only conserve water but also minimize the risk of spreading contaminants. Targeted water delivery becomes an art, optimizing both water use efficiency and crop health.

6. Innovations and Technologies in Agricultural Water Use

In the ever-evolving landscape of agriculture, innovations, and technologies play a pivotal role in shaping the efficient and sustainable use of water. realm of agricultural water use.

6.1 Smart Irrigation Technologies

Harnessing Intelligence: The Dawn of Smart Irrigation

The marriage of agriculture and technology has given rise to smart irrigation, a transformative approach that transcends traditional watering methods. This section unravels the layers of innovation woven into smart irrigation technologies, reshaping the way farmers interact with water.

Sensor-based Precision

1. **Soil Moisture Sensors:** At the heart of smart irrigation lies the ability to discern the moisture content of the soil. Soil moisture sensors, strategically placed in fields, provide real-time data on the water levels in the soil. This intelligence enables farmers to tailor irrigation schedules precisely to the needs of crops, avoiding both water wastage and moisture-related stresses on plants.

2. **Weather-based Irrigation Controllers:** The atmosphere becomes a guiding force in smart irrigation. Controllers equipped with weather data assimilate information on temperature, humidity, and precipitation forecasts. By integrating this data, smart irrigation systems adjust watering schedules dynamically, aligning with the unique climatic conditions of each day.

Automated Precision

1. **Drip Irrigation Systems:** The precision of water delivery reaches new heights with drip irrigation. This system dispenses water directly to the root zones of plants, minimizing wastage through evaporation or runoff. By automating the delivery process, drip irrigation ensures that each drop of water serves its purpose efficiently.

2. **Sprinkler Systems with Variable Rate Technology (VRT):** Precision meets versatility in sprinkler systems enhanced by Variable Rate Technology. These systems adjust the rate of water application based on the specific needs of different areas within a field. By tailoring water distribution, VRT optimizes resource use and maximizes the impact of irrigation.

Remote Monitoring and Control

1. **Smartphone-enabled Irrigation Control:** The power to manage irrigation is at the fingertips of farmers. Smartphone-enabled irrigation control systems empower farmers to monitor and control irrigation remotely. This not only enhances convenience but also allows for timely adjustments, fostering proactive water management.

2. **Satellite-based Irrigation Monitoring:** The vantage point shifts beyond the fields to the vastness of space. Satellite-based monitoring provides a bird's-eye view of agricultural landscapes. By analyzing satellite imagery, farmers gain insights into crop health, water usage, and irrigation effectiveness on a large scale.

6.2 Precision Agriculture for Water Efficiency

Fine-Tuning Farming: Precision Agriculture's Water Symphony

Precision agriculture, a transformative approach that harmonizes technology with agricultural practices, takes center stage in enhancing water efficiency. This section navigates through the dimensions of precision agriculture, exploring how it fine-tunes farming operations to achieve optimal water use.

Advanced Mapping and Monitoring

1. **GIS (Geographic Information System) Mapping:** The canvas of precision agriculture is painted with the strokes of GIS mapping. This technology integrates spatial data to create detailed maps of agricultural landscapes. By understanding the variability in soil types, topography, and water availability, farmers can make informed decisions on water allocation and management.

2. **Remote Sensing Technologies:** Satellites become the eyes in the sky for precision agriculture. Remote sensing technologies capture data on crop health, soil moisture, and water stress. This real-time information equips farmers with insights that guide targeted interventions, ensuring water is applied where and when it is most needed.

Variable Rate Application

1. **Variable Rate Fertilization:** Precision agriculture extends its finesse to nutrient application. Variable rate fertilization systems analyze soil nutrient levels and apply fertilizers in varying quantities across a field. This not only optimizes nutrient use but also indirectly influences water efficiency by promoting healthy and resilient crops.

2. **Variable Rate Seeding:** The precision mantra echoes in the act of planting itself. Variable rate seeding adjusts the density of seeds based on soil

conditions, optimizing crop establishment. By aligning seeding rates with soil moisture levels, farmers enhance water efficiency while promoting uniform crop growth.

Automated Machinery

1. **Autonomous Tractors and Equipment:** The agricultural workforce expands to include autonomous allies. Autonomous tractors equipped with GPS and AI technologies navigate fields with precision, optimizing the use of water resources. These machines not only perform tasks efficiently but also contribute to minimizing soil compaction, fostering healthier soil structure.

2. **Robotics for Crop Management:** The micro-level precision of robotics comes to the fore in crop management. Robotics assist in tasks such as weed control, disease monitoring, and even targeted application of water and nutrients. By precisely addressing specific areas of need, robotic systems contribute to water savings while promoting crop health.

6.3 Emerging Trends in Agricultural Water Management

Navigating the Future: Trends Shaping Tomorrow's Agriculture

The horizon of agricultural water management is marked by emerging trends that promise to redefine the future. This section peers into the crystal ball, unveiling the trends that hold the potential to revolutionize how water is utilized in agriculture.

Data-driven Decision Support Systems

1. **Artificial Intelligence (AI) and Machine Learning:** The era of intelligent agriculture unfolds with the integration of AI and machine learning. These technologies analyze vast datasets, offering predictive insights into crop behavior, weather patterns, and water requirements. The result is a proactive approach to decision-making that optimizes water use efficiency.

2. **Blockchain for Water Transactions:** The digital ledger technology of blockchain steps into the realm of water management. By creating transparent and secure platforms for water transactions, blockchain facilitates efficient water trading among farmers. This not only enhances water access but also incentivizes water conservation.

Sustainable Irrigation Practices

1. **Regenerative Agriculture:** A holistic approach to farming, regenerative agriculture, gains momentum. This practice focuses on building soil health, increasing water retention, and promoting biodiversity. By adopting regenerative principles, farmers contribute to sustainable water management while fostering resilient and productive agroecosystems.

2. **Closed-loop Irrigation Systems:** The concept of circularity permeates agricultural water management. Closed-loop irrigation systems capture and recycle water within a controlled environment. This closed-cycle approach minimizes water losses and creates a sustainable loop that aligns with the principles of circular agriculture.

Climate-resilient Technologies

1. **Climate-smart Crop Varieties:** The seeds of resilience are sown in the development of climate-smart crop varieties. These varieties are tailored to withstand the challenges posed by climate change, including altered precipitation patterns and increased temperatures. By cultivating crops that thrive in changing conditions, farmers enhance water use efficiency.

2. **Aerial and Satellite-based Monitoring:** The eyes in the sky evolve with advanced aerial and satellite-based monitoring. High-resolution imagery and real-time data from drones and satellites contribute to precise monitoring of water usage, crop health, and overall farm management. This technology-driven approach empowers farmers with actionable insights for efficient water management.

7. Future Prospects and Challenges

7.1 Sustainable Water Use for Future Generations

The quest for sustainable water use transcends the boundaries of the present, casting a legacy for the generations that follow. This section delves into the principles and practices that pave the way for a future where water is not just a resource but a testament to responsible stewardship.

Holistic Water Management

1. **Integrated Water Resource Management:** The silos of water management dissipate in favor of an integrated approach. Integrated Water Resource Management (IWRM) emerges as a holistic framework that considers the interconnectedness of water sources, encompassing rivers, lakes, aquifers, and precipitation. By harmonizing human needs with ecological sustainability, IWRM becomes a cornerstone for securing water for the future.

2. **Nature-based Solutions:** The wisdom of nature becomes a guide in sustainable water use. Nature-based solutions, including the restoration of wetlands, afforestation, and soil conservation, harness natural processes to enhance water availability and quality. By emulating nature's resilience, these solutions offer a blueprint for coexisting with water ecosystems.

Technological Innovation

1. **Advanced Water Recycling Technologies:** Embracing the circular economy, advanced water recycling technologies step into the limelight. From decentralized wastewater treatment to the purification of agricultural runoff, these innovations reimagine water as a reusable resource. By closing the loop, these technologies contribute to a sustainable water cycle that transcends conventional boundaries.

2. **Precision Agriculture 2.0:** The precision revolution in agriculture undergoes an evolution. Precision Agriculture 2.0 integrates advanced technologies, including AI, robotics, and nanosensors, to further enhance water efficiency. By fine-tuning irrigation, nutrient application, and crop management with unprecedented precision, this evolution heralds a future where every drop of water maximizes its impact.

7.2 Addressing Water Scarcity in a Changing Climate

Adaptive Water Management

1. **Climate-resilient Infrastructure:** The blueprint for water infrastructure undergoes a transformation to withstand the shocks of climate change. Climate-resilient infrastructure, including resilient dams, water storage systems, and irrigation networks, becomes a linchpin in ensuring water availability even in the face of extreme weather events.

2. **Drought-resistant Crop Varieties:** The fields become a battleground against drought. Drought-resistant crop varieties, bred through advanced genetic techniques, offer a frontline defense against water scarcity. By cultivating crops that can thrive with minimal water inputs, agriculture becomes more resilient to prolonged dry spells.

Water-Energy-Food Nexus

1. **Nexus Thinking for Sustainability:** The interconnected web of water, energy, and food takes center stage in sustainability discourse. Nexus thinking recognizes the interdependencies between these essential elements, emphasizing the need for integrated planning. By optimizing the use of water in agriculture, energy production, and food systems, nexus thinking becomes a strategic approach to mitigate water scarcity.

2. **Renewable Energy for Desalination:** The synergy between water and energy unfolds in the realm of desalination. Renewable energy sources, such as solar and wind power, become integral to desalination processes. By harnessing sustainable energy for water desalination, this approach addresses both water scarcity and energy sustainability.

7.3 Collaborative Approaches for Global Water Security

In a world bound by interconnected water systems, collaborative approaches emerge as imperative for ensuring global water security. This section explores the dynamics of collaboration, transcending borders and ideologies to forge pathways that secure the future of water on a global scale.

International Cooperation

1. **Transboundary Water Management:** Rivers and aquifers seldom recognize geopolitical boundaries. Transboundary water management becomes a diplomatic imperative, fostering cooperation among neighboring nations to sustainably manage shared water resources. By transcending political divides, this approach promotes regional stability and resilience.

2. **International Water Treaties and Agreements:** The ink on international agreements becomes a seal of cooperation. Treaties and agreements, such as the United Nations Watercourses Convention and the Ramsar Convention on Wetlands, serve as frameworks for collaborative water management. By establishing shared principles and commitments, these agreements contribute to the harmonious use of global water resources.

Public-Private Partnerships

1. **Private Sector Engagement in Water Management:** The private sector steps onto the stage of water stewardship. Public-private partnerships leverage the efficiency and innovation of private entities to address water challenges. From sustainable agricultural practices to water infrastructure development, these partnerships create synergies that amplify the impact of water management initiatives.

2. **Corporate Water Stewardship:** Businesses become custodians of water resources. Corporate water stewardship extends beyond compliance to proactive initiatives that promote water conservation and community engagement. By aligning business practices with sustainable water use, corporations contribute to the broader goal of global water security.

3

Adaptation Strategies for Climate -Resilient Wasteland Development and Management

[1]Savita Jangde, [2]Wajid Hasan, [3]Sheetanshu Gupta, [4]Niranjan B.N. and [4]Kh. Chandrakumar Singh

[1]Department of Plant Physiology, Institute of Agricultural Sciences B.H.U. Varanasi, U.P
[2]KVK Jehanabad, Bihar Agricultural University Sabour, Bihar
[3]Department of Biotechnology, Institute of Technology and Management BKT, Lucknow, U.P.
[4]Dakshina Kannada Milk Union (KMF), Mangalore, Karnataka, India
[5]VCSGU University of Horticulture and Forestry, Bharsar, Uttarakhand

1. Wasteland

Introduction

In the vast tapestry of Earth's landscapes, certain areas have been relegated to the sidelines, deemed unproductive and often dismissed as wastelands. However, the term "wasteland" carries a complexity that extends beyond mere barrenness; it encapsulates a narrative of neglect, degradation, and, crucially, the latent potential for transformation.

The Dynamic Nature of Wasteland

At its core, a wasteland is a geographical space characterized by low productivity, often manifesting in barren, unproductive landscapes. The term, however, fails to encapsulate the dynamic and evolving nature of these areas. Wastelands can result from a confluence of natural processes and human activities, making them heterogeneous entities requiring careful examination (fig 1).

Fig. 1: Wasteland

Natural Forces at Play

Nature, in its unforgiving dynamics, can render landscapes inhospitable. Soil erosion, salinity, aridity, and other environmental factors contribute to the creation of wastelands. For instance, a region subjected to persistent drought may transform from a once-fertile land into a desolate expanse, its productivity stripped away by the relentless forces of nature.

Human Induced Wastelands

Equally significant in the genesis of wastelands are human activities. Deforestation, improper land management, industrialization, and mining are anthropogenic actions that leave lasting imprints on the land. The consequences can range from soil degradation to the pollution of water sources, creating landscapes labeled as wastelands due to their compromised ecological integrity.

Neglect and Marginalization

Crucial to the definition of wastelands is the societal perception that relegates these areas to the periphery of attention and resource allocation. Often overlooked and marginalized, wastelands become symbolic of regions deemed unworthy of investment or efforts towards rejuvenation. This societal neglect exacerbates the challenges faced by these landscapes, creating a self-perpetuating cycle of degradation.

Potential for Regeneration

While the term "wasteland" implies desolation, it also harbors an inherent potential for regeneration. The very designation suggests an opportunity for transformation, a prospect to breathe life back into landscapes that have been marginalized. It is this potential that renders wastelands as canvases awaiting the brushstrokes of strategic interventions and innovative approaches.

Strategic Interventions for Wasteland Regeneration

Wasteland development is not merely about restoring ecological balance; it is about harnessing the potential of these landscapes for sustainable development. Thus, the definition of wasteland is not static; it evolves with the acknowledgment that, given the right interventions, these areas can transcend their barren state.

Rejuvenating Ecological Wastelands

One facet of wastelands includes areas where natural ecosystems have been significantly altered or degraded. Deforested regions, habitats affected by soil erosion, and spaces suffering from habitat destruction fall into this category. The definition expands to recognize that ecological wastelands are not irreversible; they can be rejuvenated through strategic reforestation, soil conservation, and habitat restoration efforts.

Abandoned Agricultural Lands

Abandoned agricultural lands represent a substantial portion of wastelands. These areas often bear the scars of intensive farming practices, leading to soil degradation, nutrient depletion, and loss of biodiversity. The definition acknowledges that abandoned agricultural lands, though currently unproductive, hold the potential for transformation through sustainable farming practices, agroforestry, and initiatives aimed at soil improvement.

Industrial Wastelands

The imprint of industrial activities often marks landscapes as wastelands. Pollution, soil contamination, and altered ecosystems characterize these industrial wastelands. The definition recognizes that while industrialization is essential for progress, it must be accompanied by strategies for remediation, pollution control, and the integration of green technologies to mitigate ecological impacts.

Mining Sites

Mining activities, while vital for resource extraction, contribute significantly to the creation of wastelands. Soil degradation, habitat destruction, and water pollution are common consequences. The definition extends to acknowledge that mining sites, with careful reclamation initiatives, can become symbols of responsible resource extraction, emphasizing the need for reforestation, water management, and sustainable land use practices.

2. Concept and Development of Wasteland

The concept and development of wasteland stand as an intricate dance between the forces of nature, human interventions, and societal perspectives.

Conceptualizing Wasteland

At its essence, the concept of wasteland is not a static classification but a reflection of the dynamic processes shaping our planet. These areas, often characterized by barrenness, are in constant flux due to a myriad of natural and anthropogenic factors. Recognizing wastelands necessitates an understanding that goes beyond the surface, acknowledging the fluid nature of ecosystems responding to environmental and human pressures.

Natural Processes

Nature, in its relentless dynamics, plays a significant role in the creation of wastelands. Soil erosion, salinity, aridity, and other natural forces can transform once-fertile lands into seemingly desolate expanses. The concept underscores the impact of these unyielding forces, highlighting the need for adaptive strategies that align with the evolving ecological dynamics. Understanding the intricate dance between natural processes and the landscape's response forms the foundation for effective wasteland development.

Human-induced Modifications

Wastelands are, in part, a testament to human activities that disrupt ecosystems and alter landscapes. Deforestation, industrialization, and improper land management contribute significantly to the creation of wastelands. The concept acknowledges that wastelands are not solely the outcome of natural forces but are often exacerbated and accelerated by human interventions. This recognition is crucial for formulating targeted development strategies that address the root causes of ecological imbalances (Fig 2).

Fig. 2: Human Induced Wasteland

Societal Perspectives

Beyond the ecological dimensions, the concept of wasteland encompasses societal perceptions that often perpetuate neglect and marginalization. These landscapes labeled as wastelands, carry a societal narrative that reinforces their barren image. The concept delves into the need for a paradigm shift, recognizing that wastelands are not irredeemable spaces but potential arenas for regeneration. Transforming societal perspectives is intrinsic to wasteland development, turning narratives of neglect into stories of opportunity and restoration.

Wasteland Development

Wasteland development is not merely an environmental endeavor; it is a visionary pursuit that seeks to regenerate landscapes for sustained ecological, social, and economic benefits. It involves strategic approaches that go beyond restoration, aiming for a comprehensive transformation of once-neglected spaces.

Revitalizing Ecosystems: Nurturing Biodiversity

Central to wasteland development is the strategic rejuvenation of natural ecosystems. Degraded areas, including deforested regions and habitats affected by soil erosion, become focal points for targeted efforts. Reforestation, soil conservation, and habitat restoration are integral components of the vision. The concept aligns with the idea of breathing life back into these ecosystems, fostering biodiversity and ecological resilience. It recognizes that the regeneration of natural habitats is not only ecologically significant but also a cornerstone of sustainable development.

Agricultural Renaissance: Sustainable Land Use Practices

Abandoned agricultural lands, representing a substantial portion of wastelands, become key arenas for development. The concept involves revitalizing these lands through sustainable farming practices, agroforestry initiatives, and soil improvement strategies. It envisions a renaissance in agricultural practices that not only restores productivity but also promotes long-term sustainability. The integration of traditional wisdom with modern agricultural techniques forms a crucial aspect of this vision, ensuring that agricultural development aligns with the principles of ecological balance.

Industrial Adaptation

Industrial wastelands, marked by pollution and altered ecosystems, present unique challenges and opportunities for development. The concept emphasizes the integration of green technologies, pollution control measures, and remediation efforts. Wasteland development in industrial areas aims to strike a balance between economic progress and environmental preservation. The vision is to create sustainable industrial zones where growth is harmonized with ecological well-being. This involves a paradigm shift in industrial practices, with a focus on resource efficiency, circular economy principles, and the integration of environmentally friendly technologies.

Mining Site Restoration

Mining activities contribute significantly to wasteland creation, necessitating innovative reclamation strategies. The concept revolves around transforming mining sites into models of responsible resource extraction. Wasteland development in mining areas involves reforestation, water management, and sustainable land use practices. The vision is to showcase a harmonious resource extraction and ecological preservation coexistence. It recognizes the imperative of responsible mining practices that not only mitigate environmental impacts but also contribute positively to the surrounding ecosystems and communities.

Technological Integration

Wasteland development leverages technological innovations as indispensable tools for informed decision-making and efficient execution. The integration of advanced technologies provides a comprehensive understanding of wasteland dynamics, enabling adaptive strategies and precise management.

Remote Sensing and GIS for Precision Management

The integration of remote sensing technologies and Geographic Information Systems (GIS) becomes crucial for monitoring and managing wastelands. Real-time data on vegetation health, soil quality, and land use changes provide insights for precision management. These technologies offer a bird's eye view of the evolving landscape, facilitating data-driven decision-making. The concept embraces technology as a means to enhance situational awareness, allowing for adaptive strategies that respond to the dynamic nature of wastelands.

Cutting-edge Reclamation Technologies

In the realm of industrial and mining wastelands, advanced reclamation technologies inspired by nature play a pivotal role. Phytoremediation, bioremediation, and other innovative approaches leverage natural processes for rehabilitation. The concept embraces these technologies as integral components of wasteland development, aligning with the vision of nature-inspired and sustainable solutions. It recognizes the potential of bio-inspired technologies to efficiently remediate contaminated areas, contributing to the overall success of wasteland regeneration.

Community Engagement: Catalysts for Sustainable Development

Wasteland development is inherently linked to community engagement, where residents become catalysts for sustainable and inclusive development. The concept acknowledges that communities are not merely stakeholders but essential partners in the development process.

Empowerment of Local Communities

Empowering local communities involves providing knowledge, skills, and resources, fostering a sense of ownership in the development process. The concept recognizes that sustainable development is contingent on active community participation and engagement. It envisions a shift from top-down approaches to collaborative models where local communities are active contributors to decision-making and implementation. Empowered communities become stewards of the rehabilitated landscapes, ensuring the longevity and sustainability of development initiatives.

Stakeholder Collaboration: Holistic Approaches

Effective wasteland development requires collaboration across diverse stakeholders. Government agencies, non-governmental organizations, businesses, and local communities must collaborate to devise holistic approaches. The concept emphasizes that inclusive stakeholder engagement is fundamental for addressing the multifaceted challenges of wasteland development. It recognizes the need for a shared vision and collective action, where each stakeholder brings unique perspectives and resources to the table. Holistic approaches involve breaking down silos and fostering partnerships that transcend traditional boundaries, leading to integrated and sustainable development outcomes.

3. Classification of Wasteland

The classification of wasteland is a pivotal step in formulating effective and targeted development strategies. Wastelands, often perceived as homogenous barren landscapes, encompass a diverse array of ecological, climatic, and anthropogenic factors. In this exploration, we delve into the nuanced classifications of wastelands, identifying distinct categories that demand bespoke approaches for regeneration and sustainable development.

Ecological Wastelands

Ecological wastelands represent areas where natural ecosystems have been significantly altered or degraded. This category encompasses a variety of landscapes, including deforested regions, habitats affected by soil erosion, and areas subjected to habitat destruction. Understanding the dynamics of ecological wastelands involves recognizing the need for restoration efforts such as reforestation, soil conservation, and habitat rehabilitation. The classification underscores the importance of ecological sensitivity in developing strategies tailored to the unique challenges posed by altered natural landscapes.

Abandoned Agricultural Lands

A substantial portion of wastelands falls under the category of abandoned agricultural lands. These areas often suffer from soil degradation, nutrient depletion, and loss of biodiversity due to unsustainable farming practices or agricultural abandonment. The classification emphasizes the significance of agricultural reclamation, promoting sustainable farming practices, agroforestry, and initiatives aimed at soil improvement. Strategies for abandoned agricultural lands focus on revitalizing soil health, ensuring long-term productivity, and fostering sustainable land use practices.

Industrial Wastelands

Industrial activities leave a lasting imprint on the landscape, resulting in the creation of industrial wastelands. Pollution, soil contamination, and altered ecosystems characterize these areas. The classification of industrial wastelands underscores the need for comprehensive remediation efforts, pollution control measures, and the incorporation of green technologies. The development strategies aim not only to rectify ecological damage but also to transform these areas into sustainable industrial zones, harmonizing economic progress with environmental preservation.

Mining Sites

Mining activities contribute significantly to the creation of wastelands. Mining sites face challenges such as soil degradation, habitat destruction, and water pollution. The classification of mining wastelands highlights the need for reclamation initiatives, including reforestation, water management, and sustainable land use practices. The development strategies for mining sites envision transforming these areas into models of responsible resource extraction, emphasizing the delicate balance between economic growth and ecological preservation.

Urban Wastelands

Urban areas also host wastelands, often neglected or abandoned due to urbanization processes. These urban wastelands may include brownfields, vacant lots, or areas with degraded infrastructure. The classification recognizes the need for urban regeneration initiatives, including brownfield redevelopment, green space creation, and revitalization projects. Urban wasteland development strategies aim to transform these neglected spaces into vibrant, sustainable components of the urban landscape.

Saline and Alkaline Wastelands

Some wastelands are characterized by soil salinity or alkalinity, rendering them inhospitable for conventional agriculture. These saline and alkaline wastelands require specialized development strategies focusing on soil reclamation, salt-tolerant crop cultivation, and the implementation of innovative agricultural practices. The classification emphasizes the importance of understanding and addressing specific soil quality challenges for the successful development of these landscapes.

Forest Degradation Areas

Wastelands within forested regions often arise from activities such as illegal logging, uncontrolled wildfires, or habitat destruction. The classification of forest degradation areas recognizes the importance of restoring these vital ecosystems. Development strategies involve afforestation, anti-deforestation measures, and habitat restoration efforts to bring back the ecological balance and biodiversity of these degraded forest landscapes.

Coastal and Mangrove Wastelands

Coastal areas and mangrove ecosystems are susceptible to degradation due to human activities, climate change, and industrial development. The classification of coastal and mangrove wastelands highlights the need for conservation and restoration efforts. Development strategies include mangrove reforestation, coastal habitat rehabilitation, and sustainable coastal management practices to protect these ecologically significant areas.

Classification Based on Climatic Zones

Wastelands are also classified based on climatic zones, considering factors such as aridity, desertification, or vulnerability to climate change. This classification recognizes the diverse impacts of climate variability on landscapes and necessitates adaptation strategies specific to each climatic zone. Development efforts aim to enhance climate resilience, promote water conservation, and address the unique challenges posed by different climatic conditions.

Challenges in Wasteland Classification

While the classification of wastelands is essential for targeted development, it is not without challenges. The dynamic nature of ecosystems, the interconnectedness of various factors, and the evolving landscape conditions pose difficulties in creating rigid classifications. Additionally, the intersectionality of ecological, social, and economic factors in wastelands makes it challenging to fit certain areas into a single category. Overcoming these challenges involves continuous monitoring, flexibility in classification frameworks, and adaptive management strategies that respond to the evolving nature of wastelands.

Opportunities in Tailored Development Strategies

The classification of wastelands provides a foundation for tailored development strategies that capitalize on the unique characteristics of each category. Recognizing the opportunities inherent in specific wasteland types allows for more efficient resource allocation, targeted interventions, and the maximization of positive ecological, social, and economic outcomes. Tailored development strategies also present opportunities for innovation, community engagement, and sustainable land use practices that can serve as models for broader environmental conservation efforts.

4. Estimates of Wastelands in India

Estimating the extent of wastelands in India is a multifaceted task that involves considering various factors such as land use changes, environmental policies, and ongoing reclamation efforts. As of my last knowledge update in January 2022, specific estimates might have evolved, and therefore, the most accurate and recent information should be sourced from authoritative bodies. In this exploration, we'll delve into the significance of estimating wastelands, the key factors influencing these estimates, and potential sources for obtaining up-to-date information.

Significance of Estimating Wastelands

Understanding the extent and distribution of wastelands is crucial for sustainable land management, environmental conservation, and informed policymaking. Wastelands, often characterized by low productivity and ecological degradation, represent areas with the potential for reclamation and rejuvenation. Accurate estimates serve as a baseline for developing targeted strategies aimed at transforming these lands into productive, ecologically vibrant spaces. Moreover, such estimates contribute to broader environmental monitoring efforts and the formulation of policies geared towards achieving land use sustainability goals.

Key Factors Influencing Wasteland Estimates

1. **Land Use Changes:** Dynamic changes in land use patterns, influenced by factors such as urbanization, industrialization, and agricultural practices, impact the extent of wastelands. Monitoring these changes is essential for understanding the evolving landscape.

2. **Reclamation Efforts:** Ongoing efforts to reclaim wastelands, through initiatives like afforestation, soil conservation, and sustainable land management, directly influence the estimates. Successful reclamation reduces the extent of wastelands over time.

3. **Environmental Policies:** Policies aimed at environmental conservation, sustainable development, and natural resource management play a significant role in shaping wasteland estimates. Changes in policy frameworks can impact land use patterns and contribute to variations in wasteland extent.

4. **Climate Change:** The influence of climate change on land degradation and soil quality is a crucial factor. Climate-induced alterations may lead to the expansion or contraction of wasteland areas.

5. **Technological Advancements:** Advances in remote sensing technologies, satellite imagery, and Geographic Information Systems (GIS) contribute to more accurate and detailed assessments of wasteland areas. These technological tools enable precise monitoring and mapping of land use changes.

Potential Sources for Obtaining Information on Wasteland Estimates

1. **Ministry of Environment, Forest and Climate Change (MoEFCC):** The MoEFCC is a key governmental body overseeing environmental policies in India. They often release reports and data related to land use, including estimates of wastelands. The official website and publications from MoEFCC are valuable resources.

2. **National Remote Sensing Centre (NRSC):** As part of the Indian Space Research Organisation (ISRO), NRSC plays a pivotal role in satellite-based remote sensing. They provide data and insights into land cover changes, contributing to estimates of wastelands.

3. **Indian Council of Agricultural Research (ICAR):** ICAR focuses on agricultural research and may have data related to changes in agricultural land use, which is pertinent to wasteland estimates.

4. **National Institute of Rural Development and Panchayati Raj (NIRD&PR):** NIRD&PR, being involved in rural development, may have information on wastelands, especially in the context of rural areas. Research publications and reports from NIRD&PR can provide valuable insights.

5. **State Remote Sensing Agencies:** Many Indian states have their remote sensing agencies responsible for monitoring and managing land use within their jurisdictions. These agencies often release reports and data specific to their regions.

6. **National Land Records Modernization Programme (NLRMP):** NLRMP focuses on modernizing land records and may provide information on changes in land use patterns, contributing to estimates of wastelands.

7. **Research Institutions and Universities:** Academic institutions, including the Indian Institutes of Technology (IITs) and agricultural universities, conduct research on various aspects of land use. Research papers and studies from these institutions may contain relevant information.

Challenges and Considerations

While estimates of wastelands are critical for informed decision-making, there are challenges and considerations to be aware of:

1. **Dynamic Nature:** The landscape is dynamic, and estimates may change over time due to natural processes, human interventions, and policy shifts.

2. **Interconnected Factors:** Wasteland estimates are influenced by a myriad of factors, making it challenging to attribute changes solely to one cause. Understanding the interconnectedness of these factors is crucial.

3. **Data Accuracy:** The accuracy of estimates depends on the quality of data collection methods and technologies employed. Regular updates and validations are essential to ensure data accuracy.
4. **Local Context:** Wasteland characteristics can vary significantly based on the local context. Local knowledge and community involvement are valuable for a comprehensive understanding.

5. Degraded Forest Lands: A Comprehensive Exploration of Extent, History, and Causes of Degradation

Degraded forest lands pose a significant environmental challenge globally, manifesting as a decline in the quality and health of forest ecosystems. This exploration aims to deepen our understanding of the extent, historical context, and multifaceted causes contributing to the degradation of these vital ecosystems.

Extent of Degraded Forests

Assessing the extent of degraded forest lands requires a nuanced approach, considering both spatial and ecological dimensions. Globally, it is estimated that a substantial percentage of the world's forests have experienced some degree of degradation. The Food and Agriculture Organization (FAO) of the United Nations plays a pivotal role in monitoring and providing insights into the scale of forest degradation. These degraded areas often result from a combination of natural processes and human activities, such as logging, agriculture expansion, infrastructure development, and climate change.

History of Forest Degradation

The history of forest degradation is deeply entwined with the evolution of human activities over centuries. This historical context provides a crucial foundation for understanding the cumulative impact of various interventions on forest ecosystems.

1. **Pre-Industrial Era:** In the pre-industrial era, forest exploitation was typically localized and driven by subsistence needs. Indigenous communities practiced sustainable resource use, maintaining a delicate balance with their ecosystems.
2. **Colonial Era:** The colonial period marked a significant intensification of forest exploitation, primarily driven by economic interests. Large-scale logging, often orchestrated by European powers, led to considerable deforestation and degradation in various regions.
3. **Post-Independence Period:** The post-independence era saw countries grappling with challenges related to population growth, economic development, and the expansion of agriculture. These factors collectively contributed to increased pressure on forest lands, resulting in widespread degradation.

4. **Industrialization:** The advent of industrialization brought about an unprecedented demand for timber, fuelwood, and other forest products. Mechanized logging and infrastructure development further accelerated forest degradation during this period.

5. **Globalization:** With globalization, markets for forest products expanded, resulting in increased extraction and trade. This global demand exerted additional stress on forests, contributing to widespread degradation.

Causes of Degradation

Forest degradation results from a complex interplay of human activities and environmental factors. Understanding these causes is crucial for formulating effective strategies for sustainable forest management and restoration.

1. **Logging and Timber Extraction:** Unsustainable logging practices, driven by the demand for timber and wood products, significantly contribute to forest degradation. Clear-cutting, selective logging, and illegal logging can alter forest structure, reduce biodiversity, and disturb ecosystem dynamics.

2. **Agricultural Expansion:** The conversion of forests into agricultural lands, fueled by the need for food production, stands as a major cause of degradation. Practices like slash-and-burn agriculture, shifting cultivation, and large-scale monoculture plantations lead to habitat loss and fragmentation.

3. **Infrastructure Development:** The construction of roads, dams, and other infrastructure projects can fragment forest landscapes, resulting in habitat loss and facilitating human access to remote areas. Increased accessibility often leads to further exploitation and degradation.

4. **Mining Activities:** Extractive industries, including mining, can cause extensive forest degradation. Deforestation, soil erosion, and water pollution are common consequences of mining activities, impacting both terrestrial and aquatic ecosystems.

5. **Fire:** Both natural and human-induced fires can contribute to forest degradation. Uncontrolled wildfires, exacerbated by factors like climate change and land-use practices, can alter ecosystems and reduce biodiversity.

6. **Climate Change:** Changes in climate patterns, such as increased temperatures and altered precipitation, can contribute to forest degradation. These changes affect the distribution of plant and animal species, disrupt ecological processes, and make forests more susceptible to pests and diseases.

7. **Overgrazing:** Uncontrolled grazing by livestock can negatively impact forest understories, preventing the regeneration of native vegetation and reducing the overall resilience of forest ecosystems.

8. **Invasive Species:** The introduction of invasive plant and animal species can outcompete native species, leading to changes in ecosystem structure and function. Invasive species contribute to the degradation of forest biodiversity.

9. **Lack of Sustainable Management:** Poor Forest management practices, including the absence of effective conservation measures and sustainable harvesting practices, can exacerbate degradation over time.

Addressing Forest Degradation

Effectively addressing forest degradation requires a holistic approach that integrates conservation, restoration, and sustainable management practices.

1. **Conservation Policies:** The implementation and enforcement of robust conservation policies, including the management of protected areas and regulations against illegal logging and hunting, are essential for preventing further degradation.

2. **Forest Restoration:** Active restoration efforts, such as afforestation and reforestation programs, play a vital role in rehabilitating degraded areas. Restoring native vegetation promotes biodiversity and enhances ecosystem resilience.

3. **Community Involvement:** Engaging local communities in forest management and restoration efforts is crucial. Indigenous knowledge and community-based conservation practices often contribute to more sustainable outcomes, fostering a sense of shared responsibility.

4. **Sustainable Logging Practices:** Implementing sustainable logging practices, such as selective logging and reduced-impact logging, helps minimize the ecological impact of timber extraction. Balancing economic interests with environmental considerations is key to ensuring long-term forest health.

5. **Agroforestry:** Promoting agroforestry practices that integrate trees into agricultural landscapes provides sustainable livelihoods for communities while preserving forest ecosystems. This approach encourages a harmonious coexistence between agriculture and forestry.

6. **Integrated Land-use Planning:** Adopting integrated land-use planning that considers the ecological value of forests alongside human needs is essential. This involves balancing the competing demands on land to ensure sustainable outcomes for both communities and ecosystems.

7. **Research and Monitoring:** Investing in ongoing research and monitoring initiatives enhances our understanding of the complex dynamics of

forest ecosystems. Regular assessments help track changes, evaluate the effectiveness of interventions, and adapt strategies over time.

8. **Global Collaboration:** Recognizing that forests are interconnected on a global scale, fostering international collaboration is critical. Sharing knowledge, resources, and best practices ensures a collective effort to address the challenges of forest degradation.

6. Assessment and Criteria for Soil Degradation, Sedimentation, and Desertification

The assessment of soil degradation, sedimentation, and desertification is crucial for understanding the health and resilience of ecosystems. These processes, often interrelated, have profound implications for agriculture, biodiversity, and overall environmental sustainability.

Soil Degradation Assessment:

Criteria for Soil Degradation

1. **Soil Erosion:** Assessing soil erosion involves evaluating the extent to which the topsoil is being lost due to water or wind action. Criteria include measuring soil depth, identifying exposed subsoil, and observing gully formation. Soil erosion assessments often utilize tools such as erosion models, sediment traps, and remote sensing technologies.

2. **Loss of Soil Fertility:** The decline in soil fertility is a key indicator of degradation. Criteria include assessing nutrient levels, organic matter content, and soil structure. Soil sampling and laboratory analysis are common methods for evaluating fertility changes.

3. **Compaction:** Soil compaction reduces porosity and limits root growth. Criteria for assessing compaction include bulk density measurements, penetration resistance tests, and observation of stunted plant growth. Compaction assessments are crucial for understanding the impact of agricultural machinery and land use practices.

4. **Salinization and Alkalization:** Accumulation of salts in the soil can lead to salinization, affecting plant growth. Criteria involve measuring electrical conductivity and assessing sodium levels. Additionally, visual observations of salt crust formation and plant damage contribute to salinization assessments.

5. **Waterlogging:** Excessive water retention in the soil can lead to waterlogging, impacting plant root health. Criteria for assessment include field observations of standing water, measuring groundwater levels, and assessing the presence of water-tolerant or anaerobic vegetation.

6. **Chemical Contamination:** Soil degradation can result from chemical contaminants, including pesticides and heavy metals. Criteria include soil sampling and laboratory analysis to identify the presence and concentration of contaminants. Field observations of impaired vegetation and wildlife can also provide qualitative indicators.

7. **Loss of Biodiversity:** Soil is a habitat for diverse microorganisms and invertebrates. Assessing biodiversity loss involves studying changes in species composition and abundance. Criteria include soil fauna counts, microbial diversity analysis, and observation of changes in vegetation structure.

8. **Land Use Changes:** Changes in land use, such as deforestation or conversion to agriculture, contribute to soil degradation. Criteria involve monitoring land cover changes through satellite imagery, historical land use records, and on-the-ground observations.

Sedimentation Assessment: Monitoring Erosion Impacts

Criteria for Sedimentation Assessment

1. **Sediment Yield:** Sediment yield is a fundamental criterion for assessing sedimentation. It involves measuring the quantity of sediment transported by rivers or runoff. Techniques include sediment samplers, sediment traps, and monitoring sediment deposition in reservoirs.

2. **Sediment Particle Size Distribution:** The particle size of transported sediment provides insights into erosion processes. Criteria involve sieve analysis or laser diffraction methods to determine the distribution of sediment particles, aiding in identifying sediment sources.

3. **Channel Morphology Changes:** Sedimentation often alters river channels. Criteria include monitoring changes in channel width, depth, and bed elevation. Remote sensing, aerial photography, and field surveys contribute to assessing channel morphology changes.

4. **Landform Changes:** Sedimentation influences the formation of landforms. Criteria involve observing changes in land topography, such as the development of alluvial fans, river deltas, or sediment deposition in valleys. Geospatial tools aid in mapping and analyzing landform changes.

5. **Impact on Aquatic Ecosystems:** Sedimentation negatively impacts aquatic ecosystems. Criteria include assessing changes in water quality, habitat degradation, and the abundance of aquatic species. Field surveys and water quality monitoring contribute to understanding sedimentation impacts.

Desertification Assessment

Criteria for Desertification Assessment

1. **Vegetative Cover Change:** Declines in vegetation cover are indicative of desertification. Criteria involve satellite imagery analysis, vegetation indices, and ground surveys to assess changes in plant density and species composition.

2. **Soil Moisture Content:** Desertification often leads to decreased soil moisture. Criteria include measuring soil moisture levels using instruments such as soil moisture meters or remote sensing technologies. Observations of reduced plant water stress also contribute to assessments.

3. **Wind and Water Erosion:** Erosion is a key process in desertification. Criteria involve assessing wind and water erosion through monitoring soil loss, sediment deposition, and changes in landscape features such as sand dunes. Remote sensing and field surveys are commonly employed.

4. **Land Productivity:** Declining land productivity is a criterion for desertification assessment. Criteria include evaluating crop yields, vegetation biomass, and carrying capacity for livestock. Long-term monitoring helps identify trends in land productivity.

5. **Sand Encroachment:** Expansion of sand dunes into previously vegetated areas is a clear sign of desertification. Criteria involve mapping changes in sand dune distribution, observing shifts in land cover, and assessing the impacts on infrastructure.

6. **Groundwater Depletion:** Over-extraction of groundwater contributes to desertification. Criteria include monitoring groundwater levels, assessing well yields, and evaluating the sustainability of water extraction practices. Groundwater modeling and hydrogeological studies aid in understanding depletion trends.

7. **Human Migration and Livelihood Changes:** Desertification often leads to migration and changes in livelihoods. Criteria involve socio-economic indicators, such as population displacement, altered agricultural practices, and shifts in traditional livelihoods. Surveys and interviews provide qualitative data.

8. **Climate Variability:** Desertification is influenced by climate variability. Criteria include analyzing long-term climate data, assessing trends in temperature and precipitation, and understanding the impact of climate change on arid ecosystems.

Integrated Assessment Approaches

Integrated assessment approaches are gaining prominence to comprehensively address soil degradation, sedimentation, and desertification. These approaches involve combining multiple criteria and employing diverse data sources to obtain a holistic understanding of environmental changes. Participatory methods, involving local communities and stakeholders, enhance the relevance and effectiveness of assessments.

Assessing soil degradation, sedimentation, and desertification requires a nuanced understanding of diverse criteria, employing a combination of field surveys, remote sensing technologies, and socio-economic indicators. These assessments form the foundation for designing effective conservation and restoration strategies, promoting sustainable land management practices, and mitigating the impacts of environmental degradation on ecosystems and human livelihoods. Continuous monitoring and adaptive management are essential for addressing the dynamic nature of these processes and ensuring the long-term health of the planet.

7. Role of Climatic Factors (Water and Temperature) in Soil Degradation

Climatic factors, particularly water and temperature, play a pivotal role in shaping soil health and influencing processes that contribute to soil degradation. Understanding the intricate relationship between climate and soil is essential for developing strategies to mitigate degradation and promote sustainable land management.

Water

Impact of Water Scarcity on Soil Degradation

1. **Drought Stress:** Prolonged water scarcity, often exacerbated by climate change, leads to drought stress in soils. Lack of moisture affects plant growth, reduces vegetation cover, and increases the susceptibility of soils to erosion. Drought-induced stress compromises the stability of soil structure, making it more prone to degradation processes (Fig 3).

Fig. 3: Impact of Water Scarcity on Soil-Drought Stress

2. **Increased Erosion Risk:** Water scarcity amplifies the risk of soil erosion. Without sufficient vegetation cover to shield the soil, raindrops directly impact the soil surface, leading to increased detachment and transportation of soil particles. This process contributes to the loss of topsoil, reducing fertility and promoting degradation.

3. **Salinization and Alkalization:** Water scarcity can intensify soil salinization and alkalization. Reduced water availability limits leaching, allowing salts to accumulate in the soil. Salinization negatively impacts plant growth and contributes to soil degradation, particularly in arid and semi-arid regions.

4. **Groundwater Depletion:** Excessive extraction of groundwater for agricultural irrigation can lead to soil degradation. Over-irrigation contributes to waterlogging, disrupting soil structure and nutrient balance. The subsequent rise in groundwater salinity further exacerbates degradation.

5. **Desertification:** Prolonged water scarcity is a key driver of desertification. Reduced precipitation and prolonged drought conditions contribute to soil acidification, leading to decreased vegetation cover, increased wind and water erosion, and the expansion of arid landscapes.

Impact of Water Excess on Soil Degradation

1. **Waterlogging:** Excessive water, either from heavy rainfall or poor drainage, results in waterlogged soils. Waterlogging negatively impacts soil aeration, root health, and microbial activity. The lack of oxygen in waterlogged soils leads to anaerobic conditions, promoting the release of harmful compounds and nutrient imbalances.

2. **Erosion and Sedimentation:** Intense rainfall can lead to soil erosion and sedimentation. Surface runoff, especially in areas with inadequate ground cover, carries away topsoil, reducing soil fertility. Sedimentation in water bodies negatively impacts aquatic ecosystems, contributing to soil and water degradation.

3. **Nutrient Leaching:** Excess water can leach nutrients from the soil profile. Nutrient leaching, particularly in sandy soils, leads to nutrient imbalances and compromises soil fertility. Leached nutrients may also contribute to water pollution, further impacting the environment.

4. **Soil Structure Breakdown:** Water excess can result in soil compaction and structure breakdown. Heavy rainfalls can lead to surface sealing and crusting, reducing water infiltration and exacerbating erosion. Compacted soils are less resilient to degradation processes, affecting plant root development.

5. **Flooding:** Severe flooding events have detrimental effects on soil health. Floodwaters carry sediments, contaminants, and nutrients away, leaving behind degraded soils. Flood-induced erosion and deposition contribute to the alteration of soil properties and landscape features.

Temperature

Impact of Temperature on Soil Degradation

1. **Accelerated Decomposition:** Warmer temperatures can accelerate microbial activity, leading to faster decomposition of organic matter in the soil. This process reduces soil organic carbon content, impacting soil structure, water retention, and nutrient availability.

2. **Increased Evaporation:** Higher temperatures contribute to increased evaporation rates, leading to soil drying. Drier soils are more susceptible to erosion, compaction, and reduced fertility. This drying effect is particularly pronounced in arid and semi-arid regions.

3. **Changes in Soil Moisture Regimes:** Rising temperatures influence soil moisture regimes. Increased evapotranspiration and altered precipitation patterns contribute to shifts in soil moisture availability. These changes can affect plant growth, exacerbate water stress, and contribute to soil degradation.

4. **Thermal Expansion and Contraction:** Temperature fluctuations induce thermal expansion and contraction in soils. This process can lead to the development of soil cracks, reducing soil stability and increasing vulnerability to erosion. In clayey soils, repeated expansion and contraction can lead to surface crusting.

5. **Impact on Microbial Communities:** Temperature influences the composition and activity of soil microbial communities. Changes in temperature regimes can alter microbial diversity, affecting nutrient cycling, organic matter decomposition, and overall soil health.

6. **Influence on Soil Structure:** Temperature variations impact soil structure dynamics. Freeze-thaw cycles in colder climates can lead to soil heaving, disrupting soil structure. In warmer climates, persistent high temperatures can contribute to the breakdown of soil aggregates, increasing susceptibility to erosion.

Integrated Approaches to Mitigate Climate-Induced Soil Degradation

1. **Climate-Smart Agriculture:** Implementing climate-smart agricultural practices involves adapting farming techniques to changing climatic conditions. This includes optimizing irrigation, adopting conservation tillage, and promoting agroforestry to enhance soil resilience.

2. **Water Management Strategies:** Sustainable water management is critical for mitigating the impacts of water scarcity and excess. Implementing efficient irrigation practices, enhancing water retention through soil conservation methods, and optimizing drainage systems are key strategies.

3. **Cover Cropping:** Planting cover crops helps protect the soil surface, preventing erosion and promoting water retention. Cover crops also contribute organic matter to the soil, enhancing its structure and fertility.

4. **Agroecological Practices:** Embracing agroecological principles involves designing farming systems that mimic natural ecosystems. This includes diversified cropping systems, agroforestry, and integrated pest management, fostering soil health and resilience.

5. **Reforestation and Afforestation:** Planting trees and restoring forest ecosystems contribute to mitigating the impacts of climate change on soil health. Trees enhance water infiltration, stabilize soils, and promote biodiversity.

6. **Soil Conservation Measures:** Implementing soil conservation measures, such as contour ploughing, terracing, and windbreaks, helps reduce water and wind erosion. These practices contribute to maintaining soil structure and fertility.

7. **Adaptive Land Management:** Implementing adaptive land management practices involves continuously monitoring climate and soil conditions. This allows for timely adjustments in land use planning, crop selection, and water management to mitigate the impacts of climate-induced soil degradation.

8. **Research and Innovation:** Investing in research and innovation is crucial for developing new technologies and practices that enhance soil resilience to climate change. This includes the development of drought-resistant crops, climate-resilient soil amendments, and precision agriculture tools.

The role of climatic factors, specifically water and temperature, in soil degradation is intricate and multifaceted. Climate change intensifies the frequency and severity of extreme weather events, placing additional stress on soil ecosystems. Mitigating climate-induced soil degradation requires a holistic approach that integrates sustainable land management practices, adaptive strategies, and ongoing research. By understanding the dynamic interactions between climate and soil, societies can develop resilience to the challenges posed by changing environmental conditions, ensuring the long-term health and productivity of soils for future generations.

8. Technological Innovations for the Management of Various Types of Wastelands

The effective management of diverse wastelands, including those in hot deserts, saline-alkali areas, ravines, coastal soils, laterites, rocky eroded hill slopes, and coal mine areas, requires innovative technological solutions.

1. Hot Desert Wastelands

Technological Solutions

1. **Drip Irrigation and Soil Moisture Sensors:** Efficient water management is crucial in hot deserts. Drip irrigation systems, combined with soil moisture sensors, enable precise control over water distribution, minimizing wastage and promoting plant growth in arid conditions.
2. **Solar Desalination:** Hot deserts often face salinity challenges. Solar desalination technologies harness sunlight to desalinate saline water, providing a sustainable source of irrigation water for crops in these regions.
3. **Drones for Monitoring and Precision Agriculture:** Drones equipped with remote sensing technology facilitate the monitoring of vast desert landscapes. They provide valuable data on soil quality, vegetation health, and water distribution, enabling precision agriculture practices in challenging environments.
4. **Aeroponics and Hydroponics:** Soilless cultivation methods like aeroponics and hydroponics are ideal for hot desert areas. These technologies deliver nutrient-rich solutions directly to plant roots, reducing the dependence on soil and optimizing resource use.
5. **Windbreak Technologies:** Wind erosion is a common issue in hot deserts. Innovative windbreak technologies, such as strategically placed barriers and vegetation cover, help mitigate soil erosion and create microenvironments suitable for plant growth.

2. Saline-Alkali Wastelands

Technological Solutions

1. **Phytoremediation:** Certain plant species have the ability to tolerate and accumulate salts. Phytoremediation involves cultivating these salt-tolerant plants to extract salts from the soil, gradually improving its quality over time.
2. **Subsurface Drainage Systems:** Installing subsurface drainage systems helps manage the water table in saline-alkali soils. Controlled drainage and sub-irrigation techniques prevent waterlogging and reduce the accumulation of salts in the root zone.
3. **Biochar Application:** Biochar, a form of charcoal produced from organic materials, can enhance soil structure and water retention in saline-alkali soils. It also acts as a carbon sink, promoting microbial activity and nutrient availability.
4. **Electrokinetic Soil Decontamination:** This innovative technology involves applying an electric field to the soil, mobilizing ions and contaminants. It aids in the removal of salts and alkaline substances, contributing to soil reclamation.
5. **Halophyte Farming:** Cultivating halophytes, plants adapted to saline conditions, provides a sustainable way to utilize saline-alkali wastelands. These plants can be used for food, fodder, or bioenergy production, contributing to economic and environmental sustainability.

3. Ravine Wastelands

Technological Solutions

1. **Check Dams and Contour Trenching:** Check dams help control erosion in ravine areas by slowing down water flow and trapping sediments. Contour trenching involves creating trenches along the contours of the land to capture rainwater, preventing soil erosion.
2. **Vegetative Cover:** Establishing vegetative cover through afforestation and reforestation helps stabilize ravine slopes. Planting native species with deep roots enhances soil structure, reduces erosion, and promotes biodiversity.
3. **Biotechnical Measures:** Implementing biotechnical measures, such as the use of live fascines (bundles of live branches) and brush layers, provides natural stabilization for ravine slopes. These measures enhance soil structure and prevent further erosion.
4. **Geosynthetic Reinforcements:** Geosynthetic materials, such as geotextiles and geogrids, can be used to reinforce ravine slopes. These materials

add stability to the soil and mitigate erosion, especially in areas prone to landslides.

5. **Hydroseeding:** Hydroseeding involves spraying a mixture of seeds, mulch, and stabilizing agents on ravine slopes. This accelerates vegetation growth, controls erosion, and promotes rapid soil stabilization.

4. Coastal Soil Wastelands

Technological Solutions

1. **Salt-Tolerant Crop Varieties:** Developing and cultivating crops that are tolerant to saline conditions is crucial for coastal soil management. Biotechnological advancements in breeding programs can enhance the resilience of crops in these environments.

2. **Managed Aquifer Recharge:** Coastal areas often suffer from saltwater intrusion into aquifers. Managed aquifer recharge involves intentional injection of freshwater into aquifers, creating a barrier against saltwater intrusion and improving soil and water quality.

3. **Sea Water Greenhouses:** These greenhouses utilize seawater for cooling and create a controlled environment for crop cultivation. The evaporation and condensation processes help desalinate water, making it suitable for irrigating crops in coastal regions.

4. **Floating Agriculture:** In areas with saline-affected soils and limited space, floating agriculture systems can be implemented. Floating gardens or rafts support the growth of crops, providing an innovative solution for utilizing coastal water bodies.

5. **Coastal Afforestation:** Planting salt-tolerant trees and mangroves along coastal areas helps stabilize soil, reduce erosion, and create a buffer against storm surges. Coastal afforestation also enhances biodiversity and provides valuable ecosystem services.

5. Laterite Wastelands

Technological Solutions

1. **Laterite Stabilization Techniques:** Stabilizing laterite soils using techniques such as lime stabilization, cement stabilization, or the addition of fly ash helps improve their engineering properties. This is particularly important for construction purposes in laterite-rich regions.

2. **Agroforestry with Nitrogen-Fixing Trees:** Agroforestry systems incorporating nitrogen-fixing trees help enhance soil fertility in laterite

areas. These trees contribute nitrogen to the soil, supporting agricultural productivity and improving overall soil health.

3. **Microbial Bioremediation:** Microbial bioremediation techniques involve the use of microorganisms to enhance nutrient availability and soil structure in lateritic soils. Biofertilizers and soil inoculants can be applied to improve plant growth and soil quality.
4. **Terracing and Contour Plowing:** Implementing terracing and contour ploughing helps control erosion in laterite-rich landscapes. These measures slow down water runoff, promote water infiltration, and prevent the loss of valuable topsoil.
5. **Precision Agriculture Practices:** Precision agriculture, including the use of precision nutrient management and site-specific crop management, helps optimize resource use in laterite areas. This approach improves agricultural productivity while minimizing environmental impact.

6. Rocky Eroded Hill Slopes

Technological Solutions

1. **Bioengineering Techniques:** Bioengineering involves using living plants and vegetation to stabilize rocky slopes. Planting erosion-control species with deep roots helps bind soil particles and prevent further erosion.
2. **Rock Terracing:** Constructing rock terraces on hill slopes helps create level surfaces, reducing the impact of water runoff and preventing soil erosion. These terraces also provide platforms for agriculture or afforestation.
3. **Hydraulic Erosion Control:** Hydraulic erosion control measures, such as the use of erosion control blankets and sediment control structures, help manage water runoff and minimize soil loss on rocky hill slopes.
4. **Hydroseeding with Native Species:** Hydroseeding with a mixture of native seeds, mulch, and stabilizers promotes the establishment of vegetation on rocky slopes. This accelerates the natural process of soil stabilization and erosion control.
5. **Geosynthetic Nets and Mats:** Deploying geosynthetic nets and mats on rocky hill slopes provides additional stability. These materials help prevent soil erosion by securing loose rocks and stabilizing the surface, facilitating the establishment of vegetation.

7. Coal Mine Areas

Technological Solutions

1. **Land Reclamation through Grading and Recontouring:** Grading and recontouring the land in coal mine areas help restore the topography, reduce soil erosion, and create a suitable foundation for subsequent land uses, including agriculture or afforestation.
2. **Soil Amendments and Fertilization:** Applying soil amendments such as organic matter, lime, and gypsum helps ameliorate the soil quality in coal mine areas. Fertilization with appropriate nutrients supports the establishment of vegetation and accelerates the rehabilitation process.
3. **Phytoremediation with Native Plants:** Phytoremediation involves using native plants with metal-accumulating capabilities to remediate soils contaminated with heavy metals from mining activities. These plants absorb and concentrate metals, facilitating their removal.
4. **Waste Utilization:** Utilizing waste materials from coal mining processes, such as fly ash, for soil improvement is a sustainable approach. Fly ash can be used as a soil amendment to enhance fertility and structure.
5. **Hydroseeding and Erosion Control Mats:** Hydroseeding combined with erosion control mats provides an effective method for establishing vegetation and preventing soil erosion in coal mine areas. This approach accelerates the rehabilitation of disturbed landscapes.

The management of diverse wastelands demands a tailored and technologically advanced approach, addressing the specific challenges posed by each type of degraded land. From hot deserts to coal mine areas, the integration of innovative solutions such as precision agriculture, phytoremediation, and sustainable land reclamation techniques is essential. By harnessing the power of technology, societies can not only rehabilitate wastelands but also promote sustainable land use practices that balance environmental conservation with human development. Continuous research and the application of cutting-edge technologies are crucial for mitigating the impact of land degradation and ensuring the long-term health of ecosystems.

8. Environmental Considerations for Wasteland Development

Wasteland development necessitates a nuanced and environmentally conscious strategy to guarantee sustainable land use practices. This involves thorough consideration of various ecological factors, the promotion of biodiversity, and the mitigation of adverse impacts on the environment.

1. Biodiversity Conservation

1.1 **Native Species Selection:** Native plant species play a crucial role in maintaining ecological balance. Beyond their adaptability to local conditions, they often have symbiotic relationships with native fauna, forming interconnected ecosystems. The selection of native species should involve a comprehensive understanding of the region's biodiversity, ensuring that the reintroduction or establishment of vegetation aligns with the broader ecological context.

1.2 **Habitat Restoration:** Habitat restoration is a multifaceted process involving the re-establishment of native flora and the creation of conducive environments for fauna. This can include reintroducing keystone species, restoring natural water bodies, and implementing measures to attract pollinators. Well-executed habitat restoration not only enhances biodiversity but also contributes to the resilience and sustainability of the entire ecosystem.

1.3 **Avoiding Invasive Species:** The introduction of invasive species poses a significant threat to existing ecosystems. These species can outcompete native flora, alter soil compositions, and disrupt natural nutrient cycles. Robust screening processes and environmental impact assessments are vital to identify and mitigate the risks associated with invasive species, ensuring that wasteland development contributes positively to biodiversity.

1.4 **Biodiversity Assessments:** Comprehensive biodiversity assessments are essential for understanding the baseline ecological diversity of a region. These assessments should encompass flora, fauna, and microbial life, providing a holistic view of the ecosystem. Continuous monitoring post-development is equally crucial to evaluate the effectiveness of conservation measures and adapt strategies based on emerging ecological dynamics.

Biodiversity conservation in wasteland development can benefit from innovative practices. Incorporating technology, such as DNA barcoding for species identification and remote sensing for monitoring, can provide valuable insights. Engaging local communities as citizen scientists in biodiversity monitoring fosters a sense of ownership and responsibility. Furthermore, establishing biodiversity corridors connecting fragmented habitats promotes genetic diversity and facilitates species migration, contributing to the long-term viability of ecosystems.

2. Soil Health and Conservation

2.1 **Soil Erosion Control:** Soil erosion is a pervasive issue, and its control is integral to sustainable wasteland development. Beyond traditional methods, modern erosion control techniques involve the use of geosynthetics, erosion control blankets, and check dams. These technologies provide effective erosion prevention while allowing for natural processes of vegetation establishment, ensuring long-term soil health.

2.2 **Soil Amendments:** Sustainable soil management involves not only preventing degradation but actively improving soil health. Integrating agroecological practices, such as crop rotation and cover cropping, helps enhance soil structure and fertility. Additionally, the use of organic amendments, like biochar, holds promise in sequestering carbon, promoting microbial activity, and improving nutrient availability.

2.3 **Avoiding Chemical Contamination:** The reduction of chemical inputs in agriculture is paramount for preventing soil contamination. Integrated pest management (IPM), organic farming methods, and precision agriculture technologies minimize the reliance on synthetic pesticides and fertilizers. These practices not only protect soil health but also contribute to overall environmental sustainability.

2.4 **Soil Monitoring:** Advanced soil monitoring techniques, including sensor networks and satellite-based observations, offer real-time data on soil health. Continuous monitoring allows for adaptive management strategies, enabling prompt responses to changes in soil conditions. Precision agriculture technologies, such as variable-rate fertilization, optimize nutrient application, minimizing environmental impacts.

Incorporating agroforestry practices into wasteland development provides a multifaceted approach to soil health. The strategic integration of trees with crops enhances nutrient cycling, improves water retention, and contributes to overall soil fertility. Furthermore, the promotion of mycorrhizal fungi, through practices like mycorrhizal inoculation, supports plant nutrient uptake and fosters symbiotic relationships between plants and soil microorganisms.

3. Water Management and Conservation

3.1 Water Harvesting: Advancements in water harvesting technologies offer innovative solutions for wasteland development. Rooftop rainwater harvesting, check dam construction, and subsurface water storage systems can be augmented with smart technologies. Internet of Things (IoT)-based sensors can optimize water collection and storage, ensuring efficient use for agricultural or ecological purposes.

3.2 Efficient Irrigation Practices: Precision agriculture technologies, such as sensor-based irrigation and automated drip systems, revolutionize water use efficiency. These technologies enable targeted irrigation, minimizing water wastage and ensuring that crops receive the necessary moisture. Adoption of these practices in wasteland development promotes sustainable water management.

3.3 Water Quality Protection: Beyond preventing physical erosion, measures to protect water quality are critical. Constructed wetlands, vegetative buffer strips, and sedimentation ponds act as natural filters, trapping sediments and contaminants before water enters water bodies. These ecological engineering approaches safeguard downstream water quality.

3.4 Wetland Restoration: Restoration of degraded wetlands involves not only recreating the physical habitat but also enhancing water quality. Phytoremediation, using wetland plants to absorb pollutants, can be integrated into wetland restoration projects. Additionally, smart monitoring systems can track water quality parameters in real time, ensuring the success of wetland rehabilitation efforts.

Innovative approaches to water conservation include the use of moisture-absorbing polymers in soil, which can improve water retention. Smart irrigation systems powered by artificial intelligence algorithms can dynamically adjust irrigation schedules based on weather forecasts and soil moisture levels. Integrating decentralized water treatment technologies, such as solar-powered desalination units, enhances water availability in coastal wasteland areas.

4. Climate Resilience

4.1 Climate-Smart Land Use Planning: Climate-smart land use planning involves considering future climate scenarios. Geographic information system (GIS) technologies, coupled with climate modeling, enable the identification of areas vulnerable to climate change impacts. This information guides the selection of suitable crops, infrastructure placement, and overall land use strategies.

4.2 Drought-Resistant Species: Bioengineering approaches, such as genetic modification for drought resistance, offer potential solutions for cultivating crops in water-stressed environments. Traditional breeding methods, guided by genomic studies, can identify and enhance the resilience of crops to climatic extremes, contributing to sustainable agriculture in arid and semi-arid regions.

4.3 Green Infrastructure: The development of green infrastructure, including urban green spaces and vegetated roofs, contributes to climate resilience in urban areas. Implementing cool roof technologies, which reflect sunlight and absorb less heat, can mitigate the urban heat island effect. These nature-based solutions enhance the adaptability of urban environments to rising temperatures.

4.4 Adaptive Management: The integration of satellite-based monitoring systems and climate modeling tools supports adaptive land management. Predictive analytics can help anticipate climate-induced changes, allowing for proactive adjustments in land use planning, water management, and agricultural practices. Adaptive management strategies ensure the resilience of wasteland development projects to evolving climate conditions.

Nature-based solutions, such as reforestation and afforestation, contribute not only to carbon sequestration but also to climate resilience. Forest ecosystems act as buffers against extreme weather events, stabilize soils, and regulate local climates. Implementing climate-smart agricultural practices, such as agroforestry and precision farming, enhances the adaptive capacity of agricultural systems to changing climate patterns.

5. Social and Community Engagement

5.1 Community Involvement: The engagement of local communities in biodiversity monitoring and environmental conservation programs fosters a sense of ownership. Citizen science initiatives, facilitated through mobile applications and community workshops, empower residents to actively participate in monitoring and conservation efforts, promoting environmental stewardship.

5.2 Traditional Ecological Practices: Recognizing and integrating traditional ecological practices into wasteland development is crucial. Indigenous knowledge often holds valuable insights into sustainable land management, including traditional agricultural techniques, water conservation methods, and plant-based remedies for soil health. Collaborative learning sessions facilitate the exchange of knowledge between traditional practices and modern technologies.

5.3 Livelihood Enhancement: Beyond the ecological aspects, wasteland development should aim to enhance local livelihoods. The establishment of community-based enterprises, such as eco-friendly tourism, organic farming cooperatives, or artisanal ventures, provides sustainable economic opportunities. Engaging local communities in decision-making processes ensures that development aligns with their needs and aspirations.

5.4 Capacity Building: Capacity-building programs for local communities should encompass a range of skills, from sustainable agriculture practices to environmental monitoring techniques. Training sessions, workshops, and educational campaigns empower community members to actively participate in and sustain the benefits of wasteland development initiatives. Knowledge dissemination through local schools and community centers ensures widespread awareness and understanding.

Innovative community engagement strategies include the use of virtual reality (VR) and augmented reality (AR) technologies to simulate the potential impacts of wasteland development. These technologies facilitate immersive learning experiences, allowing community members to visualize proposed changes and actively participate in the decision-making process. Furthermore, incorporating gamification elements into environmental education programs can enhance community involvement and knowledge retention.

6. Waste Management and Recycling

6.1 Waste Reduction: Adopting a circular economy approach involves minimizing waste generation at the source. Innovative packaging solutions, such as biodegradable materials and reusable containers, contribute to waste reduction. Additionally, incorporating principles of the sharing economy, such as community tool libraries and resource-sharing platforms, minimizes the demand for new products and reduces overall waste production.

6.2 Construction Materials: Sustainable construction materials play a pivotal role in reducing the environmental impact of infrastructure development. The use of recycled materials, such as reclaimed wood, recycled concrete, and repurposed metals, aligns with the principles of sustainable construction. Advanced engineering techniques, including 3D printing with recycled materials, offer innovative solutions for sustainable building practices.

6.3 Hazardous Waste Management: The proper management of hazardous waste involves employing technologies such as bioremediation and phytoremediation. These biological methods utilize microorganisms or plants to degrade or absorb contaminants, promoting environmentally friendly waste disposal. Collaboration with specialized waste management companies ensures the safe handling and disposal of hazardous materials.

6.4 Circular Economy Principles: The principles of a circular economy extend beyond waste reduction to promote the reuse and recycling of materials. Implementing closed-loop systems for product manufacturing and establishing reverse logistics channels for the collection and recycling of used goods contribute to the circular economy. This holistic approach minimizes resource depletion and environmental pollution.

Incorporating decentralized waste-to-energy systems, such as anaerobic digesters and biomass gasification, addresses both waste management and energy generation. These technologies convert organic waste into biogas or bioenergy, providing a sustainable solution for waste disposal while contributing to renewable energy production. Embracing innovative packaging designs, such as edible packaging or compostable materials, reduces the environmental impact of packaging waste.

Conclusion

A holistic and environmentally sensitive approach to wasteland development is essential for the success and sustainability of such initiatives. By extending our understanding of biodiversity conservation, soil health promotion, water management, climate resilience, community engagement, and sustainable waste practices, we contribute to the broader goals of ecological integrity and environmental health. Continuous monitoring, adaptive management, and a commitment to sustainable development principles ensure that the benefits of wasteland reclamation endure over time, fostering thriving ecosystems and resilient communities. As technology advances and our understanding of ecological systems deepens, integrating these innovations into wasteland development becomes crucial for a harmonious coexistence between human activities and the environment.

4

Biological Aspects of Soil and Water Conservation Techniques

[1]B.N., Niranjan, [2]Kh. Chandrakumar Singh, [3]Shubham, [4]Jeetendra Kumar, and [5]Sheetanshu Gupta

[1]Dakshina Kannada Milk Union (KMF), Mangalore, Karnataka
[2]VCSGU University of Horticulture and Forestry, Bharsar, Uttarakhand
[3]Department of Entomology, CCS Haryana Agricultural University Hisar, Haryana
[4]Krishi Vigyan Kendra, Jehanabad, Bihar Agricultural University Bihar
[5]Department of Biotechnology, Institute of Technology and Management BKT, Lucknow, U.P

1. Concept of Conservation Farming

Conservation farming represents a paradigm shift in agriculture, moving away from conventional practices that often degrade the environment and compromise long-term productivity. Expanding on the principles of conservation farming provides deeper insights into the multifaceted benefits and challenges associated with this approach.

Conservation farming aligns closely with the principles of agroecology, emphasizing the integration of ecological concepts into agricultural systems. The concept recognizes that agriculture is not separate from the natural environment but rather an integral part of it. By minimizing soil disturbance, conservation farming promotes the preservation of soil structure and the intricate network of microorganisms that contribute to nutrient cycling and plant health.

The practice of keeping soil covered with crop residues or cover crops extends beyond erosion control. This protective cover serves as a microhabitat for beneficial insects, fungi, and bacteria. The decomposition of plant residues enriches the soil with organic matter, enhancing its water-holding capacity and nutrient content. Cover crops, such as legumes, contribute to nitrogen fixation, reducing the reliance on synthetic fertilizers and promoting a more sustainable nutrient management approach.

Crop rotation and diversification, core principles of conservation farming, are essential for pest and disease management. Planting different crops in a sequence disrupts the life cycles of pests and pathogens, reducing the need for chemical interventions. Moreover, diverse cropping systems enhance ecosystem resilience, making agricultural landscapes less susceptible to the negative impacts of extreme weather events and climate variability.

Integrated Pest Management (IPM) in conservation farming goes beyond chemical control methods. Farmers practicing conservation agriculture often employ agroforestry techniques, intercropping, and the use of trap crops to naturally manage pest populations. This integrated approach not only reduces the environmental impact of pest control but also fosters a more balanced and biodiverse agroecosystem.

Water management in conservation farming involves a nuanced understanding of the soil's hydrological properties. The emphasis on no-till or reduced tillage practices helps maintain soil structure, allowing for better water infiltration and retention. Efficient water use is further achieved through practices like rainwater harvesting and the adoption of precision agriculture technologies. These technologies, including soil moisture sensors and automated irrigation systems, enable farmers to tailor water application to the specific needs of crops, promoting resource-use efficiency.

The principles of conservation farming extend to precision agriculture, where technologies like GPS-guided equipment and variable-rate fertilization contribute to sustainable resource management. This technological integration allows farmers to optimize inputs such as seeds, fertilizers, and pesticides, reducing waste and minimizing the environmental footprint of agriculture.

Key Principles of Conservation Farming

1. **Minimal Soil Disturbance:** Conservation farming discourages intensive tillage, opting for minimal soil disturbance. This practice helps maintain soil structure, reduces erosion, and preserves the habitat for beneficial soil organisms. No-till and reduced tillage techniques are commonly employed to achieve these objectives.

2. **Permanent Soil Cover:** Keeping the soil covered with crop residues or cover crops is a fundamental principle of conservation farming. This protective cover shields the soil from erosion caused by wind and water, enhances water retention, and contributes organic matter to the soil, promoting its fertility.

3. **Crop Rotation and Diversification:** Conservation farming encourages the rotation of crops and the diversification of plant species. Crop rotation disrupts pest and disease cycles, improves soil structure, and optimizes nutrient utilization. Diversification also enhances biodiversity, promoting the presence of beneficial organisms in the agricultural ecosystem.
4. **Integrated Pest Management (IPM):** Rather than relying heavily on chemical pesticides, conservation farming integrates IPM strategies. This involves the use of natural predators, crop rotation, resistant crop varieties, and other eco-friendly methods to manage pests, minimizing the environmental impact of pest control.
5. **Water Management:** Efficient water use is a crucial aspect of conservation farming. Practices such as rainwater harvesting, drip irrigation, and the use of soil moisture sensors help optimize water utilization, ensuring that crops receive adequate moisture while minimizing water wastage.
6. **Precision Agriculture:** Conservation farming embraces precision agriculture technologies, allowing farmers to optimize input use based on site-specific conditions. GPS-guided equipment, variable-rate fertilization, and automated machinery contribute to resource efficiency and reduced environmental impact.

Benefits of Conservation Farming

1. **Soil Health Improvement:** By minimizing soil disturbance and maintaining permanent soil cover, conservation farming promotes soil health. Improved soil structure, increased organic matter, and enhanced microbial activity contribute to sustainable and productive agriculture.
2. **Erosion Control:** One of the primary goals of conservation farming is to control soil erosion. Practices such as no-till farming and cover cropping prevent the loss of topsoil due to water and wind erosion, preserving the fertile layer essential for crop growth.
3. **Biodiversity Enhancement:** The incorporation of diverse crops, cover crops, and natural habitats in and around farmland fosters biodiversity. Conservation farming systems create a more resilient and balanced ecosystem, supporting a variety of plant and animal species.
4. **Water Conservation:** Efficient water management practices, including precision irrigation and rainwater harvesting, contribute to water conservation. By optimizing water use, conservation farming helps farmers adapt to water scarcity challenges while maintaining crop productivity.
5. **Reduced Chemical Inputs:** Conservation farming reduces reliance on chemical inputs such as pesticides and synthetic fertilizers. This not only

minimizes the environmental impact but also promotes healthier soils and ecosystems.

In lacuna, conservation farming embodies a holistic and ecologically conscious approach to agriculture. It is a dynamic system that adapts to local environmental conditions, emphasizing resilience and sustainability. As the global agricultural landscape faces challenges from climate change, soil degradation, and resource constraints, the principles of conservation farming are increasingly recognized as a cornerstone for building a resilient and productive food system.

Concept of Irrigation

Irrigation, as a fundamental component of agriculture, requires a more detailed exploration to understand its diverse applications, challenges, and innovations. Elaborating on the key aspects of irrigation sheds light on the intricacies of managing water resources in agriculture.

Water sourcing for irrigation systems is a critical consideration that varies significantly across regions. In some areas, rivers and surface water reservoirs serve as primary sources, while others rely on groundwater extracted from wells or aquifers. The sustainability of these water sources is of paramount importance, and careful management is necessary to prevent over-extraction and depletion.

Different types of irrigation systems cater to varying needs and conditions. Surface irrigation methods, including furrow, basin, and flood irrigation, are traditional approaches with broad applications. Drip irrigation, with its precise water delivery directly to the plant roots, is especially effective in water-scarce regions. Sprinkler irrigation mimics natural rainfall and is suitable for a range of crops. Sub-surface irrigation involves delivering water below the soil surface, minimizing evaporation losses. The choice of system depends on factors such as crop type, soil characteristics, and available water resources.

Water conveyance plays a crucial role in ensuring that water reaches fields efficiently. Canals, pipes, and pumps are integral components of irrigation infrastructure. The design and maintenance of these conveyance systems impact the equitable distribution of water across fields. Technological advancements in water conveyance, such as automated canal gates and sensor-based irrigation scheduling, contribute to improved efficiency and reduced water wastage.

Efficiency and water conservation have become central themes in modern irrigation practices. Drip and sprinkler irrigation systems, characterized by their precision in water application, significantly reduce water use compared to traditional methods. Smart irrigation technologies leverage data from soil moisture sensors, weather forecasts, and crop models to optimize irrigation scheduling, ensuring that crops receive the right amount of water at the right time.

Effective irrigation management involves careful scheduling based on crop water requirements, soil moisture levels, and climatic conditions. Farmers utilizing modern tools and technologies make informed decisions about when and how much water to apply. The integration of remote sensing and satellite imagery further enhances the ability to monitor crop health and water stress, allowing for timely interventions.

Challenges in irrigation, such as water scarcity and environmental impacts, necessitate innovative solutions. Sustainable irrigation practices prioritize water-use efficiency, minimize environmental degradation, and consider the social and economic implications of water distribution. Small-scale irrigation projects, community-driven water management initiatives, and the adoption of agroecological approaches contribute to more sustainable and equitable irrigation practices.

Key Aspects of Irrigation

1. **Water Source:** Irrigation systems draw water from various sources, including rivers, lakes, reservoirs, wells, and underground aquifers. The availability and sustainability of the water source are critical considerations in designing effective irrigation strategies.

2. **Types of Irrigation Systems:** There are several types of irrigation systems, each designed to deliver water to crops in specific ways. Common irrigation methods include surface irrigation (furrow, basin, and flood irrigation), drip irrigation, sprinkler irrigation, and sub-surface irrigation. The choice of system depends on factors such as crop type, soil characteristics, and water availability.

3. **Water Conveyance:** Irrigation systems involve the conveyance of water from its source to the fields. Canals, pipes, pumps, and open ditches are commonly used to transport water efficiently. Proper water conveyance ensures uniform distribution across the fields.

4. **Efficiency and Water Conservation:** Modern irrigation practices focus on improving efficiency to conserve water. Technologies such as drip irrigation and precision irrigation enable precise control over water application, reducing wastage and optimizing water use efficiency.

5. **Scheduling and Management:** Effective irrigation management involves scheduling water applications based on crop needs, soil moisture levels, and weather conditions. Monitoring tools, such as soil moisture sensors and weather forecasts, aid farmers in making informed decisions about when and how much water to apply.

Challenges and Considerations in Irrigation

1. **Water Scarcity:** In many regions, water scarcity poses a significant challenge to irrigation. Sustainable water management practices, conservation measures, and the adoption of water-efficient irrigation technologies are essential to address this issue.

2. **Environmental Impact:** Poorly managed irrigation systems can lead to environmental issues such as waterlogging, salinization, and depletion of aquifers. Sustainable irrigation practices aim to minimize these negative environmental impacts through careful water resource management.

3. **Energy Consumption:** Some irrigation methods, particularly those involving pumping water over long distances, can be energy-intensive. The choice of irrigation system and the use of energy-efficient technologies contribute to reducing the overall energy footprint of irrigation.

4. **Social and Economic Impacts:** Access to irrigation can have significant social and economic implications for communities. Ensuring equitable distribution of water resources, promoting small-scale irrigation projects, and considering the needs of local farmers are crucial for sustainable irrigation practices.

In conclusion, irrigation is a dynamic and evolving field that intersects with environmental, social, and technological dimensions. The efficient and sustainable management of water resources in agriculture is crucial for global food security, and ongoing research and innovation in irrigation practices are essential for meeting the challenges of a changing climate and growing population.

2. Role of Vegetation, Conservation Tillage, and Mulch in Various Land and Climate Conditions

1. Vegetation

Arid and Semi-Arid Regions: In arid and semi-arid regions, where water scarcity is a constant challenge, the strategic use of vegetation becomes paramount. Native plants with deep root systems play a dual role in stabilizing soil structure and tapping into groundwater resources. The conservation of water through the employment of cover crops, specifically chosen for their drought-resistant qualities, becomes an integral part of sustainable land management. These practices not only prevent soil erosion but also contribute to the replenishment of groundwater tables.

Tropical Rainforests: The lush vegetation in tropical rainforests, characterized by high rainfall, serves as a natural shield against soil erosion. The dense canopy intercepts rainfall, preventing the impact of heavy rains on the forest floor. This not only maintains the integrity of the soil structure but also prevents nutrient leaching. In agricultural settings, agroforestry practices that mirror the structure of natural forests create resilient ecosystems. Such systems promote biodiversity, protect against extreme weather events, and provide microclimates conducive to various crops.

Temperate Climates: In temperate climates, the role of vegetation is pivotal for both erosion control and biodiversity support. Introducing cover crops during non-cropping seasons protects against soil erosion. The roots of these cover crops help bind the soil, preventing it from being washed away during rainfall. Additionally, agroforestry practices, such as windbreaks and alley cropping, contribute to microclimate regulation. Windbreaks reduce wind erosion, and diverse plantings create habitats for beneficial insects, enhancing overall ecosystem resilience.

2. Conservation Tillage

Erosion-Prone Areas: Conservation tillage practices are indispensable in erosion-prone areas. Implementing techniques like contour ploughing or strip tillage minimizes soil disturbance, preventing water runoff and erosion. The combination of conservation tillage with cover crops acts as a formidable defense against soil erosion. The residue left on the field surface acts as a protective layer, shielding the soil from the impact of heavy rains and promoting water infiltration.

Dryland Farming: In regions relying on dryland farming, where water is a limiting factor, conservation tillage assumes a critical role. Adopting minimal tillage practices or transitioning to no-till systems helps conserve precious soil moisture. The residue from previous crops left on the field surface functions as a natural mulch, reducing water evaporation. This promotes improved water-use efficiency, allowing crops to endure dry periods with greater resilience.

Highly Productive Soils: Conservation tillage practices are equally beneficial in areas with naturally fertile soils. Excessive tillage can disrupt the soil structure and deplete organic matter. Adopting reduced tillage or direct seeding methods helps maintain the inherent productivity of the soil. Preservation of microbial activity and nutrient cycling contributes to sustained soil health and long-term agricultural productivity.

3. Mulch

Desert and Arid Regions: In desert and arid regions, where water conservation is critical, mulching emerges as a vital strategy. Organic mulches, such as straw or compost, create a protective layer that significantly reduces water evaporation from the soil surface. This serves as a crucial water-saving technique, promoting moisture retention and enhancing the efficiency of water use in agriculture.

Cold Climates: In cold climates, where freezing temperatures pose challenges to plant health, mulching becomes an indispensable practice. Mulch serves as insulation, preventing frost heaving and protecting plant roots from the detrimental effects of repeated freezing and thawing. Additionally, mulch provides a gradual thaw in spring, reducing abrupt temperature fluctuations that can harm emerging plants.

Intensive Agriculture: In intensively cultivated areas, where high-value crops are grown, mulching is widely employed to optimize crop productivity. The use of plastic or synthetic mulches suppresses weed growth, conserves soil moisture, and maintains warmer soil temperatures. This is particularly crucial for crops like vegetables, where precise control over environmental conditions is essential. Organic mulches, such as straw or bark, contribute not only to weed control and moisture retention but also enhance soil fertility over time.

3. Biological Measures in Different Climatic Regions

Sustainable agriculture necessitates tailored biological measures that harmonize with the unique challenges presented by varied climatic conditions. The implementation of these measures contributes to soil health, water conservation, and overall ecosystem resilience.

1. Dryland Agriculture

Agroforestry Systems: In dryland areas, where water is a precious resource, agroforestry systems play a pivotal role. Introducing trees and shrubs into the landscape helps conserve soil moisture. The deep root systems of trees can access groundwater, providing stability to the ecosystem. The tree canopy acts as a shield against the harsh sun, reducing evaporation from the soil surface.

Cover Cropping: Dryland agriculture benefits from cover cropping practices. Utilizing drought-tolerant cover crops, such as legumes or grasses, prevents soil erosion and enhances soil fertility. The ground cover formed by these crops shields the soil, reducing evaporation and improving water retention.

Water Harvesting: Biological measures in drylands include water harvesting techniques that optimize rainfall utilization. Practices like contour ploughing and planting create micro-catchments, capturing rainwater and promoting its infiltration into the soil. This method helps enhance soil moisture availability for crops during dry periods.

2. Rainfed Agriculture

Agroecological Farming Practices: Rainfed agriculture thrives with the implementation of agroecological practices. Crop diversification and intercropping create resilient ecosystems. Diverse plantings enhance biodiversity, fostering natural pest control and nutrient cycling. Companion planting exploits symbiotic relationships between different crops for maximum resource utilization.

Water Conservation through Organic Mulching: Organic mulching is a valuable biological measure in rainfed regions. Applying materials like straw or crop residues as mulch reduces water evaporation from the soil surface. This protective layer conserves soil moisture, providing a buffer against fluctuations in rainfall patterns.

Conservation Tillage: Practicing conservation tillage in rainfed regions helps retain soil moisture. Techniques like reduced tillage or no-till minimize soil disturbance, preserving soil structure and reducing the risk of water runoff. This approach contributes to sustained soil health and improved water-use efficiency.

3. Arid and Semi-Arid Regions

Drought-Resistant Crop Varieties: Selecting crop varieties adapted to water scarcity is imperative in arid and semi-arid regions. Drought-resistant crops can endure periods of water stress, maintaining productivity in challenging conditions. Breeding programs that emphasize drought tolerance contribute to the resilience of agricultural systems.

Water-Efficient Irrigation Systems: Biological measures in arid regions include the adoption of water-efficient irrigation systems. Drip or sprinkler irrigation delivers water precisely to the root zone, minimizing wastage through evaporation. These systems optimize water use, a critical consideration in regions where water resources are limited.

Agroforestry for Microclimate Regulation: Agroforestry practices, incorporating drought-resistant tree species, contribute to microclimate regulation. The presence of trees helps create a more favorable environment for crops. Reduced wind velocity, shaded areas, and improved soil moisture retention are benefits of agroforestry in arid and semi-arid regions.

4. Humid Lands

Integrated Pest Management (IPM): Humid regions often face heightened pest and disease pressures. IPM involves biological control measures to mitigate the impact of pests. Introducing natural predators, using pest-resistant crop varieties, and employing crop rotation are strategies that reduce reliance on chemical pesticides, promoting environmental sustainability.

Crop Rotation and Polyculture: Diverse cropping systems, including crop rotation and polyculture, are essential in humid lands. These practices manage soil fertility effectively and mitigate the risk of diseases. The rotation of crops disrupts pest and disease cycles, while polyculture promotes biodiversity and resilience.

Wetland Conservation for Water Quality: Preserving wetlands is a biological measure that significantly contributes to water quality in humid regions. Wetlands act as natural filters, trapping pollutants and sediment from water runoff. Conserving these ecosystems ensures the maintenance of water quality and supports overall environmental health.

Implementing these biological measures requires a nuanced understanding of local conditions and a commitment to sustainable agricultural practices. Tailoring these measures to the specific challenges posed by each climatic region fosters resilient ecosystems and ensures the long-term viability of agricultural systems.

4. Water Use Efficiency and Soil Fertility

The dynamic interaction between water use efficiency (WUE) and soil fertility is fundamental to the success of sustainable agriculture. As farmers navigate the challenges of water scarcity, climate variability, and the need for increased food production, understanding the symbiotic relationship between WUE and soil fertility becomes paramount. Let's delve deeper into the multifaceted aspects of this crucial alliance:

1. Water Use Efficiency (WUE)

Definition and Importance: Water use efficiency is not merely a measure; it is a comprehensive approach to optimizing water resources in agriculture. Maximizing WUE is essential for mitigating the impact of water scarcity and ensuring sustainable agricultural practices. As water availability becomes increasingly unpredictable, improving WUE becomes a strategic imperative for farmers globally.

Factors Influencing WUE

a. Irrigation Practices: Efficient irrigation practices play a central role in enhancing WUE. Drip or sprinkler irrigation systems deliver water precisely to the root zone, minimizing wastage through evaporation. Furthermore, adopting modern precision irrigation technologies allows for intelligent water management based on real-time data.

b. Crop Selection: The choice of crop varieties profoundly influences WUE. Selecting drought-resistant or water-efficient cultivars ensures that crops can thrive under limited water conditions. Research and breeding programs that emphasize WUE contribute to the development of resilient agricultural systems.

c. Soil Management: Soil conditions significantly impact WUE. Practices like conservation tillage, cover cropping, and organic matter incorporation improve soil structure, enhancing water infiltration and retention. Well-structured soils enable effective root development, allowing plants to access water throughout the growing season.

d. Climate and Weather Patterns: Adapting agricultural practices to local climate and weather conditions is crucial for optimizing WUE. Strategies such as adjusting planting times, selecting suitable crops for specific seasons, and incorporating climate-smart practices ensure efficient water utilization under varying climatic scenarios.

2. Soil Fertility

Definition and Importance: Soil fertility is the bedrock of agricultural productivity. It encompasses the soil's ability to provide essential nutrients for plant growth. Fertile soils support vigorous plant development, contribute to high yields, and play a vital role in sustaining the overall health of agricultural ecosystems.

Factors Influencing Soil Fertility (Table 1)

Table 1: Factors influencing Soil Fertility

S.No	Factor	Description
1	Soil pH	pH level affects nutrient availability; most crops prefer a slightly acidic to neutral pH for optimal nutrient uptake.
2	Soil Organic Matter	Decomposed plant and animal material contribute to soil structure, water retention, and nutrient availability.
3	Nutrient Content	Levels of essential nutrients such as nitrogen, phosphorus, potassium, calcium, magnesium, and trace elements influence fertility.
4	Texture and Structure	Soil composition (sand, silt, clay) and structure impact water retention, aeration, and root penetration, affecting nutrient availability.
5	Cation Exchange Capacity (CEC)	CEC represents the soil's ability to hold and exchange cations (positively charged ions), influencing nutrient retention and availability.
6	Microbial Activity	Microorganisms play a vital role in nutrient cycling, organic matter decomposition, and the release of nutrients for plant uptake.
7	Climate and Weathering	Climate influences the rate of weathering, affecting the breakdown of rocks and minerals into essential nutrients for plants.
8	Topography	Slope and landscape characteristics impact water drainage, erosion, and nutrient distribution within the soil profile.
9	Parent Material	The geological material from which the soil is formed influences its initial nutrient content and mineral composition.
10	Crop Rotation and Cover Crops	Diverse plant species contribute different nutrients to the soil; crop rotation and cover crops help maintain and enhance soil fertility.
11	Irrigation and Drainage	Proper water management influences nutrient leaching, availability, and overall soil health.
12	Human Activities	Agricultural practices, fertilization, and land use impact soil fertility; improper management can lead to nutrient depletion.

a. Nutrient Content: Balanced nutrient levels are essential for soil fertility. Nitrogen, phosphorus, potassium, and micronutrients must be present in adequate quantities and proportions to support optimal plant growth. Soil testing and precision nutrient management contribute to maintaining nutrient balance.

b. Organic Matter: Organic matter is a cornerstone of soil fertility. It improves soil structure, water retention, and nutrient-holding capacity. Practices such as cover cropping, green manuring, and the application of compost contribute to increasing organic matter content.

c. Microbial Activity: The intricate web of soil microorganisms influences nutrient cycling and organic matter decomposition. Healthy microbial communities enhance nutrient availability, supporting plant growth and contributing to soil fertility. Practices like avoiding excessive use of chemical fertilizers and incorporating organic amendments promote microbial diversity.

d. pH and Soil Structure: Optimal soil pH and well-structured soils are critical for soil fertility. Adjusting soil pH to suitable levels ensures nutrient availability, while practices like conservation tillage and organic matter incorporation contribute to favorable soil structure.

Symbiotic Relationship

1. Nutrient Availability and WUE

- Adequate soil fertility ensures a ready supply of essential nutrients, promoting robust plant development. This, in turn, enhances the physiological processes within plants, optimizing water use efficiency.
- Nutrient-rich soils support extensive root systems, enabling efficient water uptake by plants, particularly during critical growth stages.

2. Soil Structure and Water Retention

- Soil fertility, influenced by nutrient content and organic matter, contributes to improved soil structure. Well-structured soils enhance water retention, promoting efficient water use by plants.
- The enhanced water retention capacity of fertile soils ensures a continuous water supply to plants, particularly during periods of water stress.

3. Organic Matter and Microbial Activity

- Increasing soil fertility through organic matter addition boosts microbial activity. Microorganisms play a pivotal role in nutrient cycling and the breakdown of organic materials.

- Enhanced microbial activity results in improved nutrient availability for plants, fostering efficient nutrient uptake and contributing to overall soil fertility.

4. Irrigation Management

- Understanding soil fertility informs irrigation management decisions. Nutrient-rich soils may require adjustments in irrigation scheduling to match crop water needs and optimize water use efficiency.
- Efficient irrigation practices contribute to maintaining soil fertility by avoiding waterlogging or leaching of nutrients, ensuring a balanced nutrient supply to plants.

Challenges and Opportunities

1. Climate Change Impacts

- Changing climate patterns pose challenges to both WUE and soil fertility. Erratic rainfall, prolonged droughts, or extreme weather events can disrupt water availability and nutrient cycling.
- Opportunities lie in adopting climate-smart agricultural practices, such as precision irrigation, resilient crop varieties, and sustainable soil management, to mitigate the impacts of climate change.

2. Technological Innovations

- Advancements in technology, such as sensor-based irrigation systems, remote sensing for nutrient management, and precision agriculture, offer opportunities to enhance both WUE and soil fertility.
- Integrating these technologies allows for real-time monitoring and data-driven decision-making, optimizing resource use and improving overall agricultural sustainability.

3. Integrated Farming Systems

- Embracing integrated farming systems that combine crop production with livestock, agroforestry, and aquaculture provides a holistic approach to enhancing both WUE and soil fertility.
- Diversified farming systems contribute to nutrient cycling, organic matter addition, and improved water management through synergies between different components.

3. Selection of Drought-Tolerant Plants

Amidst the growing challenges posed by climate change and water scarcity, the meticulous selection of drought-tolerant plants stands as a critical strategy for ensuring agricultural sustainability. The process of choosing and cultivating plants that can withstand water scarcity not only safeguards agricultural systems but also contributes to the overall resilience of ecosystems.

1. Definition of Drought-Tolerant Plants:

Drought-tolerant plants, often referred to as xerophytes, showcase unique adaptations that enable them to thrive in arid or semi-arid environments where water is limited. These adaptations can manifest in various ways, including the development of deep root systems, the presence of water storage tissues, reduced transpiration rates, and the activation of efficient water-use mechanisms.

2. Criteria for Selection

a. Water Use Efficiency: Choosing plants with high water use efficiency is paramount. These plants can extract and utilize water more effectively, ensuring optimal growth even in conditions of water scarcity. Crops utilizing a C4 photosynthetic pathway, such as maize and sorghum, exemplify efficient water use.

b. Deep Root Systems: Plants equipped with deep root systems have the ability to tap into soil moisture at greater depths. This adaptation allows them to access water even when surface conditions are dry. Deep-rooted perennials like alfalfa and specific grass species epitomize this characteristic.

c. Succulence and Water Storage: Succulent plants have the capacity to store water in specialized tissues, allowing them to endure extended periods of drought. Cacti and certain varieties of agave are excellent examples of succulent plants adapted to arid environments.

d. Drought-Responsive Genes: Selecting plant varieties with genes that respond positively to drought stress is crucial. Modern breeding techniques, including molecular breeding and genetic modification, offer avenues to enhance drought tolerance in crops, ensuring resilience in the face of changing climatic conditions.

3. Role of Grasses and Legumes in Conservation, Pasture, and Rangeland Management

a. Grasses

Soil Conservation: Grasses play a pivotal role in preventing soil erosion. Their extensive root systems help bind soil particles, reducing the risk of water runoff and erosion. Perennial grasses, such as switchgrass and fescue, are particularly effective in soil conservation, stabilizing landscapes and protecting against degradation (Fig 1).

Fig 1: Role of Grasses in Soil and Water Conservation

Forage Production: Grasses are foundational components of forage production systems. They provide valuable grazing resources for livestock, contributing to sustainable pastoral management. Well-managed grasslands support both biodiversity and livestock productivity, creating a harmonious balance in agroecosystems.

Carbon Sequestration: Certain grass species contribute to carbon sequestration in soils, aiding in climate change mitigation. Practices like rotational grazing and improved pasture management enhance carbon storage in grassland ecosystems, fostering environmental sustainability.

b. Legumes

Nitrogen Fixation: Legumes, including clover and alfalfa, possess the unique ability to fix atmospheric nitrogen through symbiotic relationships with nitrogen-fixing bacteria. Introducing legumes into pasture and rangeland systems enhances soil fertility by providing a natural source of nitrogen, reducing the dependence on external inputs.

Biodiversity Support: Legumes contribute to biodiversity in agroecosystems by providing habitat and food for various organisms. They play a vital role in supporting pollinators and beneficial insects, fostering ecological balance in pasture and rangeland environments.

Improved Forage Quality: In mixed pastures, the inclusion of legumes improves overall forage quality. Legumes often have higher protein content than grasses, enhancing the nutritional value of forage for grazing livestock. This not only benefits animal health but also contributes to sustainable livestock production.

4. Conservation, Pasture, and Rangeland Management Improvement

a. Rotational Grazing: Implementing rotational grazing systems is a sustainable practice that prevents overgrazing and allows forage plants to recover. This management approach enhances pasture sustainability by promoting uniform forage growth and reducing soil compaction, ensuring the long-term health of grazing lands.

b. Diversification of Species: Diversifying plant species in pasture and rangeland management promotes ecosystem resilience. A diverse mix of grasses, legumes, and other forbs ensures a balanced and robust forage base, reducing vulnerability to pests and diseases. Biodiverse systems are more adaptable to changing environmental conditions.

c. Water Management: Efficient water management practices, such as rainwater harvesting and improved irrigation techniques, contribute to sustainable pasture and rangeland ecosystems. These practices enhance water availability for forage plants, supporting overall system productivity and reducing reliance on external water sources.

d. Soil Health Promotion: Conservation practices, including cover cropping and minimal tillage, contribute to improved soil health in pasture and rangeland ecosystems. Healthy soils support diverse plant communities, enhancing the resilience of these systems to environmental stresses. Additionally, maintaining soil health ensures the long-term productivity of grazing lands.

e. Adaptive Grazing Strategies: Adopting adaptive grazing strategies, such as mob grazing or holistic planned grazing, allows for better forage utilization and ecosystem regeneration. These strategies mimic natural grazing patterns, promoting soil health, enhancing biodiversity, and contributing to improved pasture resilience over time.

The careful selection of drought-tolerant plants and the strategic roles of grasses and legumes in conservation, pasture, and rangeland management underscore the holistic nature of sustainable agriculture. By embracing resilient plant varieties and implementing effective management strategies, farmers can not only mitigate the impacts of climate variability but also foster ecosystems that are adaptable and robust in the face of changing environmental conditions. This comprehensive approach not only ensures the productivity of agricultural lands but also contributes to the long-term health and sustainability of pastoral landscapes.

4. Management of Waterways, Canal Banks, and Bench Terraces

Amidst the intricate web of sustainable agriculture and land management, the strategic integration of biological means emerges as a fundamental approach to safeguarding the integrity of waterways, canal banks, and bench terraces. These components, crucial to ecosystem health and agricultural productivity, demand a nuanced blend of biological strategies for erosion control, soil conservation, and overall ecological resilience.

1. Waterways Management

Biological Erosion Control: Waterways, with their dynamic flow, require robust biological measures to combat erosion and sedimentation. The introduction of riparian vegetation stands as a key strategy. Native grasses, shrubs, and trees, carefully selected for their adaptive qualities, form a natural buffer along watercourses. Their extensive root systems act as stabilizers, anchoring the soil and reducing erosion. Complementary bioengineering techniques, such as live fascine and brush mattress installations, leverage living plant materials to create erosion-resistant structures, enhancing water quality by preventing sedimentation.

Vegetative Buffer Strips: Establishing vegetative buffer strips along waterways serves as a natural filtration system. Native vegetation, carefully chosen to suit the ecosystem, forms a green barrier that intercepts sediment, nutrients, and pollutants. This biological filter not only enhances water quality but also fosters a healthy aquatic ecosystem. Wetland restoration further contributes to water purification and sediment trapping, acting as a biological powerhouse that supports biodiversity and aids flood control.

2. Canal Bank Management

Grass Cover and Turf: The biological stabilization of canal banks involves the strategic introduction of grass cover and turf. Grasses with strong root systems are meticulously planted to stabilize the soil and prevent erosion. The intricate network of grass roots binds soil particles, reducing the risk of bank collapse. Turf reinforcement mats provide additional support, facilitating the establishment of vegetation and enhancing the overall stability of canal banks.

Biotechnical Solutions: Biotechnical solutions offer a symbiotic blend of living plant materials and traditional engineering techniques. Brush layering and live crib walls exemplify this approach, utilizing vegetation to reinforce canal banks. These living structures mitigate erosion, enhance stability, and contribute to the overall health of the canal ecosystem. An integrated approach that merges biological elements with engineering ingenuity ensures a resilient and sustainable solution.

Integrated Pest Management (IPM): Canal banks, as part of a larger ecosystem, benefit from adopting Integrated Pest Management (IPM) practices. Biological control methods, such as the introduction of natural predators or the use of pest-resistant plants, mitigate the need for chemical interventions. This approach ensures the health of canal bank vegetation, fostering a balanced and thriving ecosystem.

3. Bench Terrace Management

Cover Cropping: Bench terraces, critical for soil conservation on sloping landscapes, find biological support through cover cropping. Carefully selected cover crops, including legumes and grasses, serve as a protective shield against soil erosion. Their robust root systems enhance soil structure, reducing surface runoff and promoting overall terrace stability.

Agroforestry on Terraces: Integrating agroforestry practices on bench terraces introduces a harmonious blend of trees or shrubs alongside crops. These woody plants act as natural windbreaks, reducing soil erosion caused by wind forces. Their intricate root systems contribute to terrace stability, preventing landslides and facilitating water infiltration. Agroforestry transforms bench terraces into dynamic and resilient agroecosystems.

Compost and Organic Amendments: Biological means extend to the incorporation of compost and organic amendments on bench terraces. This practice not only improves soil structure but also enhances water retention. The organic matter fosters a thriving soil microbiome, contributing to terrace health and reducing the risk of erosion. This biological approach ensures a productive and sustainable terrace environment.

4. Common Principles for Biological Management

Native Plant Selection: Central to effective biological management is the selection of native plant species. These plants, adapted to the local ecosystem, promote biodiversity, resilience, and successful integration into the existing landscape.

Biodiversity Enhancement: Encouraging biodiversity within and around waterways, canal banks, and bench terraces is paramount. Diverse ecosystems demonstrate greater resilience to environmental stressors and contribute to the overall health and sustainability of the landscape.

Comprehensive Vegetative Cover: Striving for comprehensive vegetative cover stands as a cornerstone of biological management. A diverse mix of grasses, shrubs, and trees provides varying degrees of soil stabilization, erosion control, and habitat for beneficial organisms. This comprehensive approach ensures a balanced and resilient ecosystem.

Education and Community Involvement: Promoting education and community involvement is essential for the successful implementation of biological management practices. Engaging local communities fosters a sense of ownership and encourages sustainable practices. Knowledge dissemination ensures that biological solutions become an integral part of landscape management.

5. Advanced Soil and Water Conservation Techniques: Sustaining Agricultural Resilience

In the ever-evolving landscape of agriculture, the adoption of advanced soil and water conservation techniques stands as a beacon of progress toward sustainability and resilience. These techniques, leveraging cutting-edge technologies and innovative practices, go beyond conventional methods, addressing the challenges posed by climate change, soil degradation, and water scarcity.

1. Conservation Tillage

Definition and Principles: Conservation tillage represents a paradigm shift in soil management, emphasizing minimal disturbance compared to traditional tillage. The principles include leaving crop residues on the field surface, reducing soil erosion, enhancing water retention, and fostering soil health. Different methods, such as no-till and reduced tillage, offer flexibility in implementation based on crop requirements and soil characteristics.

Benefits

- **Erosion Control:** The primary benefit of conservation tillage is evident in erosion control. By minimizing soil disturbance, the risk of water and wind erosion is significantly reduced, preserving the topsoil and maintaining its structural integrity.
- **Water Retention:** Crop residues left on the field surface act as a protective layer, reducing water evaporation and enhancing water retention. This contributes to increased soil moisture levels, especially during dry periods.
- **Soil Structure Improvement:** Reduced tillage promotes the formation of stable soil aggregates, fostering improved aeration and root growth. This enhances the overall health and resilience of the soil.

2. Precision Agriculture

Technological Integration: Precision agriculture represents a revolutionary approach to farming, integrating advanced technologies for optimized resource management. GPS, sensors, and data analytics play pivotal roles in tailoring field-level management to the specific needs of crops, allowing for precise resource allocation.

Components and Applications

- **GPS-guided Equipment:** Tractors and machinery equipped with GPS technology enable precise control over planting, fertilizing, and harvesting operations. This accuracy ensures optimal use of resources and minimizes waste.

- **Remote Sensing:** Satellite and drone-based remote sensing provide real-time data on crop health, moisture levels, and nutrient status. This information empowers farmers to make informed decisions and implement targeted interventions.
- **Variable Rate Technology (VRT):** VRT allows for the variable application of inputs like fertilizers and pesticides based on the specific needs of different areas within a field. This approach optimizes resource use and reduces environmental impact.

Benefits

- **Resource Efficiency:** Precision agriculture optimizes the use of resources, reducing inputs such as water, fertilizers, and pesticides while maximizing yields. This efficiency contributes to both economic and environmental sustainability.
- **Cost Savings:** Targeting inputs where they are most needed results in cost savings for farmers. Precision agriculture ensures that resources are allocated efficiently, minimizing waste.
- **Environmental Impact:** The precise application of inputs reduces the environmental footprint of agriculture by minimizing nutrient runoff, soil erosion, and the use of agrochemicals.

3. Cover Cropping

Principles and Implementation: Cover cropping is a sustainable agricultural practice involving the planting of non-commercial crops during periods when the primary cash crop is not actively growing. These cover crops serve multiple purposes, including erosion prevention, weed suppression, and enhancement of soil fertility. Common cover crops include legumes, grasses, and brassicas.

Benefits

- **Erosion Control:** One of the primary functions of cover crops is to protect the soil from water and wind erosion, particularly during the offseason when fields may be left bare.
- **Weed Suppression:** The dense growth of cover crops inhibits the growth of weeds, reducing the need for herbicides. This natural weed suppression contributes to sustainable weed management practices.
- **Nutrient Cycling:** Cover crops, especially legumes, have the ability to fix atmospheric nitrogen, enriching the soil with this essential nutrient. When incorporated into the soil, cover crops contribute to improved soil fertility.

4. Agroforestry Practices

Integration of Trees in Agriculture: Agroforestry involves intentionally integrating trees and shrubs into agricultural systems, creating synergies between crop farming and sustainable tree management (Fig 2)

Fig 2: Soil Conservation through Agroforestry

Types of Agroforestry

- **Alley Cropping:** This involves planting rows of trees with alleys of crops in between. The trees provide shade and additional organic matter, benefiting both crops and soil.
- **Silvopasture:** Integrating trees with pasture or forage production for livestock provides shade and diversified forage options, promoting animal well-being and ecosystem resilience.
- **Windbreaks:** Planting rows of trees along field borders protects crops from wind erosion and provides habitat for beneficial organisms.

Benefits

- **Soil Conservation:** Tree roots stabilize the soil, preventing erosion and promoting water infiltration. The presence of trees contributes to the overall health of the soil.
- **Biodiversity Enhancement:** Agroforestry systems support diverse ecosystems, fostering increased biodiversity. The coexistence of trees, crops, and livestock creates a balanced and resilient agricultural landscape.
- **Carbon Sequestration:** Trees in agroforestry systems act as carbon sinks, sequestering carbon from the atmosphere. This contributes to climate change mitigation and overall environmental sustainability.

5. Rainwater Harvesting

Harvesting and Storing Rainwater: Rainwater harvesting involves collecting and storing rainwater for various agricultural purposes. From simple rain barrels to more sophisticated structures like check dams and rooftop harvesting systems, this technique offers a sustainable approach to water management.

Applications

- **Irrigation:** Harvested rainwater can be used for supplemental irrigation during dry periods, reducing reliance on conventional water sources. This is particularly crucial in regions with irregular rainfall patterns.
- **Livestock Watering:** Rainwater harvesting provides a reliable and sustainable water source for livestock, addressing water scarcity concerns in arid and semi-arid areas.
- **Recharge of Aquifers:** In regions experiencing declining groundwater levels, rainwater harvesting contributes to the recharge of aquifers, supporting long-term water availability.

Benefits

- **Water Security:** Rainwater harvesting provides a reliable and decentralized water source, especially in areas with erratic rainfall. It enhances water security for agricultural activities.
- **Reduced Erosion:** By capturing rainwater, soil erosion is minimized, preserving topsoil and maintaining soil fertility. This contributes to overall soil health and agricultural productivity.
- **Cost-Effective:** Rainwater harvesting systems can be cost-effective and relatively simple to implement, making them accessible to a wide range of farmers.

Conclusion

The integration of advanced soil and water conservation techniques represents a transformative shift towards sustainable and resilient agriculture. Conservation tillage, precision agriculture, cover cropping, agroforestry practices, and rainwater harvesting contribute to a holistic approach that addresses the multifaceted challenges faced by agricultural landscapes. These advanced techniques not only mitigate environmental impacts but also foster long-term soil health, water management, and overall agricultural sustainability. As farmers and land managers continue to embrace innovation, these techniques serve as pillars for building a resilient and sustainable future for global agriculture.

5

Precision Agriculture and Technology Integration - Restoration of Chemically Degraded Soils

[1]*Kh. Chandrakumar Singh,* [2]*Savita Jangde,* [3]*Niranjan B.N.*
[4]*Jeetendra Kumar and* [5]*Wajid Hasan*

[1]*VCSGU University of Horticulture and Forestry, Bharsar, Uttarakhand*
[2]*Department of Plant Physiology, Institute of Agricultural Sciences B.H.U. Varanasi, U.P.*
[3]*Dakshina Kannada Milk Union (KMF), Mangalore, Karnataka*
[4]*Krishi Vigyan Kendra, Jehanabad, Bihar Agricultural University Bihar*
[5]*KVK Jehanabad, Bihar Agricultural University Sabour, Bihar*

1. Land Degradation- Type, Factors, Distribution, Processes, and Impacts on Soil Productivity

A. Types of Land Degradation

1. Soil Erosion: Soil erosion, a pervasive form of land degradation, manifests through the detachment and transportation of soil particles by various agents like wind and water. Wind erosion is common in arid regions, where dry, loose soil is susceptible to being carried away by the wind. Water erosion, on the other hand, occurs through surface runoff and can lead to gully formation and the loss of fertile topsoil. This loss of topsoil directly impacts agricultural productivity by reducing nutrient availability and water retention capacity.

Implementing erosion control measures such as contour ploughing, cover cropping, and windbreaks can help mitigate the effects of soil erosion and promote sustainable land management.

2. Deforestation: Deforestation is a major contributor to land degradation, particularly in tropical and subtropical regions. The clearing of forests for agriculture, logging, and urbanization not only results in the direct loss of biodiversity but also exposes the soil to erosion. The intricate root systems of trees play a crucial role in stabilizing soil, and their removal can lead to increased vulnerability to landslides and decreased soil fertility.

Sustainable forestry practices, afforestation initiatives, and reforestation projects are essential to counteract the impacts of deforestation and restore ecosystem balance.

3. Desertification: Desertification is the transformation of once-productive land into arid and desert-like conditions. This process is often driven by a combination of climatic factors, such as prolonged droughts, and human activities like overgrazing and inappropriate agricultural practices. As fertile land turns into desert, soil productivity declines significantly, jeopardizing the livelihoods of communities dependent on agriculture.

Combatting desertification requires a holistic approach involving water management, afforestation, and sustainable agricultural practices tailored to the specific needs of arid regions.

4. Salinization: Salinization occurs when the concentration of salts in the soil becomes elevated, negatively impacting plant growth. This is commonly associated with irrigation practices in arid and semi-arid regions. As water evaporates, salts are left behind in the soil, leading to reduced soil fertility and hampered agricultural productivity. Crops become less tolerant to salinity, further exacerbating the problem.

Implementing proper irrigation techniques, drainage systems, and salt-tolerant crop varieties are crucial strategies to manage and prevent salinization.

5. Urbanization: Rapid urbanization results in the conversion of agricultural land into urban areas, causing soil sealing and fragmentation. Soil sealing prevents water infiltration and disrupts natural drainage systems, leading to increased surface runoff and erosion. The loss of arable land to urban expansion reduces the overall land available for agriculture, impacting food security.

Urban planning strategies that prioritize green spaces, sustainable infrastructure, and brownfield redevelopment can help minimize the negative impacts of urbanization on soil quality.

B. Factors Contributing to Land Degradation

1. Human Activities: Human activities play a pivotal role in land degradation. Overgrazing by livestock, unsustainable agricultural practices such as monoculture and excessive use of agrochemicals, and improper land management contribute to soil erosion, nutrient depletion, and loss of biodiversity. The expansion of urban areas and infrastructure further intensifies the pressure on land resources.

Adopting sustainable agricultural practices, promoting agroecology, and implementing effective land-use planning are essential in mitigating the adverse impacts of human activities on land.

2. Climate Change: Climate change exacerbates land degradation through alterations in temperature, precipitation patterns, and the frequency of extreme weather events. Increased temperatures and changing rainfall patterns can lead to more frequent droughts and floods, further stressing ecosystems and soil resilience. These climatic shifts contribute to desertification, soil erosion, and reduced agricultural productivity.

Mitigating climate change impacts on land involves global efforts to reduce greenhouse gas emissions, adapt agricultural practices to changing conditions, and implement strategies for sustainable resource management.

3. Poor Land Management: Improper land management practices, such as excessive use of irrigation, inadequate waste disposal, and deforestation without proper reforestation, contribute significantly to land degradation. Poorly planned irrigation systems can lead to waterlogging and salinization, while deforestation disrupts the natural balance of ecosystems, impacting soil stability and fertility.

Implementing sustainable land management practices, precision agriculture, and integrated watershed management can help address the challenges associated with poor land management.

4. Population Pressure: The increasing global population exerts pressure on land resources, leading to intensified land use for agriculture, housing, and infrastructure. The conversion of natural landscapes into agricultural fields and urban areas results in habitat loss, fragmentation, and soil degradation.

Sustainable population management strategies, coupled with responsible land-use planning, are crucial in addressing the challenges posed by population pressure on land resources.

C. Distribution of Land Degradation

Land degradation is a global phenomenon, affecting diverse regions with varying intensity. While arid and semi-arid areas are particularly susceptible due to their fragile ecosystems and limited water availability, no geographical region is immune. In fact, some temperate and tropical regions also experience severe land degradation due to a combination of human activities and climatic factors.

In arid and semi-arid regions, the lack of vegetation cover, coupled with irregular precipitation, creates conditions conducive to desertification. However, humid tropical regions are not exempt, as deforestation, unsustainable agriculture, and infrastructure development contribute to soil erosion and loss of biodiversity. In temperate zones, intensive agriculture and urbanization can lead to soil compaction and nutrient depletion.

Addressing land degradation requires a nuanced understanding of regional variations and tailoring intervention strategies to specific environmental contexts. International cooperation and information exchange are vital for implementing effective measures to combat land degradation globally.

Processes of Land Degradation

1. Erosion: Erosion is a dynamic process involving the detachment, transport, and deposition of soil particles. Water erosion occurs in several forms, including sheet erosion (thin layers of soil being removed uniformly), rill erosion (small channels forming on the surface), and gully erosion (larger, deeper channels). Wind erosion is characterized by the lifting and movement of soil particles by wind, often leading to the formation of sand dunes.

Sustainable land management practices, such as contour ploughing, cover cropping, and agroforestry, are effective measures to minimize erosion and maintain soil health.

2. Compaction: Soil compaction results from the compression of soil particles, reducing pore spaces and limiting water infiltration. Heavy machinery, intensive agricultural activities, and frequent foot traffic contribute to soil compaction. Compacted soils have decreased porosity, impeding root growth and nutrient uptake by plants.

Implementing reduced tillage practices, rotating crops, and using cover crops can alleviate soil compaction and improve overall soil structure.

3. Salinization: Salinization occurs when the concentration of salts in the soil exceeds the tolerance levels of crops. This process is often associated with improper irrigation practices in arid and semi-arid regions. As water evaporates, salts accumulate in the root zone, hindering plant growth and reducing agricultural productivity.

Sustainable irrigation methods, such as drip irrigation and proper drainage systems, are essential in preventing and managing salinization. Additionally, selecting salt-tolerant crop varieties can mitigate the impacts of elevated soil salinity.

4. Pollution: Pollution of soil by industrial and agricultural runoff introduces contaminants such as heavy metals, pesticides, and fertilizers. These pollutants adversely affect soil health, microbial activity, and the overall ecosystem. Soil pollution can lead to the contamination of groundwater, posing risks to human health and the environment.

Implementing pollution control measures, promoting organic farming practices, and adopting sustainable waste management strategies are crucial in preventing soil pollution and its detrimental effects.

Understanding the processes driving land degradation is essential for developing targeted and effective mitigation strategies. Sustainable land management practices that promote soil conservation, water management, and biodiversity conservation are key components of efforts to combat the various processes contributing to land degradation.

D. Impacts on Soil Productivity

1. Loss of Fertility: The loss of fertile topsoil through erosion and depletion of essential nutrients significantly reduces soil fertility. Fertile topsoil is crucial for plant growth, providing nutrients and supporting microbial activity. As soil fertility declines, agricultural productivity is compromised, leading to decreased crop yields and food insecurity.

Implementing soil conservation practices, such as cover cropping, crop rotation, and organic farming, helps maintain soil fertility and sustains agricultural productivity.

2. Reduced Water Retention: Eroded and degraded soils have reduced water retention capacity, exacerbating the impacts of drought. Soil erosion removes the finer, more water-absorbent particles, leaving behind a coarser, less porous substrate that struggles to retain moisture. This reduction in water retention makes crops more vulnerable to water stress and negatively affects overall water availability.

Adopting water conservation techniques, such as mulching, agroforestry, and rainwater harvesting, is essential to enhance soil water retention and mitigate the impacts of reduced water availability.

3. Decline in Biodiversity: Land degradation contributes to habitat loss and fragmentation, leading to a decline in biodiversity. The destruction of natural habitats and the degradation of ecosystems disrupt the balance of plant and animal species. This loss of biodiversity not only affects the resilience of ecosystems but also hampers their ability to provide essential services, such as pollination and pest control.

Biodiversity conservation measures, including protected areas, habitat restoration, and sustainable land-use planning, are crucial for maintaining ecological balance and supporting soil productivity.

4. Food Insecurity: Land degradation directly threatens global food security by reducing the productivity of agricultural lands. As soil fertility declines and water retention capacity diminishes, farmers face challenges in achieving optimal crop yields. This, in turn, contributes to food shortages, price volatility, and increased dependence on external sources for food supply.

Sustainable agricultural practices, precision farming, and agroecological approaches are key strategies to enhance food security and build resilience against the impacts of land degradation.

5. Increased Vulnerability to Climate Change: Degraded land is more susceptible to the adverse impacts of climate change, including extreme weather events such as floods and droughts. The loss of vegetation cover, disruption of natural drainage systems, and compromised soil structure amplify the effects of climate-induced challenges. This increased vulnerability further exacerbates the negative consequences on soil productivity and agricultural sustainability.

Integrated approaches that address both land degradation and climate change are essential for building resilience in agricultural systems and ensuring the long-term sustainability of food production.

Land degradation is a complex and multifaceted challenge that requires concerted global efforts to address its various dimensions. From understanding the different types and factors contributing to land degradation to examining its distribution, processes, and impacts on soil productivity, a holistic approach is necessary. Implementing sustainable land management practices, promoting conservation and restoration initiatives, and fostering international cooperation are pivotal in mitigating the adverse effects of land degradation and ensuring the long-term health of our land resources.

2. Formation, nature, and properties of problem soils

A. Formation of Problem Soils

1. Parent Material: The parent material plays a crucial role in shaping the characteristics of soils. For problem soils, the parent material may contain high concentrations of specific minerals that contribute to challenging soil properties. In some instances, the weathering of rocks rich in clay minerals can lead to the formation of soils with high clay content. This characteristic makes the soil prone to compaction, reducing its permeability and hindering root growth.

Additionally, parent materials with elevated levels of certain salts can contribute to the development of saline or sodic soils. The mineral composition of the parent material sets the stage for the potential challenges that the soil may pose to agriculture and other land uses.

2. Climate: Climate influences the formation of problem soils through its impact on weathering, precipitation, and evaporation. In arid and semi-arid regions, where evaporation rates exceed precipitation, the concentration of salts in the soil tends to increase. This leads to the formation of saline soils, which can be detrimental to plant growth due to the high salinity levels inhibiting water uptake.

In contrast, humid regions may experience intense weathering, resulting in soils with high aluminum or iron content. The nature of problem soils is thus intimately connected with the prevailing climatic conditions, highlighting the need for region-specific approaches to soil management.

3. Topography: The topography of an area contributes to soil formation and influences the nature of problem soils. Sloping areas are more susceptible to erosion, leading to the development of soils with poor structure and reduced fertility. The erosion of topsoil not only affects nutrient availability but also compromises the overall stability of the soil.

Conversely, flat or depressional areas may be prone to waterlogging, especially if drainage is inadequate. Poorly drained soils can adversely impact root health, reduce oxygen availability, and create anaerobic conditions that hinder nutrient uptake.

4. Vegetation: The type of vegetation covering an area significantly influences soil properties. Different plant species contribute to the accumulation or depletion of nutrients, organic matter, and allelopathic compounds. For instance, certain leguminous plants have nitrogen-fixing capabilities, enhancing soil fertility. Conversely, some plants may release compounds that inhibit the growth of other plants, affecting overall biodiversity.

Understanding the interaction between vegetation and soil is vital for sustainable land management. The removal of native vegetation, such as deforestation, can disrupt the delicate balance of soil ecosystems and contribute to the development of problem soils.

5. Anthropogenic Activities: Human activities can accelerate the formation of problem soils. Improper land management practices, including excessive use of agrochemicals, deforestation, and improper waste disposal, can lead to soil degradation. For example, the misuse of fertilizers can result in nutrient imbalances and soil acidification. Deforestation may expose soils to erosion, compromising their structure and fertility.

Mining activities introduce disturbances to the soil structure, altering its composition and potentially leading to the formation of soils with unfavorable characteristics. Recognizing the role of human activities in soil formation is crucial for developing sustainable practices that mitigate the impact of anthropogenic influences.

B. Nature of Problem Soils

1. Saline Soils: Saline soils are characterized by high concentrations of soluble salts, posing challenges to agriculture. The accumulation of salts can result from natural processes, such as weathering of minerals, or human activities, particularly irrigation with salty water. The presence of excess salts in the soil creates an osmotic imbalance, making it difficult for plants to absorb water.

Management strategies for saline soils often involve leaching with low-salt water to flush out accumulated salts. Additionally, incorporating organic matter can improve soil structure and enhance nutrient availability, facilitating the growth of salt-tolerant crops.

2. Sodic Soils: Sodic soils are characterized by high levels of sodium, which can adversely affect soil structure. The excessive sodium content disrupts the formation of stable aggregates, leading to poor drainage and reduced water infiltration. Sodic soils often exhibit a high pH, making them unsuitable for many crops.

Gypsum application is a common strategy for managing sodic soils. Gypsum displaces sodium ions, allowing for the formation of stable soil aggregates. This process improves soil structure and promotes better water penetration. Additionally, incorporating organic amendments helps enhance cation exchange capacity and mitigate the detrimental effects of high sodium levels.

3. Acidic Soils: Acidic soils have a low pH, primarily caused by leaching of basic cations, acid rain, or the decomposition of organic matter. Low pH can lead to nutrient imbalances and toxicities, particularly with aluminum and manganese. Plant roots may struggle to absorb essential nutrients under acidic conditions.

Liming is a common practice to raise the pH of acidic soils. Adding lime materials, such as calcium carbonate, helps neutralize acidity and improve nutrient availability. Soil testing is essential for determining the appropriate amount of lime needed for effective pH adjustment.

4. Alkaline Soils: Alkaline soils have a high pH, often resulting from the presence of excess basic cations or carbonates in the soil. High pH can limit the availability of certain nutrients, hindering plant growth. Sodium carbonate-rich soils, also known as sodic soils, fall into the alkaline category.

Soil amendments, such as elemental sulfur or acidifying agents, are used to lower the pH of alkaline soils. Acidification helps enhance nutrient availability and promotes the growth of plants adapted to slightly acidic to neutral pH ranges.

5. Erosive Soils: Soils prone to erosion often have poor structure and low organic matter content. Factors contributing to soil erosion include deforestation, improper agricultural practices, and overgrazing. Erosion leads to the loss of topsoil, which is rich in nutrients and crucial for plant growth.

Implementing soil conservation practices is essential for managing erosive soils. Techniques like contour ploughing, cover cropping, and agroforestry help stabilize the soil, reduce surface runoff, and prevent erosion. These practices contribute to the restoration of soil structure and fertility.

C. Properties of Problem Soils

1. Texture: The texture of problem soils can vary, but many exhibit fine textures such as clay. Fine-textured soils have good nutrient-holding capacity but may suffer from issues like poor drainage and compaction. Clay soils have smaller particles, leading to higher water retention but reduced aeration.

Management strategies for fine-textured soils often involve incorporating organic matter to improve soil structure and drainage. Cover cropping and crop rotation can also contribute to the enhancement of soil tilth.

2. Structure: Problem soils often exhibit poor soil structure, affecting water infiltration and root penetration. Compaction, caused by factors like heavy machinery or intense foot traffic, results in decreased pore spaces and limited oxygen availability to plant roots.

Soil aeration practices, such as deep tillage or aeration equipment, can help alleviate compaction issues. Additionally, incorporating cover crops with deep root systems can contribute to soil structure improvement.

3. Nutrient Content: Nutrient deficiencies or imbalances are common in problem soils, impacting plant growth. Saline and sodic soils may have impaired nutrient uptake due to high levels of salts or sodium. Acidic soils can have low calcium and magnesium availability, while alkaline soils may limit the uptake of certain micronutrients.

Soil testing is a critical step in understanding nutrient deficiencies and formulating appropriate fertilization strategies. Targeted nutrient applications, crop rotation, and organic amendments can help address nutrient imbalances in problem soils.

4. Organic Matter: The organic matter content in problem soils can vary based on land use and management practices. Erosive soils may have low organic matter due to topsoil loss, while poorly managed agricultural soils may suffer from organic matter depletion. Organic matter is crucial for soil structure, water retention, and nutrient availability.

Incorporating organic amendments, such as compost or cover crops, is a fundamental practice for improving organic matter content. These additions enhance microbial activity, contribute to soil aggregation, and promote overall soil health.

5. Water-Holding Capacity: The water-holding capacity of problem soils is often compromised. Saline soils may have poor water retention due to high salt concentrations, while poorly structured soils may suffer from waterlogging or inadequate drainage. The ability of the soil to hold water is critical for sustaining plant growth, especially during dry periods.

Implementing water management practices, such as mulching and controlled drainage systems, helps regulate soil moisture levels. Techniques like rainwater harvesting can also contribute to optimizing water availability for plants.

6. pH Levels: The pH of problem soils can range from highly acidic to highly alkaline. This variation significantly influences nutrient availability and microbial activity. Certain plants thrive in specific pH ranges, and the suitability of crops for a particular soil depends on its pH.

Regular soil testing is essential for monitoring pH levels and determining the need for corrective measures. Liming or acidifying agents can be applied to adjust pH and create an environment conducive to optimal plant growth.

D. Management and Rehabilitation

1. Amelioration Techniques: Amelioration techniques aim to improve the physical, chemical, and biological properties of problem soils. For saline soils, leaching with low-salt water is a common practice to remove excess salts. Incorporating organic matter, such as compost or manure, enhances soil structure and nutrient availability. Planting salt-tolerant crops, like halophytes, can also contribute to soil rehabilitation in saline areas.

Sodic soils benefit from gypsum application, as it displaces sodium ions and promotes the formation of stable soil aggregates. This improves soil structure and drainage. Combining gypsum application with organic amendments further enhances the effectiveness of rehabilitation efforts.

2. Soil Conservation Practices: Soil conservation practices are crucial for preventing erosion and maintaining soil health. Contour ploughing helps reduce water runoff on sloping terrain, minimizing soil erosion. Cover cropping involves planting vegetation that protects the soil surface, reduces erosion, and adds organic matter when incorporated. Agroforestry integrates trees with crops, providing additional root systems to stabilize the soil.

Terracing is another effective soil conservation practice for managing hilly landscapes. These techniques collectively contribute to the restoration of soil structure, preservation of topsoil, and prevention of further degradation.

3. Precision Agriculture: Precision agriculture utilizes technology to optimize resource use and improve crop production. Soil sensors and mapping tools help farmers understand the variability in soil properties within their fields. This information guides precise applications of fertilizers and irrigation, reducing input waste and minimizing environmental impact.

Precision agriculture promotes sustainable land management by tailoring interventions to the specific needs of problem soils. It enhances resource efficiency and supports the long-term health of the soil and surrounding ecosystems.

4. Afforestation and Green Manure: Afforestation involves planting trees on degraded lands to restore soil health and biodiversity. Trees contribute organic matter, stabilize the soil, and enhance nutrient cycling. Green manure crops, such as legumes, are grown specifically to improve soil fertility. When incorporated into the soil, green manure adds organic matter, fixes nitrogen, and enhances nutrient availability.

Afforestation and green manure practices play a vital role in restoring the ecological balance of problem soils. These approaches contribute to long-term soil improvement and support sustainable land use.

5. Soil pH Management: Managing soil pH is essential for optimizing nutrient availability and supporting plant growth. For acidic soils, liming materials like calcium carbonate are applied to raise pH. Soil testing helps determine the appropriate amount of lime needed for effective pH adjustment. In alkaline soils, elemental sulfur or acidifying agents are used to lower pH and improve nutrient availability.

Regular monitoring of soil pH and targeted amendments ensure that the soil environment remains suitable for the crops grown. Proper pH management is a foundational aspect of sustainable soil fertility.

6. Water Management: Efficient water management is critical for problem soils prone to waterlogging or drought. Installing drainage systems, such as subsurface drains, helps alleviate waterlogging issues by facilitating the removal of excess water. Drip irrigation systems provide precise water application, optimizing water use and minimizing the risk of overwatering or waterlogging.

Rainwater harvesting techniques capture and store rainwater for later use, especially in regions with erratic precipitation patterns. These practices contribute to effective water management, supporting plant growth and minimizing the impact of water-related challenges on problem soils.

3. Land restoration and conservation techniques of reclamation of chemically degraded and problem soils

Land restoration and conservation techniques are crucial for reclaiming chemically degraded and problem soils, ensuring their sustainable use for agriculture, forestry, and other purposes. These techniques aim to improve soil structure, fertility, and overall health. Here are comprehensive explanations of various strategies:

1. Cover Cropping and Green Manure

Cover Cropping: Cover cropping is a versatile and effective technique for enhancing soil health and preventing erosion. The choice of cover crops depends on the specific needs of the soil and climate. Leguminous cover crops, such as clover or vetch, fix nitrogen from the atmosphere, enriching the soil with this essential nutrient. Grass-cover crops, like rye or oats, provide biomass, which, when incorporated into the soil, enhances organic matter content.

In regions with chemically degraded soils, cover cropping plays a vital role in preventing further nutrient leaching, erosion, and weed growth. Cover crops act as living mulch, protecting the soil from the impact of raindrops and minimizing surface runoff. Their root systems contribute to soil structure improvement by breaking up compacted layers.

Green Manure: Green manure involves the incorporation of fresh, green plant material into the soil, providing an immediate nutrient boost. Leguminous green manure, such as crimson clover or winter peas, not only enriches the soil with nitrogen but also acts as a natural weed suppressant. These crops are often grown during fallow periods or before planting the main cash crop.

The incorporation of green manure crops into chemically degraded soils aids in nutrient recycling and fosters microbial activity. This practice accelerates the decomposition of organic material, releasing nutrients that become readily available for subsequent crops. Additionally, green manure improves water retention in the soil, crucial for areas prone to drought or erratic precipitation.

2. Agroforestry

Agroforestry is a sustainable land management approach that integrates trees or shrubs with agricultural crops or livestock. This technique is particularly relevant for the reclamation of chemically degraded soils, providing multifaceted benefits:

- **Soil Structure Improvement:** The root systems of trees contribute to soil aeration and help break up compacted layers. This enhances water infiltration and nutrient absorption by plant roots.
- **Nutrient Cycling:** Trees contribute organic matter through fallen leaves, enhancing nutrient cycling in the soil. This organic material improves soil structure and fosters beneficial microbial activity.
- **Erosion Control:** Tree canopies act as natural windbreaks, preventing soil erosion in windy areas. The roots stabilize the soil, reducing the risk of water erosion during heavy rainfall.
- **Diversification of Yield:** Agroforestry systems provide diverse products, including timber, fruits, and fodder. This diversification ensures a more stable income for farmers and promotes sustainable land use.

Implementing agroforestry practices in chemically degraded soils involves selecting tree species that are well-adapted to the local climate and soil conditions. Integration may include alley cropping, where rows of trees are planted between rows of crops, or silvopasture, combining trees with pasture for livestock.

3. Contour Plowing and Terracing

Contour Plowing: Contour ploughing is an effective technique for preventing soil erosion on sloping terrain. It involves ploughing perpendicular to the slope, creating ridges and furrows that follow the contour lines of the land. This method slows down water runoff, allowing it to infiltrate the soil rather than causing erosion.

For chemically degraded soils, contour ploughing helps in preventing the loss of topsoil, which is often rich in essential nutrients. It minimizes the impact of rainfall on the soil surface, reducing the risk of nutrient leaching. This technique is particularly relevant in areas with intense rainfall or where irrigation is applied.

Terracing: Terracing is a more structured approach to preventing soil erosion on steep slopes. It involves constructing steps or platforms on the slope to create level surfaces for cultivation. Each terrace acts as a mini-contour, slowing down water flow and preventing the concentration of erosive forces.

In chemically degraded soils, terracing helps in retaining soil moisture, mitigating the impacts of drought. It provides a stable platform for cultivation and allows for more efficient water use. Terracing is often combined with cover cropping or agroforestry to further enhance soil stability and fertility.

4. Mulching

Mulching is a practice that involves covering the soil surface with organic or inorganic materials to conserve moisture, suppress weeds, and regulate soil temperature. The application of mulch offers several benefits for chemically degraded soils:

- **Moisture Conservation:** Mulch reduces water evaporation from the soil surface, helping to maintain adequate soil moisture. This is particularly crucial in arid or semi-arid regions where water scarcity is a significant concern.
- **Weed Suppression:** Mulch acts as a physical barrier, suppressing weed growth and competition for nutrients. This is important in chemically degraded soils where the presence of weeds can exacerbate nutrient imbalances.
- **Temperature Regulation:** Mulch provides insulation, moderating soil temperature fluctuations. This is beneficial for crops, especially in areas with extreme temperature variations.
- **Organic Matter Addition:** Organic mulches, such as straw or wood chips, gradually decompose, adding organic matter to the soil. This enhances soil structure and microbial activity.

Applying mulch to chemically degraded soils is a cost-effective and sustainable practice. Organic mulches contribute to the gradual improvement of soil health over time.

5. Crop Rotation and Diversification

Crop Rotation: Crop rotation involves systematically changing the types of crops grown in a particular field over time. This practice helps break pest and disease cycles, improve nutrient balance, and enhance soil structure. In chemically degraded soils, crop rotation can be tailored to address specific nutrient deficiencies or imbalances.

Crop rotation with nitrogen-fixing legumes, such as peas or beans, is beneficial for replenishing nitrogen levels in chemically degraded soils. Including deep-rooted crops, like tap-rooted radishes, helps break up compacted layers and improve soil structure.

Diversification: Diversifying crops in a given area is essential for maintaining soil health and preventing the depletion of specific nutrients. This practice can include growing different varieties of the same crop or integrating multiple crops within the same field.

In chemically degraded soils, diversification contributes to the resilience of the agroecosystem. Planting a mix of crops with different nutrient requirements helps balance nutrient availability in the soil. Additionally, diverse crops attract a variety of beneficial organisms, promoting a healthier soil ecosystem.

6. Gypsum Application

Gypsum, a calcium sulfate compound, is a valuable amendment for reclaiming sodic soils. Sodic soils have high sodium levels, leading to poor soil structure and drainage issues. Gypsum application offers several benefits:

- **Sodium Displacement:** Gypsum helps displace sodium ions in the soil, allowing for the preferential adsorption of calcium. This promotes the formation of stable soil aggregates, improving soil structure.
- **Improved Infiltration:** Enhanced soil structure facilitates better water infiltration, reducing the risk of waterlogging in sodic soils.
- **Nutrient Availability:** Gypsum contributes calcium, an essential nutrient, to the soil. It enhances cation exchange capacity, promoting nutrient availability for plants.

Applying gypsum to chemically degraded sodic soils is a targeted and effective strategy for improving soil conditions. It is essential to consider the specific requirements of the soil and conduct soil tests to determine the optimal application rates.

7. Soil Amendments

Lime Application: Lime is a fundamental soil amendment used to correct acidity in soils. In chemically degraded soils with low pH, lime application is essential for several reasons:

- **pH Adjustment:** Lime raises soil pH, neutralizing acidity and creating a more favorable environment for plant growth. This is crucial for nutrient availability and microbial activity.
- **Aluminum and Manganese Toxicity:** Liming helps alleviate issues related to aluminum and manganese toxicity, which can occur in acidic soils and negatively impact plant health.
- **Promotion of Microbial Activity:** Lime fosters microbial activity in the soil, contributing to nutrient cycling and organic matter decomposition.

Conducting soil tests is crucial to determine the appropriate type and amount of lime needed for effective pH adjustment. Over-liming should be avoided, as it may lead to nutrient imbalances.

Organic Amendments: Adding organic amendments, such as compost, manure, or crop residues, is a key practice for improving soil structure and fertility in chemically degraded soils:

- **Organic Matter Addition:** Organic amendments contribute to the buildup of organic matter, enhancing soil structure and water retention.
- **Microbial Activity:** Organic matter serves as a food source for soil microorganisms, promoting beneficial microbial activity. This, in turn, supports nutrient cycling.
- **Nutrient Supply:** Organic amendments release nutrients gradually, providing a sustained nutrient supply for plants. This is especially important in chemically degraded soils where nutrient imbalances may exist.

Incorporating organic amendments is a long-term strategy that requires regular monitoring of soil conditions. Compost, for example, can be produced on-site, utilizing agricultural residues, kitchen waste, and other organic materials.

8. No-Till and Reduced Tillage

Reducing or eliminating tillage minimizes soil disturbance, preserving soil structure and reducing erosion. No-till practices involve planting directly into untilled soil. These techniques have numerous benefits for chemically degraded soils:

- **Soil Structure Preservation:** No-till practices prevent the disruption of soil aggregates, maintaining soil structure and porosity. This is crucial for water infiltration and root penetration.

- **Erosion Control:** By leaving crop residues on the soil surface, no-till reduces water runoff and minimizes soil erosion. This is particularly important in areas prone to erosion and degradation.
- **Carbon Sequestration:** No-till systems contribute to carbon sequestration in the soil. The accumulation of organic matter enhances soil fertility and resilience.

Implementing no-till or reduced tillage practices requires adapting machinery and adjusting planting techniques. Crop residues left on the soil surface act as natural mulch, further improving water retention.

9. Bio-Drainage

Bio-drainage involves planting deep-rooted vegetation in areas prone to waterlogging. This technique is valuable for chemically degraded soils experiencing poor drainage:

- **Excess Water Absorption:** Deep-rooted plants, such as willows or poplars, absorb excess water from the soil, lowering the water table and preventing waterlogging.
- **Improved Drainage:** The presence of deep-rooted plants enhances soil structure and creates channels for water movement, improving overall drainage.
- **Nutrient Uptake:** Bio-drainage plants actively take up nutrients, contributing to nutrient cycling in the soil. This is particularly beneficial in soils with nutrient imbalances.

Bio-drainage is a sustainable alternative to conventional drainage systems and can be integrated into agroforestry or riparian buffer zones. Plant species selection should consider adaptability to local conditions.

10. Precision Agriculture

Precision agriculture utilizes technology to optimize resource use and improve crop production. This approach is highly relevant for chemically degraded soils, offering several advantages:

- **Site-Specific Management:** Precision agriculture employs soil sensors and mapping tools to assess variability in soil properties within a field. This information guides precise applications of fertilizers and irrigation.
- **Resource Efficiency:** By tailoring interventions to the specific needs of the soil, precision agriculture minimizes input waste, reducing environmental impact and operational costs.

- **Data-Driven Decision-Making:** Real-time data on soil moisture, nutrients, and pH enable farmers to make informed decisions about crop management practices. This improves overall land productivity.

Implementing precision agriculture involves investing in technologies such as GPS-guided tractors, drones, and soil sensors. Training farmers in data interpretation and decision-making is essential for successful adoption.

11. Phytoremediation

Phytoremediation is an innovative approach that utilizes specific plants, known as hyperaccumulators, to extract, sequester, or degrade contaminants in the soil. While primarily applied to address soil pollution, phytoremediation can also be relevant for chemically degraded soils:

- **Heavy Metal Uptake:** Hyperaccumulator plants absorb heavy metals from the soil, reducing their concentrations and toxicity.
- **Contaminant Breakdown:** Some plants can break down organic contaminants, contributing to soil detoxification.
- **Soil Improvement:** Phytoremediation plants contribute organic matter and enhance microbial activity, promoting overall soil health.

Identifying suitable hyperaccumulator species for chemically degraded soils requires thorough research and consideration of specific contaminants. Implementation should be carefully monitored to prevent unintended ecological consequences.

12. Integrated Watershed Management

Integrated watershed management considers the interactions between land and water within a specific watershed. This holistic approach addresses both upstream and downstream effects on soil and water quality:

- **Reforestation:** Planting trees in upstream areas helps prevent soil erosion, reducing sedimentation downstream.
- **Contour Bunding:** Constructing contour bunds on sloping terrain minimizes water runoff, preventing soil erosion.
- **Check Dams:** Building check dams along watercourses slows down water flow, reducing the risk of downstream flooding and erosion.

Integrated watershed management involves collaboration between various stakeholders, including farmers, local communities, and government agencies. It requires a comprehensive understanding of the watershed's hydrological dynamics.

13. Windbreaks and Shelterbelts

Planting windbreaks and shelterbelts involves establishing rows of trees or shrubs along field borders to reduce wind erosion. This technique is particularly relevant in arid regions where wind erosion is a significant concern:

- **Wind Speed Reduction:** Trees act as windbreaks, slowing down wind speed and preventing the removal of topsoil through erosion.
- **Microclimate Improvement:** Shelterbelts create a microclimate within the agricultural area, reducing temperature extremes and enhancing soil moisture retention.
- **Biodiversity Promotion:** Planting a variety of tree species contributes to biodiversity, supporting beneficial organisms that contribute to soil health.

Designing effective windbreaks requires consideration of wind direction, tree species selection, and spacing. Proper maintenance is essential for long-term effectiveness.

4. Acid Soils: Nature, Distribution, Formation, and Properties

Acid soils, also known as acidic soils, are characterized by a low pH level, indicating an elevated concentration of hydrogen ions. This acidity can significantly impact the availability of essential nutrients for plant growth, affecting agricultural productivity. Understanding the nature, distribution, formation, and properties of acid soils is crucial for implementing effective soil management practices and ensuring sustainable land use.

Nature of Acid Soils

1. **pH Levels:** Acid soils are defined by a pH level below 7.0, with values typically ranging from 4.0 to 6.5. The low pH is a result of the accumulation of hydrogen ions, often in the form of aluminum and iron ions, which can be toxic to plants at high concentrations. The acidic nature of these soils affects various soil processes, nutrient availability, and microbial activity.
2. **Soil Solution Chemistry:** The soil solution in acid soils is characterized by an abundance of protons (H^+ ions). This high concentration of hydrogen ions influences the solubility of minerals and nutrient ions, leading to leaching and the potential for nutrient deficiencies. Aluminum and iron become more soluble in acidic conditions, posing challenges to plant roots and affecting overall nutrient uptake.
3. **Soil Color:** Acid soils may exhibit certain color characteristics related to the presence of iron. In some cases, these soils may have a reddish or yellowish hue due to the oxidation of iron minerals. The color variations can provide visual indicators of soil acidity, although other factors such as organic matter content can also influence soil color.

Distribution of Acid Soils

1. **Global Distribution:** Acid soils are widespread across the globe, with varying degrees of severity. They are found in both tropical and temperate regions. In tropical areas, weathering processes contribute to soil acidity, while in temperate regions, factors such as vegetation type and rainfall patterns play a role. Acidic soils are commonly associated with specific ecosystems, including coniferous forests and heathlands.

2. **Regional Variations:** The distribution of acid soils can vary within regions due to local geological, climatic, and land-use factors. For example, regions with high rainfall may experience more leaching of basic cations, contributing to soil acidity. Human activities, such as deforestation and agriculture, can also influence the prevalence of acid soils in specific areas.

3. **Landscape Position:** Acid soils can be found in various landscape positions, including uplands, lowlands, and slopes. Sloping terrain may experience increased leaching of basic cations, leading to the development of acidic conditions. The distribution of acid soils within a landscape is influenced by factors such as topography, drainage patterns, and parent material.

Formation of Acid Soils

1. **Weathering Processes:** The primary mechanism leading to the formation of acid soils is the weathering of minerals in parent materials. During weathering, minerals containing basic cations, such as calcium and magnesium, are broken down. The released elements, including aluminum and hydrogen, contribute to soil acidity. Silicate minerals, in particular, undergo weathering processes that release acidic components.

2. **Organic Matter Decomposition:** Organic matter decomposition can contribute to soil acidity. As organic materials break down, organic acids are released, contributing to the accumulation of hydrogen ions in the soil solution. This process is more pronounced in ecosystems with high organic matter turnover, such as peatlands and certain forested areas.

3. **Anthropogenic Activities:** Human activities can accelerate the formation of acid soils. Deforestation, agricultural practices, and the use of certain fertilizers can contribute to soil acidity. For example, the application of ammonium-based fertilizers can lead to the release of hydrogen ions during nitrification, further lowering soil pH.

Properties of Acid Soils

1. **Nutrient Availability:** One of the significant properties of acid soils is their impact on nutrient availability. The low pH can lead to the increased solubility of aluminum and iron, which can be toxic to plants. Additionally,

essential nutrients such as calcium, magnesium, and phosphorus may become less available for plant uptake in acidic conditions.

2. **Aluminum Toxicity:** Acid soils often exhibit elevated levels of soluble aluminum, which can be toxic to plant roots. Aluminum interferes with root development and nutrient uptake, leading to stunted growth and reduced crop yields. Managing aluminum toxicity is a key consideration in agriculture on acid soils.
3. **Microbial Activity:** Soil microorganisms play a crucial role in nutrient cycling and organic matter decomposition. However, the acidic conditions in acid soils can impact microbial activity. Some microbial species are sensitive to low pH, affecting the overall functioning of soil ecosystems. Acid soil management strategies should consider maintaining microbial diversity and activity.
4. **Soil Structure:** Acid soils may experience changes in soil structure due to the interactions between soil particles and acidic components. In some cases, the breakdown of soil aggregates can lead to increased soil compaction and reduced water infiltration. Addressing soil structure is essential for promoting root growth and water movement in acid soils.
5. **Liming Requirement:** Liming is a common practice in managing acid soils. The application of agricultural lime (calcium carbonate) helps neutralize soil acidity and raise pH. The liming requirement is influenced by factors such as soil texture, organic matter content, and crop selection. Proper liming is essential for improving nutrient availability and optimizing agricultural productivity.
6. **Buffering Capacity:** The buffering capacity of acid soils refers to their ability to resist changes in pH when external acidic or alkaline substances are added. Soils with low buffering capacity can experience rapid pH changes, making them more susceptible to external influences. Understanding buffering capacity is crucial for developing effective soil management strategies.
7. **Plant Adaptations:** Some plant species have developed adaptations to thrive in acidic conditions. These acid-tolerant plants often possess mechanisms to cope with aluminum toxicity and nutrient limitations. Incorporating acid-tolerant crops or crop varieties is a consideration in agricultural practices on acid soils.

Management of Acid Soils

1. **Liming:** Liming is a primary management practice for acid soils. Agricultural lime is applied to neutralize soil acidity, raise pH, and improve

nutrient availability. The optimal lime application rate depends on soil characteristics and the desired crop.

2. **Soil Amendments:** Adding organic amendments, such as compost or well-rotted manure, can enhance soil structure and microbial activity. Organic matter acts as a buffer, helping to stabilize pH and improve nutrient retention.

3. **Nutrient Management:** Implementing precise nutrient management practices is crucial on acid soils. Soil testing guides the application of fertilizers, ensuring that essential nutrients are provided in appropriate amounts for plant growth.

4. **Crop Selection:** Choosing crop varieties adapted to acid soils can enhance agricultural productivity. Some crops, such as certain varieties of maize and potatoes, exhibit better tolerance to acidic conditions.

5. **Erosion Control:** Preventing soil erosion is essential for maintaining the integrity of acid soils. Erosion control measures, including cover cropping and contour ploughing, help reduce the loss of topsoil and nutrients.

6. **Conservation Tillage:** Adopting conservation tillage practices, such as no-till or reduced tillage, helps preserve soil structure and minimize disturbance. Reduced soil disturbance can contribute to the retention of organic matter and buffering capacity.

7. **Research and Monitoring:** Continuous research and monitoring of soil conditions are essential for effective acid soil management. Regular soil testing provides insights into pH levels, nutrient status, and other relevant parameters, enabling informed decision-making.

5. The effect of acidic, halomorphic, and hydromorphic conditions on plant growth and nutrient availability

The effects of acidic, halomorphic, and hydromorphic conditions on plant growth and nutrient availability are significant and can significantly impact the productivity of ecosystems. Each condition poses distinct challenges to plants, influencing nutrient cycling, root development, and overall plant health.

1. Effect of Acidic Conditions

a. Impact on Nutrient Availability

- **Aluminum Toxicity**
  - In acidic soils, aluminum is released in a soluble form that can be toxic to plant roots.
  - Aluminum interferes with root elongation, nutrient uptake, and disrupts cellular processes.

- Plants often exhibit stunted growth and yellowing of leaves due to impaired nutrient absorption.

- **Nutrient Leaching**
    - Acidic conditions lead to increased leaching of essential cations such as calcium, magnesium, and potassium.
    - Leaching results in nutrient imbalances and deficiencies for plants, affecting overall health and productivity.

b. Microbial Activity

- **Microbial Sensitivity**
    - Soil microorganisms play a crucial role in nutrient cycling and organic matter decomposition.
    - Acidic conditions negatively impact microbial activity, reducing the breakdown of organic matter and nutrient cycling.

- **Soil Structure**
    - Disruption of soil aggregates occurs in acidic conditions, leading to increased soil compaction.
    - Poor soil structure limits root penetration, water infiltration, and nutrient movement within the soil.

- **Altered Root Zone Chemistry**
    - The pH-dependent availability of ions crucial for plant nutrition, such as phosphorus, is altered in acidic conditions.
    - Plants may experience phosphorus deficiency due to reduced solubility of phosphates in the root zone.

- **Aluminum and Manganese Interactions**
    - Increased solubility of manganese in acidic soils can lead to manganese toxicity for plants.
    - Manganese toxicity affects root function and nutrient uptake, contributing to poor plant growth.

2. Effect of Halomorphic (Saline and Alkaline) Conditions

a. Saline Conditions

- **Osmotic Stress**
    - High salinity creates osmotic stress for plants, reducing water uptake by roots.

- Osmotic stress leads to dehydration, inhibiting essential physiological processes.

- **Ion Imbalance**
  - Elevated sodium and chloride levels disrupt ion balance in plant cells.
  - Ion imbalance negatively affects nutrient uptake, leading to deficiencies and toxicity issues.

- **Reduced Germination**
  - High salinity inhibits seed germination, limiting the establishment of new plant populations.
  - Reduced germination rates affect plant diversity and ecosystem regeneration.

b. Alkaline Conditions

- **Nutrient Precipitation**
  - Alkaline soils lead to the precipitation of certain nutrients, reducing their availability for plants.
  - Essential elements like iron, phosphorus, and manganese may become less accessible to plant roots.

- **Limited Microbial Activity**
  - Alkaline conditions limit the activity of soil microorganisms.
  - Reduced microbial activity affects organic matter decomposition and nutrient cycling, impacting nutrient availability.

- **Iron Deficiency**
  - High pH induces iron deficiency in plants as iron becomes less soluble and less accessible.
  - Iron deficiency leads to chlorosis (yellowing of leaves) and affects photosynthetic processes.

3. Effect of Hydromorphic Conditions:

a. Waterlogging

- **Reduced Oxygen Availability**
  - Waterlogged conditions limit oxygen availability to plant roots, causing anaerobic conditions.
  - Anaerobic conditions hinder root respiration and nutrient uptake.

- **Microbial Processes**
    - Waterlogged soils affect soil microbial processes, leading to anaerobic microbial activities.
    - Anaerobic microorganisms may produce toxic by-products such as hydrogen sulfide, impacting plant health.
- **Nutrient Imbalances**
    - Prolonged waterlogging leads to nutrient imbalances, including nitrogen losses through denitrification.
    - Nutrient imbalances affect the nutrient status of the soil, impacting plant growth and productivity.

b. Iron Toxicity

- **Redox Reactions**
    - Waterlogged conditions create reduced (anaerobic) environments where iron becomes more soluble.
    - Increased iron solubility can lead to iron toxicity, negatively impacting root function and nutrient uptake.
- **Impaired Root Respiration**
    - Waterlogged soils limit root respiration, disrupting energy production processes.
    - Impaired root respiration affects overall plant metabolic activities and growth.

c. Altered Microbial Communities

- **Shift in Microbial Communities:**
    - Hydromorphic conditions can lead to shifts in soil microbial communities.
    - Anaerobic microorganisms become dominant, influencing nutrient cycling and organic matter decomposition.

d. Phosphorus Availability

- **Reduced Phosphorus Uptake**
    - Waterlogged conditions limit phosphorus availability to plant roots.
    - Reduced mobility of phosphorus in waterlogged soils affects critical physiological processes in plants.

e. Root Damage

- **Reduced Root Growth**
  - Prolonged waterlogging leads to reduced root growth and overall plant stunting.
  - Compromised root systems limit nutrient and water uptake, impacting plant health.

Mitigation and Management Strategies

a. Acidic Conditions

- **Liming**
  - Agricultural lime application to neutralize soil acidity and raise pH.
  - Adjusting soil pH to an optimal range for nutrient availability and plant growth.
- **Organic Amendments**
  - Addition of organic matter (compost, manure) to improve soil structure and buffer pH changes.
  - Enhanced microbial activity through organic amendments supports nutrient cycling.
- **Balanced Fertilization:**
  - Precise nutrient management to address specific nutrient deficiencies or imbalances.
  - Customized fertilization practices based on soil testing and plant requirements.

b. Halomorphic Conditions

- **Leaching**
  - Enhanced leaching practices to reduce salt concentrations in saline soils.
  - Promoting the removal of excess salts from the root zone to improve plant water uptake.
- **Gypsum Application:**
  - Application of gypsum to saline soils to improve soil structure and reduce sodium toxicity.
  - Gypsum aids in the displacement of sodium from the soil exchange complex.

- **Selecting Salt-Tolerant Crops**
    - Planting crops that exhibit tolerance to high salinity levels.
    - Choosing plant varieties adapted to saline or alkaline conditions for improved productivity.

c. Hydromorphic Conditions

- **Improved Drainage**
    - Implementing drainage systems to alleviate waterlogged conditions.
    - Installation of subsurface drainage to enhance oxygen availability for plant roots.
- **Raised Bed Planting**
    - Elevating planting beds to avoid waterlogging and enhance root oxygenation.
    - Raised beds promote better drainage and prevent water accumulation around plant roots.
- **Wetland Plants**
    - Selecting plant species adapted to hydromorphic conditions for wetland restoration.
    - Planting wetland vegetation that can thrive in waterlogged environments.

Understanding the specific conditions and their effects on plant growth is crucial for implementing targeted management strategies. Site-specific approaches, incorporating soil amendments, proper drainage, and selecting suitable crops, play key roles in mitigating the challenges posed by acidic, halomorphic, and hydromorphic conditions, ultimately fostering sustainable and productive ecosystems.

1. Effect of Acidic Conditions

a. Impact on Nutrient Availability

Aluminum Toxicity: In acidic soils, aluminum toxicity is a significant concern for plant growth. The increased solubility of aluminum ions in low pH conditions can negatively affect root development and nutrient uptake. Aluminum can interfere with the proper functioning of plant cells, leading to stunted growth and reduced yields. The toxic effects of aluminum on plant roots can result in nutrient deficiencies, particularly phosphorus, as roots struggle to take up essential nutrients.

Nutrient Leaching: Acidic conditions contribute to the leaching of essential nutrients from the soil. Calcium, magnesium, and potassium, vital for plant growth, become more soluble in acidic environments, leading to increased leaching. Nutrient imbalances and deficiencies can result, impacting plant health and reducing overall productivity. The leaching of nutrients further exacerbates the challenges faced by plants in acidic soils.

b. Microbial Activity

Microbial Sensitivity: Soil microorganisms play a crucial role in nutrient cycling, organic matter decomposition, and overall soil health. However, acidic conditions can negatively impact microbial activity. Many microorganisms are sensitive to changes in pH, and their activity may decline in acidic soils. This reduction in microbial activity affects nutrient cycling and the availability of organic matter, impacting the overall health of the soil ecosystem.

Soil Structure: Acidic conditions can disrupt soil structure by breaking down soil aggregates. This breakdown leads to increased soil compaction, limiting water infiltration and root penetration. Poor soil structure hinders the movement of water and nutrients, exacerbating the challenges faced by plants in acidic soils. Addressing soil structure becomes crucial for promoting root growth and nutrient uptake.

Altered Root Zone Chemistry: The pH-dependent availability of essential ions for plant nutrition, such as phosphorus, is altered in acidic conditions. As the solubility of phosphates decreases in acidic soils, plants may face challenges in acquiring sufficient phosphorus for optimal growth. This can result in phosphorus deficiencies, affecting critical physiological processes within the plant.

Aluminum and Manganese Interactions: Increased solubility of manganese in acidic soils can lead to manganese toxicity for plants. Manganese, at elevated levels, can interfere with root function and nutrient uptake. The presence of toxic concentrations of manganese in the soil exacerbates the challenges posed by aluminum toxicity, further impacting plant growth and productivity.

2. Effect of Halomorphic (Saline and Alkaline) Conditions

a. Saline Conditions

Osmotic Stress: High salinity in soils creates osmotic stress for plants. The elevated concentration of salts in the soil reduces water availability to plant roots, leading to dehydration. Osmotic stress inhibits essential physiological processes, affecting overall plant health and growth. Plants face challenges in maintaining proper water balance under saline conditions.

Ion Imbalance: Saline soils often contain elevated levels of sodium and chloride ions, disrupting the ion balance within plant cells. The imbalanced uptake of ions affects nutrient absorption, leading to deficiencies and toxicity issues. The disruption of ion balance contributes to further stress on plants, impacting their ability to thrive in saline environments.

Reduced Germination: High salinity levels in soils can inhibit seed germination. This limitation on germination rates affects the establishment of new plant populations. Reduced germination hampers the regeneration of plant communities, impacting biodiversity and ecosystem resilience.

b. Alkaline Conditions

Nutrient Precipitation: Alkaline soils can lead to the precipitation of certain nutrients, making them less available for plant uptake. Essential elements such as iron, phosphorus, and manganese may form insoluble compounds in alkaline conditions. This reduced availability of key nutrients can hinder plant growth and development.

Limited Microbial Activity: Alkaline conditions limit the activity of soil microorganisms. Microbial communities that thrive in neutral to acidic soils may face challenges in alkaline environments. Reduced microbial activity affects organic matter decomposition and nutrient cycling, further impacting nutrient availability for plants.

Iron Deficiency: High pH in alkaline soils induces iron deficiency in plants. Iron becomes less soluble and, therefore, less accessible to plant roots. Iron deficiency leads to chlorosis, characterized by the yellowing of leaves due to insufficient chlorophyll production. The limited availability of iron affects photosynthetic processes within the plant.

3. Effect of Hydromorphic Conditions

a. Waterlogging

Reduced Oxygen Availability: Waterlogged conditions limit the availability of oxygen to plant roots. The lack of oxygen in waterlogged soils creates anaerobic conditions, negatively impacting root respiration. Reduced oxygen availability hinders essential metabolic processes, affecting overall plant health and nutrient uptake.

Microbial Processes: Waterlogged soils disrupt microbial processes. The anaerobic conditions favor the growth of anaerobic microorganisms that produce by-products such as hydrogen sulfide. These by-products can be toxic to plant roots, adding an additional layer of stress to plants in waterlogged environments.

Nutrient Imbalances: Prolonged waterlogging can lead to nutrient imbalances, including nitrogen losses through denitrification. Denitrification, which converts nitrate to nitrogen gas, results in nutrient losses from the soil. Nutrient imbalances affect the overall nutrient status of the soil, impacting plant growth and productivity.

b. Iron Toxicity

Redox Reactions: Waterlogged conditions create reduced (anaerobic) environments where iron becomes more soluble. Increased iron solubility can lead to iron toxicity for plants. Iron toxicity negatively impacts root function, nutrient uptake, and overall plant growth. It adds an additional layer of stress to plants already facing challenges in waterlogged soils.

Impaired Root Respiration: Waterlogged soils limit root respiration due to the reduced availability of oxygen. Impaired root respiration disrupts energy production processes within the plant. The lack of oxygen further contributes to the overall stress on plants in waterlogged conditions.

c. Altered Microbial Communities

Shift in Microbial Communities: Hydromorphic conditions can lead to shifts in soil microbial communities. Anaerobic microorganisms become dominant in waterlogged soils. These shifts in microbial populations influence nutrient cycling and organic matter decomposition, affecting nutrient availability for plants.

d. Phosphorus Availability

Reduced Phosphorus Uptake: Waterlogged conditions limit phosphorus availability to plant roots. Phosphorus, an essential nutrient for plant growth, becomes less mobile in waterlogged soils. The reduced mobility of phosphorus affects critical physiological processes within the plant, impacting overall growth and development.

e. Root Damage

Reduced Root Growth: Prolonged waterlogging leads to reduced root growth and overall plant stunting. The compromised root system limits the uptake of nutrients and water, negatively impacting plant health. Reduced root growth becomes a significant factor in limiting the overall productivity of plants in waterlogged conditions.

Mitigation and Management Strategies

a. Acidic Conditions

Liming

- Agricultural lime application to neutralize soil acidity and raise pH.
- Adjusting soil pH to an optimal range for nutrient availability and plant growth.

Organic Amendments

- Addition of organic matter (compost, manure) to improve soil structure and buffer pH changes.
- Enhanced microbial activity through organic amendments supports nutrient cycling.

Balanced Fertilization

- Precise nutrient management to address specific nutrient deficiencies or imbalances.
- Customized fertilization practices based on soil testing and plant requirements.

b. Halomorphic Conditions

Leaching

- Enhanced leaching practices to reduce salt concentrations in saline soils.
- Promoting the removal of excess salts from the root zone to improve plant water uptake.

Gypsum Application

- Application of gypsum to saline soils to improve soil structure and reduce sodium toxicity.
- Gypsum aids in the displacement of sodium from the soil exchange complex.

Selecting Salt-Tolerant Crops

- Planting crops that exhibit tolerance to high salinity levels.
- Choosing plant varieties adapted to saline or alkaline conditions for improved productivity.

c. Hydromorphic Conditions

Improved Drainage

- Implementing drainage systems to alleviate waterlogged conditions.
- Installation of subsurface drainage to enhance oxygen availability for plant roots.

Raised Bed Planting

- Elevating planting beds to avoid waterlogging and enhance root oxygenation.
- Raised beds promote better drainage and prevent water accumulation around plant roots.

Wetland Plants

- Selecting plant species adapted to hydromorphic conditions for wetland restoration.
- Planting wetland vegetation that can thrive in waterlogged environments.

6. Acid Sulfate Soils: Occurrence, Distribution, Characteristics, Effects on Plant Growth, Nutrient Availability, and Reclamation Techniques

Occurrence and Distribution

Acid sulfate soils (ASS) are a global phenomenon, found in various geographic locations. They are particularly prevalent in coastal regions and areas with a history of marine influence. The occurrence of ASS is closely tied to the presence of sulfide minerals, primarily pyrite, in sediments or parent materials. Coastal plains, estuaries, and tidal flats often exhibit the characteristics of acid sulfate soils.

These soils are not limited to coastal regions, as inland areas with sulfide-rich geological formations may also host acid sulfate soils. Understanding the distribution of ASS involves considering factors such as geological history, sedimentation patterns, and the influence of past sea-level changes.

In terms of spatial distribution, the prevalence of acid sulfate soils varies across continents. Regions with a combination of geological conditions conducive to sulfide mineral formation and past or present marine influence are more likely to harbor these soils. For example, parts of Southeast Asia, Australia, North America, and Europe have reported occurrences of acid sulfate soils.

Detailed geological surveys and soil mapping are essential for accurately identifying the distribution of acid sulfate soils within a given area. This information is crucial for land-use planning, environmental management, and the development of strategies for soil reclamation.

Characteristics

The defining characteristics of acid sulfate soils extend beyond their low pH and high iron content. A comprehensive understanding of these features is essential for effectively managing and reclaiming such soils.

1. Low pH

- The low pH of acid sulfate soils results from the oxidation of sulfide minerals, primarily pyrite. This process leads to the release of sulfuric acid into the soil, causing acidity. pH levels can drop significantly, affecting the chemical and biological properties of the soil.

2. High Iron Content

- Iron is a key component of acid sulfate soils, existing in the form of iron sulfides and iron oxides. The presence of iron contributes to the soil's coloration and influences its chemical and physical properties.

3. Potential Toxicity

- The acidification process not only results in low pH but can also lead to the release of toxic elements. Aluminum, in particular, is released in elevated amounts under acidic conditions, posing a threat to plant roots and aquatic ecosystems. The potential toxicity of acid sulfate soils extends to other heavy metals as well, depending on the specific mineral composition.

4. Variability in Sulfide Mineral Content

- Acid sulfate soils exhibit variability in sulfide mineral content, with some areas containing higher concentrations of pyrite than others. This variability influences the degree of acidity and the potential environmental impact of these soils.

Effects on Plant Growth

The impact of acid sulfate soils on plant growth is multifaceted, encompassing several physiological and nutritional aspects that can hinder the development of vegetation in affected areas.

1. Aluminum Toxicity

- One of the primary challenges for plant growth in acid sulfate soils is the potential toxicity of aluminum. Under acidic conditions, aluminum ions are released into the soil solution, adversely affecting root growth and nutrient uptake. This toxicity can lead to stunted plant development and, in severe cases, result in plant mortality.

2. Nutrient Imbalance

- The low pH of acid sulfate soils can create an imbalance in nutrient availability. Essential elements such as phosphorus, calcium, and magnesium become less accessible to plants in acidic conditions. This nutrient imbalance further exacerbates the challenges faced by plants trying to establish and thrive in acid sulfate soil environments.

3. Reduced Microbial Activity

- Microbial activity in the soil is vital for nutrient cycling and organic matter decomposition. The acidic conditions of acid sulfate soils can negatively impact microbial communities, reducing their activity. This, in turn, affects nutrient mineralization and the overall health of the soil ecosystem.

4. Impacts on Mycorrhizal Associations

- Acid sulfate soils can disrupt beneficial mycorrhizal associations between plant roots and fungi. Mycorrhizae play a crucial role in nutrient uptake for many plant species, and their disruption can further limit the ability of plants to access essential nutrients.

To mitigate these effects on plant growth, interventions must address both the direct impacts of acidity and the associated challenges related to nutrient availability and microbial activity.

Nutrient Availability

Understanding how acid sulfate soils affect nutrient availability is essential for developing effective reclamation strategies and promoting sustainable land use. The impact on specific nutrients varies, and addressing these challenges requires a targeted approach.

1. Phosphorus Fixation

- Acid sulfate soils are notorious for fixing phosphorus in forms that are less available to plants. The chemical reactions in the soil result in the precipitation of phosphorus compounds, limiting its bioavailability. Overcoming phosphorus fixation is crucial for promoting plant growth in these soils.

2. Calcium and Magnesium Deficiency

- The low pH conditions of acid sulfate soils contribute to deficiencies in calcium and magnesium. These essential nutrients are vital for various physiological processes in plants, including cell wall formation and enzyme activation. Reclamation strategies must aim to address these deficiencies to create a more conducive environment for plant growth.

3. Iron Toxicity

- While iron is a necessary nutrient for plants, acid sulfate soils can lead to iron toxicity due to elevated concentrations. Managing iron levels is essential to prevent adverse effects on plant roots and overall plant health.

4. Nitrogen Dynamics

- Acidic conditions can influence nitrogen dynamics in the soil, potentially leading to the loss of nitrogen through leaching or gas emissions. Understanding these dynamics is crucial for optimizing nitrogen availability for plants and minimizing environmental impacts.

To improve nutrient availability in acid sulfate soils, reclamation efforts often involve amendments that address specific deficiencies, promote nutrient cycling, and create a more balanced soil environment for plant growth.

Reclamation Techniques

Reclaiming acid sulfate soils involves a combination of physical, chemical, and biological strategies to restore soil health, mitigate environmental risks, and create conditions conducive to sustainable land use.

1.Lime Addition

- Adding lime (calcium carbonate) is a fundamental approach to neutralize acidity in acid sulfate soils. The addition of lime raises the soil pH, counteracting the effects of sulfuric acid. This process, known as liming, not only ameliorates soil acidity but also reduces the solubility of aluminum, mitigating its toxic effects on plant roots.
- Lime application rates need to be carefully calculated based on the degree of soil acidity, and adjustments may be necessary over time as the lime reacts with the soil.

2. Drainage and Water Management

- Improving drainage is critical for managing acid sulfate soils. Waterlogged conditions contribute to the oxidation of sulfide minerals, initiating the acidification process. Implementing effective drainage systems helps prevent waterlogging and reduces the risk of acidification.
- Controlled drainage and water management practices, such as constructing levees or installing drainage tiles, allow for the regulation of water levels in affected areas.

3. Selective Breeding of Tolerant Plants

- Developing and utilizing plant varieties that are adapted to acidic and aluminum-rich conditions is an important aspect of reclamation. Selective breeding programs aim to produce crop varieties with increased tolerance to low pH and aluminum toxicity.
- This strategy involves identifying and selecting plant genotypes with inherent tolerance traits and incorporating them into breeding programs to develop improved varieties for acid sulfate soil environments.

4. Organic Amendments

- Incorporating organic materials into acid sulfate soils can have multiple benefits. Organic amendments improve soil structure, enhance nutrient availability, and stimulate microbial activity. Organic matter acts as a buffer against pH fluctuations and promotes the development of a more resilient and fertile soil.

- Common organic amendments include compost, manure, and cover crops, which contribute to soil organic carbon and provide a source of essential nutrients.

5. Buffer Strips and Vegetative Cover

- Planting buffer strips and cover crops is an effective erosion control measure for acid sulfate soils. Vegetative cover stabilizes the soil, reducing the risk of erosion and minimizing the transport of sediments and contaminants into adjacent water bodies.
- Additionally, the root systems of plants contribute to soil structure improvement and can assist in nutrient cycling, further enhancing the overall health of the reclaimed soil.

6. Monitoring and Adaptive Management

- Successful reclamation of acid sulfate soils requires ongoing monitoring and adaptive management. Regular assessments of soil pH, nutrient levels, and plant growth are essential to track the effectiveness of reclamation efforts.
- Adaptive management involves adjusting reclamation strategies based on monitoring results, ensuring that interventions are tailored to the specific needs of the site and evolving environmental conditions.

7. Public Awareness and Stakeholder Engagement

- Effective reclamation involves collaboration and engagement with local communities, government agencies, and other stakeholders. Public awareness campaigns can promote understanding of the importance of acid sulfate soil management and encourage sustainable land-use practices.
- Engaging stakeholders in the decision-making process fosters a sense of ownership and commitment to long-term soil health and environmental sustainability.

6

Organic Farming Practices for the Management of Soil and Water

[1]*Wajid Hasan*, [2]*Sheetanshu Gupta*, [3]*Deepa Rawat*, [4]*Adarsh Pandey* *and* [4]*Kh. Chandrakumar Singh*

[1]*KVK Jehanabad, Bihar Agricultural University Sabour, Bihar*
[2]*Department of Biotechnology, Institute of Technology and Management BKT, Lucknow, U.P.*
[3]*College of Forestry, Ranichauri, VCSG University of Horticulture and Forestry, Bharsar, Uttarakhand*
[4]*Department of Botany. Swami Shukdevanand College. Shahjahanpur, U.P.*
[5]*VCSGU University of Horticulture and Forestry, Bharsar, Uttarakhand*

1. Introduction to Organic Farming

Definition and Principles of Organic Farming

Organic farming is not merely a set of agricultural practices; it is a profound philosophy that seeks to redefine humanity's relationship with the land. At its core, organic farming is an intricate dance with nature, a commitment to fostering biodiversity, ecological balance, and sustainability.

Defining Organic Farming

Organic farming is a departure from conventional agricultural methods that rely heavily on synthetic inputs. It is an agricultural system that embraces natural and traditional approaches to cultivation. The rejection of synthetic pesticides, fertilizers, and genetically modified organisms is not merely a technical choice; it reflects a fundamental shift in the understanding of agriculture as an integral part of a larger ecological system (Fig 1).

Fig. 1: Practice of Organic Farming in Village

In organic farming, the soil is not just a substrate for plant growth; it is a living, dynamic entity. Practices such as composting, cover cropping, and minimal tillage are employed not just to enhance crop yields but to nourish the soil, fostering a symbiotic relationship between the land and the crops it supports.

Principles Guiding Organic Farming

1. **Biodiversity and Ecological Harmony:** The heartbeat of organic farming is biodiversity. Monoculture is replaced by a diverse tapestry of crops, each playing a unique role in the ecosystem. Companion planting, where mutually beneficial plants are grown together, exemplifies the principle of ecological harmony. This not only deters pests but also promotes a resilient and balanced ecosystem.

2. **Soil Health and Fertility:** Organic farming places a premium on the health of the soil. Rather than viewing soil as a static medium, organic farmers recognize it as a living organism. Techniques like cover cropping enhance soil structure, prevent erosion, and contribute to the vitality of the microbial community. The focus is on maintaining and enhancing soil fertility without resorting to synthetic additives.

3. **Rejecting Synthetic Inputs:** At the core of organic farming is a resolute rejection of synthetic inputs. Chemical pesticides and fertilizers are eschewed in favor of natural alternatives. This rejection is not merely a technical choice; it is a commitment to creating a self-sustaining agricultural system where the reliance on external, potentially harmful inputs is minimized.

4. **Holistic Farm Management:** Organic farming views the farm as a dynamic and interconnected system. It goes beyond individual crops and animals,

considering the interactions between various components. Holistic farm management seeks to optimize the entire farm as an ecosystem, where each element contributes to the overall health and resilience of the agricultural landscape.

Importance of Sustainable Agriculture

As global challenges such as climate change, soil degradation, and water scarcity intensify, the importance of sustainable agriculture, epitomized by organic farming, becomes increasingly pronounced. This section explores the multifaceted significance of adopting sustainable agricultural practices for the health of the environment, the well-being of communities, and the long-term resilience of our food systems.

Ecological Balance and Environmental Stewardship

1. **Mitigating Soil Erosion:** Sustainable agriculture, particularly as embodied by organic farming, prioritizes soil conservation. Techniques such as minimal tillage, cover cropping, and contour ploughing mitigate soil erosion, preserving the topsoil and ensuring its fertility for generations to come. This not only safeguards agricultural productivity but also contributes to broader environmental health.

2. **Preserving Water Quality:** Conventional agriculture often contributes to water pollution through the runoff of synthetic chemicals. Sustainable practices, such as organic farming, emphasize the judicious use of water and the prevention of contamination. This commitment to water quality extends beyond the farm, positively impacting downstream ecosystems and human communities.

3. **Promoting Biodiversity:** Sustainable agriculture recognizes the inherent value of diverse ecosystems. Organic farming actively promotes biodiversity through practices like intercropping and maintaining natural habitats on the farm. By avoiding the use of synthetic pesticides, organic farmers create environments where beneficial insects thrive, contributing to a more resilient and balanced ecosystem.

Human Health and Nutritional Benefits

1. **Reducing Exposure to Chemical Residues:** One of the primary concerns in conventional agriculture is the presence of chemical residues in food. Organic farming, by avoiding synthetic pesticides and fertilizers, significantly reduces the risk of chemical residues in produce. This is not only beneficial for consumers but also for the health of farmworkers and nearby communities (Fig 2).

Fig. 2: Minimum Exposure to Chemical Residues through Organic Farming

2. **Enhancing Nutritional Value:** Research suggests that organically grown crops may contain higher levels of certain nutrients. The absence of synthetic fertilizers encourages plants to draw a broader range of nutrients from the soil, potentially resulting in more nutrient-dense produce. While the debate on nutritional differences continues, the emphasis on diverse, nutrient-rich soils in organic farming aligns with broader health considerations.

Long-Term Viability of Food Systems

1. **Resilience to Climate Change:** Climate change poses significant challenges to agriculture, including extreme weather events and shifts in growing seasons. Sustainable agriculture, with organic farming as a pioneer, enhances the resilience of food systems. Diverse cropping systems and practices such as agroforestry contribute to climate resilience, ensuring a more secure food supply even in the face of unpredictable environmental changes.

2. **Community and Local Economy:** Sustainable agriculture often involves smaller-scale, local farming operations. Supporting local, organic farmers not only fosters community resilience but also reduces the carbon footprint associated with the transportation of food over long distances. This localization of food systems contributes to the sustainability of rural communities and the overall health of local economies.

2. Soil Management in Organic Farming

Soil Health and Its Significance

In the vast tapestry of organic farming, the soil is not merely a substrate; it is a living, breathing entity that sustains life and dictates the success of agricultural endeavors. Soil health is the heartbeat of organic farming, encompassing a dynamic interplay of physical, chemical, and biological factors that collectively define the soil's ability to function as a living system.

The Components of Soil Health:

1. **Physical Structure:** The physical structure of soil is crucial for water infiltration, root penetration, and overall plant growth. Healthy soils exhibit a well-aggregated structure, fostering aeration and drainage. Organic farming practices, such as minimal tillage and cover cropping, contribute to the development and maintenance of a resilient soil structure.

2. **Chemical Composition:** The chemical composition of the soil influences nutrient availability to plants. Soil pH, nutrient levels, and the presence of organic matter all play pivotal roles in determining soil fertility. Organic farming emphasizes the use of natural amendments to maintain a balanced and nutrient-rich soil, steering clear of synthetic inputs that can disrupt the delicate chemical equilibrium.

3. **Biological Activity:** A thriving community of microorganisms, including bacteria, fungi, and protozoa, constitutes the biological aspect of soil health. These organisms contribute to nutrient cycling, organic matter decomposition, and disease suppression. Organic farming practices prioritize the enhancement of microbial diversity, recognizing the symbiotic relationship between soil organisms and plant roots.

Significance of Soil Health

1. **Nutrient Availability:** Healthy soils provide a reservoir of nutrients essential for plant growth. The intricate web of soil microorganisms helps break down organic matter into forms that plants can absorb. Organic farmers focus on maintaining a nutrient-rich soil through natural processes, ensuring a steady supply of essential elements for their crops.

2. **Water Retention:** Well-structured soils with good organic matter content have enhanced water retention capabilities. This is particularly crucial in regions facing water scarcity or erratic precipitation patterns. Organic farming practices, such as cover cropping and minimal tillage, contribute to improved water retention, reducing the risk of drought stress for crops.

3. **Erosion Prevention:** Soil erosion poses a significant threat to agricultural sustainability. Healthy soils, with robust structure and ample ground cover, are more resistant to erosion. Organic farmers implement strategies like cover cropping and agroforestry to safeguard against soil erosion, preserving the topsoil and maintaining its fertility.

4. **Resilience to Climate Change:** As climate patterns become more unpredictable, the resilience of agricultural systems becomes paramount.

Healthy soils, with their ability to retain water and withstand extreme weather events, contribute to the overall resilience of organic farming in the face of climate change impacts.

Understanding the intricacies of soil health is foundational to the organic farming ethos. Organic farmers are not just stewards of their crops; they are stewards of the soil, nurturing an environment where the land and its inhabitants coexist in a delicate equilibrium.

Organic Soil Amendments

Central to the philosophy of organic farming is the belief that healthy soils lead to healthy crops. In contrast to conventional agriculture, which often relies on synthetic fertilizers, organic farmers turn to a rich array of organic amendments to fortify and nourish their soils.

The Role of Organic Amendments

1. **Compost:** Composting, a time-honored practice, involves the decomposition of organic materials to create nutrient-rich humus. Beyond providing essential nutrients, compost improves soil structure, water retention, and microbial activity. Organic farmers carefully create compost blends, ensuring a balanced mix of carbon and nitrogen-rich materials.

2. **Manure:** Animal manure, when properly composted, is a valuable source of organic matter and nutrients. The application of manure not only enhances soil fertility but also contributes to the overall improvement of soil structure. Organic farmers emphasize responsible manure management to prevent nutrient runoff and minimize environmental impact.

3. **Green Manure:** Green manure crops, strategically planted and later incorporated into the soil, offer a double benefit. These cover crops not only add organic matter but also contribute nitrogen through a natural process known as nitrogen fixation. Leguminous green manure crops, such as clover or vetch, play a vital role in organic soil management.

4. **Biochar:** The use of biochar, a form of charcoal produced by heating organic matter in a low-oxygen environment, is gaining traction in organic farming. Biochar improves soil structure, enhances water retention, and provides a stable habitat for beneficial microorganisms. Additionally, it sequesters carbon, contributing to climate change mitigation.

Balancing Act

Organic farmers engage in a delicate balancing act, considering the nutritional needs of their crops while respecting the intricacies of the soil ecosystem. The choice and application of organic amendments are not arbitrary but tailored to the specific requirements of the soil and the crops being cultivated.

1. **Nutrient Cycling:** Organic amendments contribute to the ongoing process of nutrient cycling in the soil. As organic matter breaks down, nutrients become available to plants in a slow-release form. This contrasts with the quick-release nature of synthetic fertilizers, aligning with the organic farming principle of fostering long-term soil health.

2. **Microbial Support:** The introduction of organic amendments nourishes and supports the diverse community of microorganisms in the soil. These microorganisms play a pivotal role in nutrient mineralization, making organic nutrients accessible to plants. The emphasis on microbial activity distinguishes organic farming as a holistic approach to soil fertility.

3. **Soil pH Management:** Certain organic amendments, such as lime, can be used to manage soil pH. Organic farmers carefully assess the pH requirements of their crops and adjust soil acidity or alkalinity using natural amendments, ensuring optimal nutrient availability.

Organic soil amendments embody the essence of sustainability and environmental stewardship. Rather than depleting the soil through the application of synthetic inputs, organic farmers work with nature, enriching the soil in a regenerative way and fostering long-term agricultural productivity.

Cover Cropping and Green Manure

In the intricate choreography of organic farming, cover cropping and green manure emerge as key players, contributing not just to soil health but to the overall sustainability and resilience of the agricultural ecosystem.

Cover Cropping

1. **Benefits Beyond Soil Protection:** While the primary purpose of cover cropping is to protect the soil during periods when the main cash crop is not growing, the benefits extend far beyond erosion prevention. Cover crops act as green blankets, shielding the soil from the impact of raindrops, reducing compaction, and promoting a favorable microclimate for soil organisms.

2. **Nitrogen Fixation:** Leguminous cover crops, such as clover or peas, have the unique ability to fix nitrogen from the atmosphere. This natural process enriches the soil with nitrogen, a vital nutrient for plant growth. Organic farmers strategically incorporate nitrogen-fixing cover crops into their rotation plans, enhancing soil fertility without resorting to synthetic inputs.

3. **Weed Suppression:** Cover crops play a vital role in weed suppression. By forming a dense canopy, they shade out weeds, reducing competition for nutrients and sunlight. This natural weed management strategy aligns with the organic farming principle of minimizing reliance on herbicides.

4. **Diverse Benefits:** The choice of cover crops is diverse and can be tailored to specific farm goals. Some cover crops, known as "trap crops," attract pests away from cash crops. Others, like radishes, penetrate compacted soil layers, improving overall soil structure.

Green Manure

1. **Nutrient Contribution:** Green manure involves growing crops specifically for incorporation into the soil while still green. This practice contributes organic matter, enhancing soil structure and promoting microbial activity. As the green manure crop decomposes, it releases nutrients in a form that is readily available to subsequent crops.

2. **Timing and Incorporation:** Green manure crops are strategically grown during periods when the main cash crop is not in the field. Incorporation methods vary, with some farmers ploughing the green manure into the soil and others employing no-till techniques. The choice depends on factors such as the specific crop, soil type, and local climate.

3. **Diverse Green Manure Crops:** The selection of green manure crops is diverse, allowing farmers to tailor their choices based on soil needs and regional conditions. Leguminous crops are often favored for their nitrogen-fixing capabilities, while others contribute unique benefits, such as biofumigation, where certain plants release compounds that suppress soil-borne pests and diseases.

Integration into Crop Rotation

1. **Crop Rotation Strategies:** Organic farmers integrate cover cropping and green manure into their broader crop rotation strategies. By carefully planning the sequence of crops and cover crops, they maximize soil health benefits, break pest and disease cycles, and optimize nutrient availability. Crop rotation, coupled with cover cropping, exemplifies the intricate dance of organic farming.

2. **Seasonal Considerations:** The choice of cover crops and green manure is influenced by seasonal considerations and the specific needs of the succeeding crops. Winter cover crops, for example, protect the soil during the fallow season, while summer cover crops provide shade and moisture retention during the hotter months.

3. **Enhanced Pest and Disease Management:** Beyond soil health benefits, the integration of cover crops and green manure into crop rotations contributes to enhanced pest and disease management. Some cover crops, such as mustard, release compounds that naturally suppress nematodes and other soil-borne pathogens.

Cover cropping and green manure are not just practices; they are integral components of a holistic approach to organic farming. By incorporating these strategies into their agricultural systems, organic farmers foster an environment where the soil is nurtured and protected, ensuring the sustainability and resilience of their farms.

3. Water Management in Organic Farming

Importance of Water Conservation

In the intricate tapestry of sustainable agriculture, water management emerges as a critical thread, weaving through the principles of organic farming. Water, a finite and vital resource, plays a pivotal role in shaping the sustainability and success of agricultural practices. Organic farmers, attuned to the delicate balance of ecosystems, prioritize water conservation as a cornerstone of their agricultural ethos.

Conserving Water Resources

1. **Preservation of Natural Ecosystems:** Organic farming is inherently linked to the preservation of natural ecosystems, including water bodies. By avoiding the use of synthetic chemicals and adopting practices that minimize soil disturbance, organic farmers contribute significantly to the protection of watersheds and aquatic habitats. This approach safeguards biodiversity and maintains the delicate balance of aquatic ecosystems.

2. **Mitigating Soil Erosion:** Healthy soils, nurtured through organic farming practices, exhibit enhanced water retention capacities. This not only reduces the risk of soil erosion but also ensures efficient water infiltration, contributing to groundwater recharge. Organic farmers employ techniques like cover cropping and minimal tillage to promote soil health, indirectly aiding in water conservation.

3. **Promoting Biodiversity:** The conservation of water resources is inherently tied to the broader goal of promoting biodiversity. Organic farmers, through practices such as cover cropping and minimal tillage, create environments conducive to diverse plant and animal species. The result is not just the conservation of water but the preservation of a thriving ecosystem where various organisms contribute to the health of the soil and water.

4. **Reduced Water Pollution:** Organic farming's avoidance of synthetic pesticides and fertilizers minimizes the risk of water pollution. Runoff from organic farms is less likely to contain harmful chemicals, contributing to the preservation of water quality in nearby streams and rivers. This reduction in water pollution aligns with the broader principles of environmental stewardship inherent in organic agriculture.

Efficient Irrigation Methods in Organic Farming

Watering crops is a critical aspect of agriculture, and organic farmers employ efficient irrigation methods to ensure judicious and sustainable water use.

Drip Irrigation

1. **Precision Watering:** Drip irrigation stands as a beacon of precision watering in organic farming. This method delivers water directly to the base of plants, minimizing water wastage through evaporation or runoff. The precision ensures that crops receive the necessary moisture without excess usage, contributing to overall water efficiency.

2. **Soil Health:** Drip irrigation systems contribute to soil health by maintaining consistent moisture levels. In organic farming, where soil structure and microbial activity are paramount, this consistency ensures optimal conditions for plant growth. The system supports both water efficiency and the health of the agricultural ecosystem.

3. **Weed Control:** Beyond efficient water usage, drip irrigation also aids in natural weed control. By delivering water directly to the root zone, it reduces moisture on the soil surface, creating an environment less conducive to weed growth. This integrated approach aligns with organic farming principles, emphasizing natural methods over synthetic interventions.

Mulching

1. **Water Retention:** Mulching, another strategic practice in water management, involves covering the soil with organic materials such as straw or compost. This protective layer significantly reduces water evaporation, helping the soil retain moisture and reducing the frequency of irrigation. Mulching is a versatile technique that complements other water conservation strategies in organic farming.

2. **Temperature Regulation:** Mulch acts as a natural temperature regulator, preventing soil overheating during hot periods. This not only benefits plant roots by maintaining a suitable soil temperature but also contributes to overall water conservation by reducing the need for additional irrigation during periods of intense heat. Mulching exemplifies the interconnectedness of soil health and water management in organic farming.

Rainwater Harvesting in Organic Systems

As organic farmers seek sustainable water sources, rainwater harvesting emerges as a pragmatic and eco-friendly solution.

Rain Barrels and Cisterns

1. **Capturing Precious Rainfall:** Rain barrels and cisterns are instrumental in capturing and storing rainwater from roofs. This harvested rainwater serves as a supplementary water source during dry periods, reducing reliance on external water supplies. The utilization of rain barrels and cisterns aligns with the ethos of organic farming, emphasizing self-sufficiency and sustainability.

2. **Drip Irrigation Integration:** The collected rainwater can be seamlessly integrated into drip irrigation systems, providing a natural and nutrient-rich water source for crops. This integration exemplifies the closed-loop approach of organic farming, where resources are thoughtfully managed within the agricultural ecosystem, promoting sustainability and resilience.

Key Components of Rainwater Harvesting

1. **Collection Surfaces:** Organic farmers strategically position collection surfaces, such as the roofs of barns or greenhouses, to maximize rainwater capture. By directing runoff to designated collection points, the risk of direct soil contact and potential contamination is minimized. This thoughtful planning ensures that the harvested rainwater remains ecologically sound.

2. **Filtration Systems:** Filtration systems are integral to rainwater harvesting, ensuring that the collected rainwater is free from debris or contaminants. Organic farmers often opt for organic filtration methods, such as sand or gravel filters, to maintain the ecological integrity of the water. This commitment to natural filtration aligns with the principles of organic farming.

3. **Storage and Distribution:** Collected rainwater is stored in tanks or cisterns and distributed as needed. Organic farmers meticulously manage the distribution of harvested rainwater, ensuring that it complements existing irrigation practices without causing waterlogged conditions in the soil. The controlled distribution maximizes the effectiveness of rainwater as a sustainable water source.

Benefits of Rainwater Harvesting in Organic Farming

1. **Sustainable Water Source:** Rainwater harvesting provides organic farmers with a sustainable and self-replenishing water source. This reduces dependence on external water supplies and contributes to the overall resilience of the farm. The cyclical nature of rainwater harvesting aligns with the closed-loop systems advocated by organic farming.

2. **Cost-Effective:** Rainwater harvesting proves to be a cost-effective solution for water needs in organic farming. While there is an initial investment

in collection systems, the long-term benefits in reduced water bills and increased sustainability make it an economically viable and environmentally friendly choice.

3. **Reduced Environmental Impact:** Harvesting rainwater minimizes the environmental impact associated with traditional water sources, such as groundwater pumping. By utilizing a natural and locally available resource, organic farmers demonstrate a commitment to minimizing the ecological footprint of their agricultural activities.

Integration with Organic Practices

1. **Closed-Loop Systems:** Rainwater harvesting seamlessly integrates with the closed-loop systems advocated by organic farming. The collected rainwater becomes an integral part of the farm's water management, creating a self-sustaining ecosystem. This closed-loop approach embodies the holistic philosophy of organic agriculture.

2. **Complementary to Sustainable Agriculture:** Rainwater harvesting is not merely a technique; it is a philosophy that complements the principles of sustainable and organic agriculture. It embodies the ethos of utilizing natural resources judiciously, minimizing environmental impact, and promoting long-term resilience in farming practices.

4. Companion Planting and Crop Rotation

Benefits of Companion Planting:

In the intricate world of organic farming, companion planting stands as a time-honored strategy, revealing a harmonious dance between different plant species. This technique involves cultivating specific plants near one another, with each contributing to the well-being of the other. The benefits of companion planting extend beyond the individual plants, encompassing pest control, enhanced pollination, and improved nutrient uptake.

Pest Control and Insect Harmony

1. **Natural Pest Deterrence:** Companion planting leverages the natural properties of certain plants to deter pests. For instance, planting aromatic herbs like basil alongside tomatoes can repel common tomato pests like aphids and hornworms. This natural pest control minimizes the need for chemical interventions, aligning with the principles of organic farming.

2. **Attracting Beneficial Insects:** Certain plants act as magnets for beneficial insects, such as ladybugs and predatory beetles, which feed on harmful pests. Marigolds, for example, emit compounds that attract these beneficial insects, creating a balanced and self-regulating ecosystem within the organic garden.

3. **Interference with Insect Life Cycles:** Companion planting disrupts the life cycles of specific pests. Planting onions or garlic alongside susceptible crops like lettuce can deter pests that would otherwise target these plants. This interference strategy adds a layer of complexity to the garden, making it less hospitable to destructive pests.

Improved Nutrient Uptake

1. **Nitrogen Fixation:** Some companion plants, particularly legumes like peas and beans, have the ability to fix nitrogen from the air into a form that plants can absorb. Planting these nitrogen-fixing companions alongside crops with high nitrogen requirements, such as corn or lettuce, enhances the overall nutrient availability in the soil.

2. **Dynamic Nutrient Exchange:** Companion plants often engage in dynamic nutrient exchange through their root systems. Certain plants release compounds that enhance the availability of specific nutrients for neighboring plants. This collaborative nutrient exchange contributes to a more balanced and resilient soil ecosystem.

3. **Biochemical Cooperation:** Companion plants can release biochemical signals into the soil that affect the growth and development of nearby plants. This cooperative biochemical communication can stimulate root growth, improve nutrient absorption, and contribute to the overall health of the plant community.

Enhanced Pollination and Yield

1. **Attracting Pollinators:** Companion planting includes the intentional cultivation of flowers that attract pollinators. Bees, butterflies, and other pollinators play a crucial role in the reproduction of many fruit and vegetable crops. Planting companion flowers like marigolds or borage alongside crops ensures a vibrant and diverse pollinator population.

2. **Increased Yield through Diversity:** Companion planting promotes biodiversity within the garden. This diversity contributes to the overall health and resilience of the ecosystem. The interplay between different plant species can result in increased yields as the garden becomes a dynamic and interconnected web of mutually beneficial relationships.

3. **Reduced Competition for Resources:** Carefully selecting companion plants that have different nutrient requirements can reduce competition for resources like water and sunlight. For instance, tall, sun-loving plants can provide shade for shorter, shade-tolerant companions. This spatial optimization ensures that each plant receives the resources it needs for optimal growth.

Crop Rotation as a Strategy for Soil and Water Management

Crop rotation, another integral practice in organic farming, involves systematically changing the location of crops within a designated area over successive seasons. This strategic rotation offers a myriad of benefits, including pest management, disease prevention, and the improvement of soil structure and fertility.

Pest and Disease Management

1. **Breaking Pest Cycles:** Crop rotation disrupts the life cycles of pests that are specific to certain crops. For example, rotating crops in the tomato family can prevent the buildup of soil-borne diseases and pests that affect tomatoes, reducing the need for chemical interventions and fostering a healthier plant environment.

2. **Minimizing Soil-Borne Diseases:** Some plant pathogens, such as fungi and nematodes, are specific to certain crops. By rotating crops, these pathogens are denied a consistent host, reducing their prevalence in the soil. This natural approach to disease management aligns with organic farming principles.

3. **Balancing Nutrient Depletion:** Different crops have varying nutrient requirements. Crop rotation helps balance nutrient depletion in the soil by alternating crops with high nutrient demands (heavy feeders) with those that require fewer nutrients (light feeders). This thoughtful approach contributes to sustained soil fertility.

Improving Soil Structure

1. **Reducing Soil Erosion:** Crop rotation, especially when combined with cover cropping, helps reduce soil erosion. The cultivation of cover crops during fallow periods protects the soil from the impact of raindrops and minimizes nutrient runoff. This erosion control is vital for maintaining soil health and preventing the loss of topsoil.

2. **Enhancing Soil Organic Matter:** Different crops contribute different types and amounts of organic matter to the soil. Crop residues and root systems from various plants enrich the soil with organic material. This, in turn, enhances soil structure, water retention, and microbial activity, creating a fertile and resilient environment for future crops.

3. **Preventing Soil Compaction:** Certain crops have deep taproots that can break up compacted soil layers. Alternating these deep-rooted crops with shallow-rooted ones helps prevent soil compaction. Improved soil structure facilitates better water infiltration, root penetration, and overall plant growth.

Water Management and Conservation

1. **Optimizing Water Use Efficiency:** Crop rotation allows for the strategic planning of water-efficient crops based on their water requirements. For example, rotating drought-tolerant crops with those that require more water ensures optimal water use efficiency in the agricultural system. This adaptability is crucial, especially in regions facing water scarcity.

2. **Minimizing Soil Moisture Imbalances:** Different crops have varying moisture needs. By rotating crops with diverse water requirements, organic farmers can minimize imbalances in soil moisture. This adaptability is crucial for regions with irregular rainfall patterns or limited water resources.

3. **Facilitating Cover Cropping:** Crop rotation can be seamlessly integrated with cover cropping, further enhancing water management. Cover crops, strategically planted during fallow periods, contribute to soil moisture retention, weed suppression, and erosion control. The synergy between crop rotation and cover cropping exemplifies a comprehensive approach to water conservation.

Examples of Effective Companion Planting Combinations

1. Tomatoes and Basil

- **Benefit:** Basil repels common tomato pests such as aphids and hornworms.
- **Complementary Traits:** Tomatoes provide a taller structure for basil, while basil offers natural pest protection.

2. Corn, Beans, and Squash (Three Sisters)

- **Benefit:** Known as the Three Sisters, this trio exemplifies mutual support.
- **Complementary Traits:** Corn provides a structure for beans to climb, beans fix nitrogen for the soil, and squash provides ground cover, suppressing weeds and conserving soil moisture.

3. Carrots and Onions

- **Benefit:** Onions deter carrot flies and other pests.
- **Complementary Traits:** Carrots grow underground, allowing onions to provide pest protection without competing for space.

4. Cucumbers and Nasturtiums

- **Benefit:** Nasturtiums act as a trap crop for aphids, protecting cucumbers.
- **Complementary Traits:** Cucumbers benefit from the pest-deterrent properties of nasturtiums.

5. Lettuce and Radishes

- **Benefit:** Radishes deter soil-borne pests that affect lettuce.
- **Complementary Traits:** Radishes grow quickly, providing an early harvest before lettuce requires more space.

6. Strawberries and Borage

- **Benefit:** Borage attracts pollinators and deters certain pests, benefiting strawberries.
- **Complementary Traits:** Borage's bright blue flowers attract bees, enhancing pollination for strawberries.

Companion planting and crop rotation are not mere gardening techniques; they are intricate strategies that epitomize the holistic philosophy of organic farming. By cultivating diverse and mutually beneficial plant relationships, organic farmers create resilient ecosystems where the principles of pest management, nutrient cycling, soil health, and water conservation converge into a sustainable and harmonious agricultural dance.

5. Organic Pest and Disease Control

Integrated Pest Management (IPM) in Organic Farming

In the realm of organic farming, Integrated Pest Management (IPM) is not just a set of practices; it's a philosophy that acknowledges the intricate web of relationships within ecosystems. This holistic approach emphasizes prevention, monitoring, and the use of various strategies to manage pests and diseases while minimizing the impact on the environment. IPM in organic agriculture is a dynamic and adaptive process that integrates cultural, biological, and mechanical controls to maintain a balanced and sustainable agricultural system.

Preventive Measures

1. **Crop Rotation:** Crop rotation is a cornerstone of preventive pest management in organic farming. The intentional rotation of crops disrupts the life cycles of pests that are specific to certain crops. This practice minimizes the risk of pest buildup in the soil, promoting a healthier and more resilient ecosystem. For example, alternating crops like tomatoes and beans with unrelated crops in subsequent seasons helps break pest cycles and reduce the need for chemical interventions.
2. **Companion Planting:** The art of companion planting involves selecting plant combinations that offer mutual benefits. Aromatic herbs like basil planted alongside susceptible crops act as natural repellents for common pests. Beyond pest control, companion planting enhances biodiversity and

promotes a harmonious coexistence of plants. This strategy exemplifies the interconnectedness of species and their ability to support each other in the face of potential threats.

3. **Healthy Soil Practices:** Healthy soil is the foundation of organic farming, and it plays a pivotal role in pest prevention. Organic farmers focus on building and maintaining nutrient-rich, well-structured soils. Practices such as composting, cover cropping, and minimal tillage contribute to soil health, creating an environment where plants are naturally more resistant to pests and diseases. Robust plants, nurtured by vibrant soils, become less susceptible to common pests.

Cultural Practices

1. **Proper Watering Techniques:** Water management is intricately linked to pest control. Organic farmers employ proper watering techniques to maintain optimal soil moisture levels. Consistent moisture reduces stress on plants, making them less susceptible to pest attacks. Drip irrigation, mulching, and careful observation of plant water needs contribute to effective water management, minimizing conditions favorable to pests.

2. **Pruning and Plant Maintenance:** Regular pruning and maintenance are vital cultural practices in organic pest control. Pruning removes diseased or infested plant parts, preventing the spread of pests and diseases. It also improves air circulation and sunlight penetration, creating an environment less favorable for pathogens. By focusing on plant health and promptly addressing any signs of stress, organic farmers mitigate potential pest and disease issues.

3. **Diversity in Planting:** Monoculture, or the cultivation of a single crop over large areas, can be a magnet for pests. Organic farmers embrace polyculture, planting a diverse array of crops. This diversity confuses pests, making it more difficult for them to establish large populations. Additionally, diverse plantings contribute to a more resilient ecosystem, where natural predators find abundant food sources, keeping pest populations in check.

Natural Predators and Beneficial Insects

1. **Encouraging Predator Habitats:** Organic farmers recognize the role of natural predators in maintaining a balanced ecosystem. Providing habitats for beneficial insects, such as ladybugs, lacewings, and predatory beetles, ensures a steady population of natural enemies for common pests. This habitat creation involves preserving natural areas, planting hedgerows, and allowing beneficial weeds to flourish.

2. **Attracting Beneficial Insects:** Companion planting extends beyond pest repellence to the intentional cultivation of flowers that attract beneficial insects. Planting flowering herbs and plants like dill, coriander, or marigolds attracts pollinators and predatory insects. These insects play a dual role in enhancing pollination and controlling pest populations, creating a symbiotic relationship within the organic farm.

3. **Biological Pest Control:** Biological control involves the deliberate introduction of natural enemies to manage pest populations. Organic farmers strategically release predators like ladybugs, parasitic wasps, or predatory nematodes to control specific pests. This targeted approach minimizes the impact on non-target organisms and promotes a self-regulating ecosystem.

Disease-Resistant Crop Varieties in Organic Systems

1. **Genetic Diversity:** Genetic diversity is a key component of disease resistance in organic farming. Planting a variety of crop varieties with different genetic traits helps create a population that is less susceptible to specific diseases. This diversity acts as a natural barrier, preventing the rapid spread of diseases through populations of genetically uniform crops.

2. **Selection of Resistant Varieties:** Organic farmers deliberately select crop varieties that exhibit natural resistance to prevalent diseases. Through traditional breeding methods, these resistant varieties are developed to withstand specific pathogens without the need for chemical treatments. This selection process aligns with the principles of organic farming, promoting resilience and sustainability.

3. **Seed Saving and Local Adaptation:** The practice of seed saving contributes to local adaptation and the development of plant varieties that thrive in specific environmental conditions. Over time, seeds adapted to local climates and disease pressures emerge. This localized adaptation enhances disease resistance, as plants have evolved to withstand the unique challenges of the region.

Cultural and Biological Controls in Conjunction

1. **Comprehensive Pest Management:** Effective pest and disease control in organic farming often involves the seamless integration of cultural and biological controls. For example, the strategic rotation of crops not only disrupts pest cycles but also creates an environment conducive to natural predators. This comprehensive approach reduces the reliance on external inputs and fosters a self-regulating ecosystem.

2. **Biopesticides and Microbial Control:** Organic farmers utilize biopesticides derived from naturally occurring substances, such as neem oil or insecticidal

soaps. Additionally, microbial agents, such as beneficial bacteria or fungi, are employed to control specific pests or diseases. These microbial controls are targeted, minimizing the impact on non-target organisms while effectively managing pest populations.

3. **Timely Intervention:** Organic farmers adopt a proactive approach to pest and disease management, intervening at the first signs of trouble. Timely intervention, whether through the release of beneficial insects, the application of biopesticides, or other cultural practices, helps prevent the escalation of pest and disease issues. Regular monitoring and swift action ensure that pest populations are kept in check without compromising the ecological balance.

6. Conservation Tillage Practices

Minimal Tillage and its Advantages

In the realm of organic farming, the shift towards conservation tillage practices is a testament to the commitment to sustainable and environmentally friendly agricultural methods. Minimal tillage, a key component of conservation tillage, represents a departure from conventional practices that often involve extensive soil disturbance. Instead, minimal tillage focuses on reducing the frequency and intensity of soil manipulation, thereby preserving the structure and health of the soil.

Reduced Soil Disturbance

1. **Preservation of Soil Structure:** One of the primary advantages of minimal tillage is the preservation of soil structure. Excessive or deep tillage can disrupt the natural arrangement of soil particles and the intricate network of soil aggregates. Minimal tillage, by contrast, disturbs the soil less, allowing for the retention of soil structure. This preserved structure enhances the soil's ability to resist erosion, promotes aeration, and creates a more conducive environment for plant roots.

2. **Conservation of Soil Microbial Communities:** Soil is a dynamic ecosystem teeming with microbial life that plays a crucial role in nutrient cycling and overall soil health. Excessive tillage can disrupt these microbial communities, affecting their diversity and functionality. Minimal tillage practices, by minimizing soil disturbance, contribute to the conservation of microbial habitats. This preservation of soil microbial communities fosters a healthy and resilient soil ecosystem.

3. **Mitigation of Soil Erosion:** Minimal tillage practices significantly reduce the risk of soil erosion. By leaving crop residues on the soil surface and minimizing soil disturbance, organic farmers create a protective layer that

shields the soil from the impact of raindrops. This protective cover reduces surface runoff, enhances water infiltration, and mitigates soil erosion. The result is a more sustainable and erosion-resistant agricultural landscape.

4. **Improved Water Infiltration:** Minimal tillage enhances water infiltration by maintaining a porous soil structure. The presence of crop residues on the soil surface acts as a natural barrier, preventing water runoff and allowing rainwater to penetrate the soil. Improved water infiltration is crucial for water conservation, groundwater recharge, and supporting plant root development. Minimal tillage practices align with the principles of sustainable water management in organic farming.

No-till Farming in Organic Agriculture

Taking the concept of minimal tillage a step further, no-till farming represents a radical departure from traditional soil cultivation practices. In no-till systems, the soil is left undisturbed as much as possible, with seeds planted directly into the residues of the previous crop. This approach has gained traction in organic agriculture due to its potential for soil conservation, water retention, and overall sustainability.

Zero or Minimal Soil Disturbance

1. **Preservation of Soil Organic Matter:** No-till farming goes beyond minimal tillage by eliminating soil disturbance altogether. This approach preserves soil organic matter by leaving crop residues on the surface. The gradual decomposition of these residues contributes to the organic content of the soil, enhancing its fertility and nutrient-holding capacity. This preservation of organic matter is especially beneficial in organic farming systems where soil health is paramount.

2. **Weed Suppression and Reduction in Herbicide Use:** The residue cover in no-till systems acts as a natural mulch, suppressing weed growth. By minimizing soil disturbance and relying on the residue cover, organic farmers can reduce the need for herbicides. This not only aligns with organic farming principles but also contributes to the overall sustainability of the agricultural system by minimizing synthetic inputs.

3. **Conservation of Soil Moisture:** No-till farming plays a crucial role in the conservation of soil moisture. The residue cover helps reduce evaporation, keeping the soil surface cooler and preventing rapid moisture loss. Conserved soil moisture is vital for plant growth, especially in regions with erratic rainfall patterns or limited water resources. No-till practices contribute to water efficiency and support the resilience of organic farming systems.

4. **Enhanced Carbon Sequestration:** The minimal disturbance of soil in no-till systems contributes to enhanced carbon sequestration. By preventing the oxidation of organic matter that occurs with traditional tillage, carbon is retained in the soil for more extended periods. This carbon sequestration not only supports soil fertility but also has broader environmental benefits by mitigating climate change through the storage of carbon in the soil.

Impacts on Soil Structure and Water Retention

1. **Maintenance of Soil Aggregates:** Conservation tillage practices, including minimal tillage and no-till farming, prioritize the maintenance of soil aggregates. Soil aggregates are clusters of soil particles bound together by organic matter, roots, and microbial activity. Preserving these aggregates enhances soil structure, providing a stable environment for plant roots and allowing for better water infiltration.

2. **Water Retention and Reduced Runoff:** The impact of conservation tillage on water retention is significant. By leaving crop residues on the soil surface, these practices create a protective layer that reduces water runoff. This protective cover minimizes soil erosion, enhances water infiltration, and improves the overall water-holding capacity of the soil. The result is a more resilient soil that can better withstand periods of drought and erratic rainfall.

3. **Promotion of Soil Health:** Conservation tillage practices contribute to overall soil health. By minimizing soil disturbance, these practices create a habitat conducive to beneficial soil organisms. Earthworms, microbes, and other soil fauna thrive in undisturbed soils, contributing to nutrient cycling, organic matter decomposition, and the enhancement of soil structure. Healthy soils, in turn, support robust plant growth and sustainable agricultural practices.

4. **Reduced Carbon Loss and Sequestration:** The reduction in soil disturbance associated with conservation tillage practices helps minimize carbon loss from the soil. Traditional tillage releases stored carbon into the atmosphere through the oxidation of organic matter. Conservation tillage, particularly no-till farming, reduces this carbon loss, contributing to the sequestration of carbon in the soil. This sequestration aligns with global efforts to mitigate climate change by reducing greenhouse gas emissions.

7. Mulching Techniques

Mulching for Weed Control

In the intricate tapestry of organic farming, mulching stands out as a versatile and effective technique, playing a pivotal role in weed control. Weeds, though resilient and persistent, meet their match in the strategic use of organic mulches, which not only suppress weed growth but also contribute to the overall health of the soil.

Natural Weed Suppression

1. **Physical Barrier to Weed Emergence:** Mulching serves as a physical barrier, preventing weed seeds from reaching the soil surface and germinating. This barrier effect inhibits the penetration of light essential for weed seed germination. By denying weeds the opportunity to establish themselves, organic farmers employing mulching techniques create a weed-suppressed environment without resorting to synthetic herbicides.

2. **Temperature Regulation:** Organic mulches, such as straw or bark, contribute to temperature regulation in the soil. By creating a cooler microclimate at the soil surface, mulches hinder the growth of warm-season weeds that thrive in higher temperatures. This temperature moderation suppresses weed emergence, making it an effective and eco-friendly weed control strategy.

3. **Competition for Resources:** Mulches play a dual role by not only physically impeding weed growth but also competing with weeds for essential resources. As organic mulches decompose, they release nutrients into the soil. This competition for nutrients makes it challenging for weeds to establish and thrive. The result is a weed-controlled environment that fosters the growth of desired crops.

Organic Mulches and Their Benefits

1. **Straw Mulch:** Straw mulch, derived from the stalks of cereal crops, is a popular organic mulch in agriculture. Its benefits extend beyond weed suppression. Straw mulch helps retain soil moisture, regulates soil temperature, and contributes to soil organic matter as it decomposes. Additionally, straw mulch provides a habitat for beneficial insects, fostering a balanced ecosystem within the agricultural landscape.

2. **Wood Chip Mulch:** Wood chip mulch, made from chipped or shredded wood, is a versatile and long-lasting organic mulch. Its benefits include weed suppression, moisture retention, and temperature regulation. Wood chip mulch also contributes to soil structure as it decomposes, enhancing the overall health of the soil. The slow decomposition of wood chips means that they provide lasting benefits, requiring less frequent reapplication.

3. **Cover Crop Mulch:** Cover crops, when grown and left to cover the soil surface, serve as living mulch. This dynamic form of mulching offers weed suppression, nutrient cycling, and improved soil structure. Cover crop mulch, which includes crops like clover or vetch, provides a continuous

cover that protects the soil from erosion, enhances water infiltration, and creates a conducive environment for beneficial microorganisms.

4. **Strawberry Mulch:** Organic farmers often use straw or hay mulch specifically for strawberry plants. This mulching technique helps suppress weeds around the delicate strawberry plants, conserves soil moisture, and prevents soil-borne diseases by creating a barrier between the fruit and the soil. Strawberry mulch is an example of targeted mulching designed to address the unique needs of specific crops.

Water Conservation through Mulching

1. **Reduced Evaporation:** One of the primary contributions of mulching to water conservation is the reduction of soil moisture evaporation. Mulch acts as a protective layer, shielding the soil from direct sunlight and wind. This coverage minimizes the exposure of the soil surface to the elements, reducing water loss through evaporation. As a result, organic farmers can optimize water use efficiency in their fields.
2. **Enhanced Water Infiltration:** Mulches facilitate improved water infiltration into the soil. By preventing surface runoff, mulches enable rainwater or irrigation to penetrate the soil more effectively. This enhanced water infiltration ensures that moisture reaches plant roots, promoting healthy plant growth and reducing the need for additional irrigation. Mulching, therefore, becomes a critical component of sustainable water management in organic farming.
3. **Conserved Soil Moisture:** The protective barrier created by mulch contributes to the conservation of soil moisture. Organic mulches, such as straw or wood chips, create a microenvironment where soil moisture is conserved by reducing the impact of temperature extremes. Conserved soil moisture is especially vital during dry periods, ensuring that plants have a steady supply of water to support their growth and development.
4. **Prevention of Soil Erosion:** Mulches serve as a protective armor against soil erosion. By covering the soil surface, organic mulches prevent raindrops from directly impacting the soil, minimizing the risk of erosion. This erosion control is crucial for maintaining soil structure, preventing nutrient runoff, and preserving the topsoil. Organic farmers adopting mulching techniques contribute to sustainable land management practices.

In the organic farming playbook, mulching emerges as a multi-faceted strategy, seamlessly integrating weed control, soil health promotion, and water conservation. Through the thoughtful selection of organic mulches and the strategic implementation of mulching techniques, organic farmers cultivate environments where crops thrive, weeds are subdued, and the delicate balance of the ecosystem is maintained. As a cornerstone of sustainable agriculture, mulching exemplifies the harmony that can be achieved when nature-inspired practices take center stage in farming systems.

8. Use of Organic Fertilizers

Types of Organic Fertilizers

Organic farming stands as a testament to the symbiotic relationship between agriculture and nature, emphasizing the use of organic fertilizers to nourish the soil and cultivate healthy crops. These natural inputs, derived from various organic sources, serve as the lifeblood of sustainable farming practices.

1. Compost

- The Art of Decomposition: Compost, often hailed as "black gold" in organic farming, embodies the art of decomposition. Comprising a harmonious blend of kitchen scraps, yard waste, and animal manure, compost undergoes a transformative process orchestrated by a myriad of microorganisms. As these microorganisms break down organic matter, they release a symphony of nutrients, creating a humus-rich concoction that elevates soil fertility.
- Microbial Marvels: Beyond its nutrient content, compost is a haven for beneficial microbes. The diverse microbial community within compost includes bacteria, fungi, and other soil-dwelling organisms. This microbial orchestra not only enhances nutrient availability but also contributes to soil structure, water retention, and disease suppression. Organic farmers, recognizing the intricate dance of microorganisms, view compost as a cornerstone in the quest for a vibrant and resilient soil ecosystem.

2. Manure

- Nature's Nutrient Package: Animal manure, a stalwart in organic farming, embodies nature's nutrient package. Rich in nitrogen, phosphorus, and potassium, manure provides a balanced blend of essential elements for plant growth. The diversity of nutrients within manure mirrors the nutritional needs of crops, making it a versatile and holistic organic fertilizer.
- The Art of Aging: While manure is a potent fertilizer, the art lies in its proper aging. Fresh manure, though nutrient-rich, can be too strong and may contain harmful pathogens. Organic farmers employ composting or aging techniques to ensure the safe and effective use of manure. This meticulous approach transforms manure into a potent yet balanced elixir that nurtures the soil without compromising its health.

3. Bone Meal

- A Phosphorus Powerhouse: Bone meal, derived from finely ground animal bones, emerges as a phosphorus powerhouse in organic fertilization. Phosphorus, a vital element for root development, flowering, and fruiting, finds its natural source in bone meal. What sets bone meal apart is its slow-release nature, providing a sustained phosphorus supply to plants over time.
- Rooting for Growth: Organic farmers often turn to bone meal when rooting for robust root systems. The gradual breakdown of bone meal aligns with the pace of root development, fostering strong and resilient roots. This nutrient pacing not only supports current plant needs but also contributes to the long-term health and vitality of the soil.

4. Blood Meal

- Nitrogen on Demand: Blood meal, a byproduct of animal slaughter, takes center stage as a nitrogen-rich organic fertilizer. Nitrogen, crucial for leafy green growth, finds a natural source in blood meal. Organic farmers, conscious of the delicate balance in nitrogen availability, turn to blood meal for a quick and accessible nitrogen boost.
- Balancing Act: While blood meal offers rapid nitrogen release, organic farmers employ it judiciously to avoid nitrogen imbalances. The careful integration of blood meal aligns with the principles of organic farming, emphasizing a holistic and balanced approach to nutrient management.

5. Fish Emulsion

- From the Depths of the Sea: Fish emulsion, a liquid fertilizer derived from fermented fish remains, embodies the richness of marine life in organic farming. Bursting with nitrogen, phosphorus, and potassium, fish emulsion provides a well-rounded nutrient profile. Its liquid form makes it a versatile option for foliar feeding, ensuring efficient nutrient absorption by plants.
- Sustainable Sourcing: Fish emulsion aligns with the sustainability ethos of organic farming, utilizing fish byproducts that might otherwise go to waste. The utilization of fish remnants in organic fertilization not only adds valuable nutrients to the soil but also minimizes environmental impact through the repurposing of marine resources.

Nutrient Management in Organic Systems

1. Soil Testing and Analysis

- Knowledge is Fertility: Nutrient management in organic systems begins with knowledge, and soil testing is the diagnostic tool that unlocks this knowledge. Organic farmers conduct thorough soil tests to understand the current nutrient levels, pH, and organic matter content. This information serves as the compass guiding nutrient management decisions, ensuring a precise and tailored approach.

- Balancing Act: Soil testing reveals the nutritional needs of the soil, allowing organic farmers to balance nutrient inputs accordingly. The careful calibration of organic fertilizers based on soil test results ensures that crops receive the nutrients they require, avoiding both deficiencies and excesses. In the organic farming playbook, soil testing is the foundation upon which nutrient management strategies are built.

2. Crop Rotation and Diversity

- Nature's Nutrient Rotation: Crop rotation, a time-honored tradition in organic farming, embodies nature's nutrient rotation. Different crops have distinct nutrient requirements, and rotating crops strategically ensures a harmonious balance in nutrient utilization. Beyond nutrient management, crop rotation disrupts pest and disease cycles, contributing to overall ecosystem health.
- The Dance of Diversity: Organic farmers view crop diversity as a choreography of nutrient exchanges. Each crop brings its unique set of nutritional needs and contributions to the soil. Legumes, for instance, participate in the nitrogen-fixing dance, enriching the soil with this vital nutrient. By embracing the dance of diversity, organic farmers choreograph a symphony of nutrients that resonates through the seasons.

3. Green Manure Cover Crops

- Cover Crop Ballet: Green manure cover crops, a form of living mulch, perform a ballet of nutrient enrichment. Sown and then incorporated into the soil before maturity, these cover crops contribute organic matter, nitrogen, and other essential nutrients. Leguminous cover crops, with their nitrogen-fixing prowess, lead the nutrient replenishment ensemble.
- Underground Nutrient Exchange: Beneath the soil surface, the roots of cover crops engage in an intricate nutrient exchange. As cover crop roots break down, they release nutrients, contributing to the soil's nutrient bank. This underground ballet enhances soil fertility, structure, and resilience, preparing the stage for the next act of organic cultivation.

4. Crop Residue Incorporation

- Nature's Recycling Bin: Crop residues, the remnants of harvested plants, become nature's recycling bin in organic farming. Organic farmers incorporate these residues back into the soil, initiating a recycling process that enriches the soil with organic matter. As crop residues decompose, they release nutrients, completing the circle of nutrient replenishment.

- The Carbon Connection: Crop residue incorporation is not just about nutrient recycling; it's also about carbon management. The carbon content in crop residues contributes to soil organic matter, enhancing soil structure and water retention. The marriage of nutrient cycling and carbon sequestration makes crop residue incorporation a fundamental practice in organic nutrient management.

Making and Applying Compost in Organic Farming

1. Compost Ingredients

- The Symphony of Ingredients: Compost-making is a symphony of ingredients, each playing a unique role in the composition. Green materials, such as kitchen scraps and fresh plant material, provide nitrogen, while brown materials, including dried leaves and straw, offer carbon. The art lies in achieving the right balance, creating a composting symphony that resonates with nutrient richness.

- Diversity in the Bin: The diversity of compost ingredients mirrors the biodiversity principles of organic farming. A varied mix of materials ensures a broad spectrum of nutrients and microbial life. Just as a diverse ecosystem thrives, a diverse compost blend creates a nutrient-rich concoction that nourishes the soil ecosystem.

2. Composting Process

- Microbial Choreography: The heart of composting lies in the microbial choreography orchestrated by bacteria, fungi, and other microorganisms. As these organisms break down organic matter, heat is generated, marking the active phase of composting. Regular turning of the compost pile ensures oxygen availability, facilitating microbial activity and accelerating the composting dance.

- Turning and Temperature: The turning of the compost pile serves a dual purpose. It introduces oxygen, an essential element for aerobic composting, and ensures even decomposition. Temperature, a key indicator of microbial activity, rises during the active phase. Organic farmers monitor the temperature to gauge the progress of composting, ensuring that the microbial dance unfolds optimally.

3. Compost Curing

- Maturation Magic: The culmination of the composting journey is the curing or maturation phase. Cured compost, akin to a fine wine, undergoes a maturation process where it stabilizes, and any remaining heat-producing reactions subside. This maturation magic results in compost that is dark, crumbly, and imbued with the earthy aroma of success.

- Ready for the Stage: The readiness of compost for application is akin to a grand performance ready for the stage. Cured compost, with its balanced nutrient profile and microbial richness, takes center stage in the organic farming production. It is a testament to the synergy between organic inputs and microbial alchemy that transforms waste into a nutrient-rich resource.

4. Application of Compost

- Top Dressing Ballet: Compost application unfolds as a top dressing ballet or a soil incorporation symphony. As a top dressing, compost is spread on the soil surface, offering a protective layer that retains moisture, suppresses weeds, and provides a slow-release of nutrients. The gradual breakdown of compost nurtures plants throughout the growing season.
- Incorporation Harmony: Incorporating compost into the soil ensures a seamless integration of organic matter and nutrients. This incorporation harmonizes with soil structure, enhancing water retention, aeration, and nutrient availability. Organic farmers choreograph the application of compost to align with the specific needs of crops, fostering a balanced and nutrient-rich soil environment.

5. Compost Tea

- Brewing Nutrient Elixirs: Compost tea, a liquid extract brewed from steeping compost in water, emerges as a nutrient elixir in organic farming. This concentrated liquid carries the essence of compost, delivering a potent blend of nutrients and beneficial microorganisms. As a foliar spray or soil drench, compost tea nourishes plants and fortifies them against environmental stressors.
- Foliar Feeding Ballet: The foliar feeding ballet, performed with compost tea, exemplifies the efficiency of nutrient uptake through plant leaves. The liquid nature of compost tea facilitates rapid nutrient absorption, providing plants with a quick nutrient boost. This foliar ballet supports overall plant health and resilience, showcasing the versatility of compost in organic farming.

6. Compost Application Timing

- Seasonal Synergy: The timing of compost application in organic farming aligns with the seasonal rhythm of crops. Some crops benefit from pre-planting incorporation of compost, enriching the soil before the growing season. Others may receive compost as a mid-season top dressing, offering a nutrient boost during critical growth stages. The strategic timing ensures that compost plays a supporting role in the symphony of crop development.

- Cultivating Nutrient Harmony: Organic farmers, attuned to the nutrient needs of their crops, cultivate a nutrient harmony through precise compost application. Whether preparing the soil for planting, nurturing plants during growth, or fortifying them during key developmental stages, compost application becomes an orchestrated practice that resonates with the natural cycles of organic cultivation.

In the organic farming, the use of organic fertilizers is not just a means to provide nutrients; it is an intricate dance with nature. From the humble beginnings of compost ingredients to the seasoned maturity of nutrient management strategies, organic farmers engage in a symphony of practices that foster soil health, crop vitality, and the sustainability of agriculture. Through the use of organic fertilizers, agriculture becomes a regenerative force, harmonizing with the principles of nature to cultivate thriving ecosystems.

9. Crop Diversity and Biodiversity

Cultivating Diversity in Organic Agriculture

Diversity is the heartbeat of organic agriculture, pulsating through the fields and orchards, enriching the soil, and fostering resilient ecosystems. In the organic farming narrative, crop diversity and biodiversity stand as guardians of sustainability, promoting harmony between cultivation and the natural world.

Benefits of Diverse Cropping Systems

1. Enhanced Nutrient Utilization

- A Nutrient Ballet: Diverse cropping systems orchestrate a nutrient ballet where each plant plays a unique role in nutrient utilization. Different crops have varied nutrient requirements, and by cultivating a diverse array of species, organic farmers ensure that the soil's nutrient profile remains balanced. This intricate dance of nutrient uptake and release contributes to long-term soil fertility, allowing the soil to serve as a reservoir of essential elements for plant growth.

2. Microbial Diversity in the Rhizosphere

- Underground Harmony: Beneath the soil surface, the rhizosphere—a zone influenced by plant roots—comes alive with microbial diversity in diverse cropping systems. Different crops exude a variety of root exudates, shaping a unique environment that supports a rich microbial community. This underground harmony enhances nutrient cycling, disease suppression, and overall soil health. The interplay between plant roots and microbes becomes a hidden symphony that sustains the vitality of the agricultural landscape.

3. Adaptive Weed Management

- Nature's Weed Suppressants: Diverse cropping systems embody a natural approach to weed management. The varied heights, growth habits, and allelopathic properties of different crops create a complex environment where weeds struggle to establish dominance. Some crops act as natural weed suppressants, shading out unwanted vegetation, while others release compounds that inhibit weed growth. This adaptive weed management strategy reduces the reliance on herbicides, aligning with the principles of organic farming.

4. Cultural Heritage Preservation

- Seeds of Culture: Diverse cropping systems often include heirloom and traditional crop varieties, preserving cultural heritage. These varieties carry the stories of communities, passed down through generations. By cultivating and maintaining diverse crops, organic farmers become stewards of cultural biodiversity. The retention of traditional varieties not only connects farmers with their roots but also safeguards agricultural traditions that may hold resilience in the face of changing climates.

Enhancing Biodiversity on Organic Farms

1. Polyculture and Ecological Harmony

- Fields of Harmony: Polyculture, the cultivation of multiple crops in proximity, becomes a canvas for ecological harmony on organic farms. Interplanting various crops mimics natural ecosystems, where different plant species coexist and complement each other. This practice encourages a diverse array of insects, birds, and beneficial microorganisms, fostering a thriving ecosystem within the agricultural landscape.

2. Habitat Restoration and Connectivity

- Linking Landscapes: Organic farmers, as ecological architects, engage in habitat restoration to enhance biodiversity. This involves creating and preserving natural habitats within and around farms. By establishing corridors of native vegetation, organic farmers contribute to landscape connectivity. These green corridors serve as wildlife highways, allowing species to move freely, supporting genetic diversity, and mitigating the isolation of populations.

3. Biological Pest Control

- Predators on Patrol: The diverse flora in organic farming systems becomes a haven for natural predators, the unsung heroes of biological pest control. Ladybugs, lacewings, and predatory beetles are attracted to diverse landscapes, where they find a variety of prey. By nurturing biodiversity, organic farmers enlist these natural allies in the battle against pests. The balance achieved in diverse ecosystems encourages a delicate interplay where predators keep pest populations in check.

4. Genetic Diversity in Seed Banks

- Seeds as Genetic Archives: Organic farmers recognize the importance of maintaining genetic diversity in their seed banks. Preserving a diverse collection of seeds, including traditional and heirloom varieties, safeguards genetic resources. This genetic diversity is a reservoir of resilience, offering options for adaptation to changing environmental conditions. It also provides a source of genetic material for ongoing crop improvement efforts.

Role of Agroforestry in Organic Agriculture

1. Silvopasture Biodiversity Hotspots

- Diversity Beneath Canopies: Silvopasture systems, where trees are integrated with livestock grazing, emerge as biodiversity hotspots. The canopy of trees provides shade for animals, creating microenvironments that support diverse plant species. The understory becomes a tapestry of grasses, herbs, and shrubs, contributing to the overall biodiversity of the system. Silvopasture showcases the potential for coexistence between trees, livestock, and a myriad of plant species.

2. Wildlife Corridors in Agroforestry

- Green Highways: Agroforestry practices contribute to the creation of wildlife corridors. Trees planted in linear arrangements, such as windbreaks and hedgerows, serve as green highways for wildlife. Birds, insects, and small mammals utilize these corridors for foraging, nesting, and movement. The establishment of wildlife corridors enhances landscape connectivity, promoting genetic exchange among populations and supporting overall biodiversity.

3. Medicinal and Aromatic Plant Integration

- Healing Aromas: Agroforestry systems provide an ideal setting for the integration of medicinal and aromatic plants. These plants not only contribute to the diversity of the agroecosystem but also offer additional economic opportunities for farmers. The synergy between trees, crops, and medicinal plants creates multifunctional landscapes that resonate with the principles of organic agriculture.

4. Carbon Sequestration and Habitat Creation

- Trees as Carbon Architects: Agroforestry, as a carbon-smart strategy, not only sequesters carbon in trees and soil but also creates habitats for diverse species. The structure of agroforestry systems—combining tree canopies, understory vegetation, and open spaces—offers niches for a variety of organisms. This diverse habitat creation aligns with organic farming's commitment to holistic and regenerative land management.

10. Soil Erosion Control

Preserving Earth's Foundation in Organic Agriculture

Soil erosion, the silent thief of fertile topsoil, poses a significant challenge to sustainable agriculture. In the organic farming paradigm, the preservation of soil health takes center stage, and effective erosion control becomes a cornerstone in nurturing the foundation of productive and resilient farmlands.

Strategies for Preventing Soil Erosion

1. Cover Cropping as a Living Blanket

- Green Armor: Cover cropping stands as an organic farmer's first line of defense against soil erosion. The strategic planting of cover crops, such as legumes, grasses, and clover, provides a living blanket over the soil. The dense foliage acts as a shield, reducing the impact of raindrops on the soil surface and preventing the dislodging of soil particles. Cover crops also contribute to root structure, binding soil particles together and enhancing soil structure.

2. Mulching for Ground Protection

- Nature's Blanket: Mulching, whether with organic materials like straw or plant residues, serves as nature's blanket for the soil. Mulch creates a protective layer that shields the soil from the impact of rainfall, preventing the splash and detachment of soil particles. Additionally, mulch conserves soil moisture, reducing the erosive force of water runoff. Organic farmers strategically apply mulch around crops and in between rows to create a resilient barrier against erosion.

3. Contour Plowing and Terracing

- Following Nature's Lines: Contour ploughing aligns with the natural contours of the land, mimicking the way water naturally flows across the landscape. By ploughing along the contour lines, organic farmers slow down the flow of water, giving it time to infiltrate the soil. Terracing, a technique that creates flat areas on sloping land, further reduces the velocity of water runoff. Both practices contribute to erosion control by minimizing the erosive force of water on slopes.

4. Windbreaks for Erosion Resistance

- Shielding Against the Wind: Wind erosion, often underestimated, is addressed through the establishment of windbreaks. Rows of trees or shrubs strategically planted along field edges create a barrier that shields the soil from wind-driven erosion. Windbreaks not only protect against soil loss but also contribute to improved microclimates within the fields, creating a more conducive environment for plant growth.

Contour Plowing and Other Erosion Control Techniques

1. Contour Farming

- Sculpting the Landscape: Contour farming is a deliberate approach to land management where crops are planted in rows following the contour lines of the land. This method creates a series of level, curved furrows that effectively capture and slow down water runoff. Contour farming minimizes the formation of straight and rapid water channels, reducing soil erosion and promoting water infiltration.

2. Terracing for Slope Stabilization

- Steps to Erosion Control: Terracing transforms steep slopes into a series of flat surfaces, resembling steps on a staircase. These terraced levels slow down water runoff, allowing it to infiltrate the soil gradually. Terracing is particularly effective on hilly or mountainous terrain where the risk of soil erosion is heightened. The step-like structure reduces the erosive force of water, preventing soil from being carried away.

3. Covered Erosion Control Mats

- Nature-Inspired Blankets: Covered erosion control mats, often made from organic materials, provide a protective layer over vulnerable soil. These mats, resembling natural blankets, stabilize the soil surface, preventing erosion caused by rainfall or wind. Organic farmers strategically place these mats in areas prone to erosion, offering an effective and environmentally friendly solution to soil loss.

4. Grassed Waterways

- Guiding the Flow: Grassed waterways are channels or strips of land with established grass cover designed to guide water flow. These natural channels slow down water runoff, allowing it to follow a vegetated path rather than creating erosive channels. Grassed waterways not only control soil erosion but also contribute to water infiltration and sediment deposition.

Importance of Ground Cover in Erosion Prevention

1. Living Roots as Soil Anchors

- Roots: Nature's Anchors: Ground cover, whether in the form of cover crops or permanent vegetation, plays a vital role in preventing soil erosion. The roots of plants act as nature's anchors, binding soil particles together and creating a stable soil structure. The living roots provide structural support, reducing the likelihood of soil being carried away by water or wind.

2. Microbial Activity for Soil Stability

- Microbial Cement: Ground cover supports a thriving soil microbial community, contributing to the stability of soil aggregates. Microorganisms, such as fungi and bacteria, produce substances that act as a natural glue, binding soil particles together. This microbial activity enhances soil structure, making it more resistant to erosion. Organic farmers prioritize practices that promote healthy microbial populations, fostering a resilient and erosion-resistant soil ecosystem.

3. Conservation Tillage Techniques

- Disturbance Minimization: Conservation tillage techniques, such as no-till or minimal tillage, contribute to ground cover preservation. By minimizing soil disturbance, these practices leave crop residues on the field surface, acting as a protective layer. The presence of crop residues shields the soil from the impact of rainfall and wind, reducing erosion risk. Conservation tillage aligns with the principles of organic farming, emphasizing minimal disruption to the soil ecosystem.

4. Permanent Vegetation for Erosion Control

- Perennial Protectors: Permanent vegetation, including grasses, legumes, and other perennial plants, offers ongoing protection against erosion. Unlike annual crops that leave the soil exposed during certain periods, permanent vegetation maintains continuous ground cover throughout the year. This sustained cover provides a reliable shield against erosion, ensuring the long-term stability of the soil.

11. Monitoring and Assessment in Organic Systems

Nurturing Organic Excellence through Vigilance

In organic farming, the commitment to sustainable practices goes hand in hand with a vigilant approach to monitoring and assessment. Organic farmers, as stewards of the land, engage in systematic evaluations to ensure the health of soil and water, assess the success of organic practices, and continuously refine farming methods for enduring ecological harmony.

Soil and Water Quality Testing

1. Strategic Soil Sampling

- Unveiling Soil Secrets: Organic farmers initiate the monitoring process with strategic soil sampling. This involves systematically collecting soil samples from various points across the farm. These samples represent the diversity of the landscape, capturing variations in soil texture, structure, and nutrient content. Soil testing laboratories analyze these samples, unveiling the intricate details of soil health, including nutrient levels, pH, and microbial activity.

2. Comprehensive Soil Health Parameters

- Beyond Nutrients: Soil quality testing in organic systems extends beyond traditional nutrient analysis. It encompasses a comprehensive set of parameters, including soil organic matter content, cation exchange capacity, and aggregate stability. The emphasis is on understanding the holistic health of the soil ecosystem. Organic farmers recognize that a vibrant soil, rich in organic matter and teeming with microbial life, forms the foundation for sustainable agriculture.

3. Water Quality Monitoring

- Guardians of Aquatic Ecosystems: Organic farmers extend their monitoring efforts to the water resources on and around their farms. Regular water quality testing ensures that runoff from fields does not compromise nearby water bodies. Parameters such as nutrient levels, sedimentation, and the presence of pollutants are assessed. The goal is to be guardians of aquatic ecosystems, preventing contamination and preserving water quality for both farm and surrounding environments.

4. Efficient Water Usage Monitoring

- Optimizing Irrigation: Organic farming emphasizes efficient water usage, and monitoring plays a key role in achieving this goal. Farmers track irrigation practices, ensuring that water is applied judiciously based on the needs of crops. Techniques such as soil moisture monitoring and the use of weather data contribute to precision irrigation. By minimizing water wastage, organic farmers align their practices with principles of conservation and sustainability.

Assessing the Success of Organic Practices

1. Crop Performance Evaluation

- Beyond Yields: The success of organic practices is gauged not only by yields but also by the overall performance of crops. Organic farmers assess factors such as crop vigor, resistance to pests and diseases, and the development of strong root systems. Healthy plants are seen as indicators of a balanced and thriving agroecosystem.

2. Biodiversity Metrics

- Diverse Ecosystems: Assessing the biodiversity on the farm becomes a metric for success in organic systems. The presence of beneficial insects, pollinators, and diverse plant species is observed. Organic farmers recognize that a vibrant biodiversity contributes to natural pest control, soil fertility, and the overall resilience of the farm.

3. Soil Structure and Aggregation

- Building Soil Resilience: The success of organic practices is closely tied to improvements in soil structure. Farmers monitor soil aggregation, a key indicator of soil resilience. Enhanced soil aggregation promotes water infiltration, root development, and resistance to erosion. Observing improvements in soil structure validates the efficacy of organic farming methods.

4. Weed and Pest Management Efficacy

- Balancing Act: Successful organic practices strike a balance in weed and pest management. Monitoring the prevalence of weeds and the impact of natural pest control measures provides insights into the effectiveness of organic approaches. Organic farmers celebrate the presence of beneficial insects and the dynamic equilibrium achieved through integrated pest management (IPM) strategies.

Continuous Improvement in Organic Farming Methods

1. Adaptive Crop Planning

- Nature-Informed Planning: Organic farmers engage in adaptive crop planning based on continuous monitoring. Observations from previous seasons guide adjustments in crop rotations, cover cropping strategies, and the selection of plant varieties. This adaptive approach ensures that organic farming methods evolve in harmony with the dynamic nature of ecosystems.

2. Feedback Loop with Soil Amendments

- Responsive Nutrient Management: Monitoring soil nutrient levels creates a feedback loop for nutrient management. Organic farmers adjust their soil amendment practices based on the results of soil tests. This responsive approach ensures that nutrient inputs align with the specific needs of crops, minimizing excesses and deficiencies.

3. Incorporating Innovative Techniques

- Experimentation and Innovation: Continuous improvement in organic farming involves a spirit of experimentation. Organic farmers explore innovative techniques, such as agroecological practices, biofertilizers, and crop diversification. Experimentation allows farmers to test the feasibility and effectiveness of new methods, contributing to the evolution of sustainable and regenerative farming practices.

4. Education and Knowledge Sharing

- Cultivating a Learning Community: Organic farmers embrace a culture of education and knowledge sharing. Monitoring results and lessons learned from on-farm experiences are shared within the farming community. Workshops, field days, and collaborative initiatives contribute to a collective learning environment where farmers benefit from each other's insights and successes.

Conclusion

In the fertile fields of organic agriculture, where the soil breathes life, and the rhythms of nature orchestrate a harmonious symphony, the journey concludes as a testament to the profound impact of sustainable and regenerative practices. This narrative, woven with the threads of organic farming principles, soil and water management, biodiversity conservation, and continuous improvement, celebrates the resilience of ecosystems and the steadfast commitment of organic farmers to cultivate a legacy of thriving landscapes.

The organic farming ethos, rooted in the principles of ecological balance, biodiversity, and soil health, emerges as a beacon illuminating a path toward sustainable agriculture. As we traverse the landscapes nurtured by organic practices, the definition and principles of organic farming come to life. It is not merely a mode of cultivation; it is a holistic philosophy that acknowledges the interdependence of soil, water, plants, and the intricate web of life. The principles of organic farming extend an invitation to view agriculture as a partnership with nature, where every element plays a vital role in the co-creation of abundance. The importance of sustainable agriculture echoes through the chapters that unfold, beginning with the foundational understanding of organic farming. The definition and principles laid bare showcase a commitment to cultivating food in a manner that respects and supports the natural systems that sustain life.

Soil, the sacred canvas upon which the story of organic agriculture is written, takes center stage in the chapters dedicated to soil management. The exploration of soil health and its significance unveils the profound role that this living entity plays in sustaining life. Organic soil amendments become the nurturing hands that replenish and invigorate the soil, fostering a dynamic environment where microbial life, nutrients, and organic matter intertwine in a dance of regeneration. Cover cropping and green manure, depicted as vibrant strokes on the canvas, showcase the organic farmer's commitment to preserving and enhancing soil fertility through nature-inspired practices.

Water, the lifeblood of the land, becomes a focal point in the narrative of organic farming. The importance of water conservation is emphasized, not merely as a practical necessity but as a profound responsibility. Efficient irrigation methods and rainwater harvesting techniques showcase the organic farmer's innovative spirit in harnessing and managing water resources sustainably. The organic landscape becomes a canvas where water flows harmoniously, nourishing crops and ecosystems alike.

Companion planting and crop rotation emerge as strategies in organic farming that go beyond mere agricultural practices; they are nuanced expressions of ecological wisdom. The benefits of companion planting, illustrated through the symbiotic relationships between different plant species, exemplify the elegance with which nature orchestrates collaborations. Crop rotation, akin to a strategic dance, becomes a methodical approach to soil and water management, breaking the cycles of pests and diseases while nurturing the vitality of the soil.

Pest and disease control in organic farming unfold as a narrative of coexistence rather than eradication. Integrated Pest Management (IPM) becomes the guiding principle, where natural predators and beneficial insects take center stage in maintaining a delicate balance. Disease-resistant crop varieties, celebrated for their resilience, become the heroes in the story of organic systems, embodying the organic farmer's commitment to sustainable and resilient agriculture.

In the chapters exploring conservation tillage practices, minimal tillage and no-till farming become the gentle strokes that preserve soil structure and enhance water retention. The impacts on soil structure and water dynamics are portrayed not as disruptions but as nuanced adaptations that align with the principles of organic agriculture. The organic landscape, with its undisturbed soil structure, becomes a testament to the farmer's reverence for the interconnectedness of soil and water.

Mulching techniques, depicted as organic blankets, unfold as powerful tools for weed control, soil protection, and water conservation. The organic mulches, carefully chosen from nature's repertoire, become the guardians that shield the soil from erosive forces and nurture a microclimate where plants flourish. Water conservation through mulching becomes a practice where every layer of organic matter contributes to the resilience of the soil ecosystem.

The use of organic fertilizers becomes a choreography, where nature's recycling bin and the carbon connection form the rhythmic dance of nutrient replenishment. Types of organic fertilizers, ranging from compost to compost tea, become the instruments that play in harmony with the needs of crops. Nutrient management in organic systems becomes a mindful endeavor, where the farmer's understanding of the soil's needs aligns with the cyclical processes of nature.

Crop diversity and biodiversity emerge as essential themes, woven into the fabric of organic agriculture. The benefits of diverse cropping systems unfold as a nutritional banquet for the soil, offering a diverse array of nutrients and fostering resilience in the face of climate variability. Enhancing biodiversity on organic farms becomes a commitment to creating habitats, preserving native plant species, and integrating agroforestry practices that harmonize with nature's intricate designs.

The narrative extends to soil erosion control, where organic farmers become custodians of the Earth's foundation. Strategies for preventing soil erosion, contour ploughing, and ground cover emerge as the guardians that protect the soil from the erosive forces of water and wind. The organic landscape becomes a haven where every undulating contour, every mulched surface, and every covered erosion control mat becomes a statement of resilience against soil loss.

Monitoring and assessment in organic systems become the vigilant eyes that ensure the integrity of the organic journey. Soil and water quality testing become the diagnostic tools that unveil the intricacies of soil health and the purity of water resources. Assessing the success of organic practices becomes a holistic evaluation that goes beyond yields, encompassing biodiversity, soil structure, and the efficacy of pest management. Continuous improvement becomes the ethos that propels organic farmers forward, where adaptive crop planning, feedback loops with soil amendments, incorporation of innovative techniques, and a culture of education and knowledge sharing become the driving forces of progress.

7

Agroforestry and Sustainable Land Use for Soil Management and Conservation

[1]Deepa Rawat, [2]Kh. Chandrakumar Singh, [3]Savita Jangde [4]Wajid Hasan and [5]Sheetanshu Gupta

[1]College of Forestry, Ranichauri, VCSG University of Horticulture and Forestry, Bharsar, Uttarakhand
[2]VCSGU University of Horticulture and Forestry, Bharsar, Uttarakhand
[3]Department of Plant Physiology, Institute of Agricultural Sciences B.H.U. Varanasi, U.P.
[4]KVK Jehanabad, Bihar Agricultural University Sabour, Bihar
[5]Department of Biotechnology, Insitute of Technology and Management BKT, Lucknow, U.P.

1. Introduction to Agroforestry

Agroforestry stands at the intersection of agriculture and forestry, embodying a holistic approach to land management that harnesses the synergies between trees, crops, and livestock. At its essence, agroforestry transcends conventional farming practices, weaving together the ecological, economic, and social dimensions of sustainable land use.

Defining Agroforestry

In the tapestry of agroforestry, the definition extends beyond the mere coexistence of trees and crops; it embodies a philosophy of intentional and symbiotic relationships. Agroforestry is a deliberate and dynamic land-use strategy where trees are integrated into agricultural systems, fostering mutually beneficial interactions. These interactions go beyond immediate yields; they contribute to the resilience of ecosystems, the improvement of soil health, and the creation of microclimates conducive to plant and animal life.

In its essence, agroforestry acknowledges that the productivity of the land is not a zero-sum game. Instead, it recognizes that the combination of trees and crops can lead to enhanced yields and environmental sustainability. The definition encapsulates a mindful and strategic approach, where the selection and placement of trees are guided by an understanding of their ecological functions and their potential to support and augment agricultural production.

Historical Perspectives

To comprehend agroforestry fully, one must delve into its historical roots, where indigenous and traditional communities cultivated wisdom that embraced the synergy between trees and agriculture. Across cultures and continents, historical perspectives reveal the innate understanding of the land's interconnectedness. Indigenous practices often incorporated agroforestry as a way to optimize resources, improve soil fertility, and create resilient landscapes.

The historical narrative of agroforestry is not confined to a specific region or era; it spans the ancient civilizations of the Mayans, Greeks, and Romans, each contributing to the rich tapestry of agroforestry practices. In Southeast Asia, the taungya system, where crops are grown in the understory of forests, exemplifies a historical tradition of harmonizing agriculture with the existing natural ecosystems.

As societies evolved, the intrinsic principles of agroforestry persisted, adapted, and found resonance in different cultural contexts. The historical lens allows us to appreciate agroforestry not merely as a contemporary agricultural strategy but as a timeless practice deeply rooted in the collective wisdom of humanity's agrarian heritage.

Principles of Agroforestry

The principles that guide agroforestry practices serve as a compass, steering the intentional integration of trees into agricultural landscapes. These principles not only define the ecological logic behind agroforestry but also underscore its economic and social significance.

Biodiversity Integration: Agroforestry's commitment to biodiversity goes beyond the mere coexistence of different species. It aims to create synergies that enhance the overall ecological diversity, promoting a thriving community of plants, animals, and microorganisms. Diverse plant species contribute to ecological resilience by providing habitat, attracting beneficial insects, and supporting natural pest control.

Synergistic Interactions: The strategic placement of trees in agroforestry systems is informed by the understanding of synergistic relationships. For example, the shade provided by trees can be harnessed to create microclimates that benefit certain crops. The root systems of trees contribute to soil structure, preventing erosion and enhancing nutrient cycling. These intentional interactions amplify the overall productivity and sustainability of the agricultural landscape.

Economic Viability: Agroforestry is designed not only with ecological resilience in mind but also with a keen eye on economic sustainability. Trees are selected based on their market value, providing farmers with additional revenue streams. This economic viability ensures that agroforestry systems are not only environmentally sustainable but also economically viable for the farmers who steward the land.

Adaptability: Agroforestry systems are crafted with adaptability at their core. Recognizing the diverse climates, soils, and ecological contexts in which they operate, these systems are designed to be flexible and responsive. This adaptability allows agroforestry to thrive in a range of environments, from tropical rainforests to arid landscapes.

Enhanced Environmental Services: Trees in agroforestry systems are not mere ornaments; they serve as active contributors to environmental services. Carbon sequestration, water conservation, and erosion control are among the many services provided by trees in these systems. Agroforestry becomes a tool for addressing broader environmental challenges, aligning with global efforts to mitigate climate change and conserve natural resources.

Community Engagement: The success of agroforestry is intricately linked to the active involvement of local communities. Traditional knowledge, cultural practices, and community participation play pivotal roles in the implementation and sustainability of agroforestry systems. By engaging local communities, agroforestry becomes a shared endeavor that fosters a sense of ownership and responsibility toward the land.

Long-Term Planning: Agroforestry is a commitment to long-term planning, looking beyond immediate gains and considering the enduring health of ecosystems. By adopting a holistic perspective, agroforestry systems are designed to withstand the test of time, providing benefits to current and future generations.

2. Benefits of Agroforestry for Soil Management

Agroforestry, with its symbiotic integration of trees into agricultural systems, unfolds a rich tapestry of benefits that weave together the intricate threads of enhanced soil fertility, improved soil structure, reduced erosion risk, and the promotion of biodiversity (Table 1).

Table 1: Benefits of Agroforestry for Soil Management

S.No	Benefits	Description
1	Soil Conservation	Tree roots help bind soil particles, preventing erosion and reducing the risk of soil loss through water runoff.
2	Improved Soil Structure	Agroforestry systems enhance soil structure by promoting the development of stable aggregates, improving water infiltration and aeration.
3	Nutrient Cycling	Trees contribute organic matter to the soil through leaf litter and root exudates, promoting nutrient cycling and enhancing soil fertility.

S.No	Benefits	Description
4	Reduced Soil Erosion	The presence of trees provides a natural windbreak and reduces the impact of rainfall on the soil surface, minimizing erosion risk.
5	Increased Water Infiltration	Tree roots create channels in the soil, enhancing water infiltration and reducing surface runoff, leading to better water availability for crops.
6	Drought Mitigation	Trees with deep root systems can access water from lower soil layers, contributing to improved water availability during periods of drought.
7	Biodiversity Support	Agroforestry promotes biodiversity, including beneficial microorganisms and insects, contributing to a healthier and more resilient soil ecosystem.
8	Carbon Sequestration	Trees absorb carbon dioxide (CO2) from the atmosphere and store carbon in their biomass and in the soil, helping mitigate climate change.
9	Windbreak and Microclimate	Trees act as windbreaks, reducing wind erosion, and create a favorable microclimate, moderating temperature and protecting crops from extreme conditions.
10	Reduced Soil Compaction	Tree roots can alleviate soil compaction by penetrating and breaking up compacted layers, improving soil structure and root growth.
11	Enhanced Agroecosystem Resilience	Agroforestry systems provide a buffer against environmental stresses, making the agroecosystem more resilient to climatic variations and disturbances.

Enhanced Soil Fertility

Agroforestry stands as a testament to the profound influence trees wield in shaping soil fertility within agricultural landscapes. The integration of trees introduces a dynamic interplay of organic matter, nutrients, and microbial activity that transforms the soil into a fertile matrix supporting robust plant growth.

The leaf litter and organic materials shed by trees act as a continuous source of nutrients, fostering a nutrient-rich environment. As these organic materials decompose, they release essential elements such as nitrogen, phosphorus, and potassium into the soil. This natural fertilization process is a cornerstone of agroforestry, providing a sustainable and self-renewing nutrient cycle that supports long-term agricultural productivity.

The nitrogen-fixing ability of certain tree species further contributes to enhanced soil fertility. These trees have specialized root nodules that house nitrogen-fixing bacteria, converting atmospheric nitrogen into a form usable by plants. This biological nitrogen fixation reduces the need for external nitrogen inputs, promoting a more self-sufficient and ecologically harmonious agricultural system.

In agroforestry systems, trees act as nutrient pumps, drawing minerals from deeper soil layers and making them available to surface-rooted crops. This dynamic nutrient cycling, facilitated by the synergy between trees and crops, creates a balanced and resilient soil ecosystem that sustains plant growth over successive cropping seasons.

Improved Soil Structure

Agroforestry emerges as a transformative force in shaping soil structure, enhancing its physical properties and creating an environment conducive to plant root development. The intricate root systems of trees play a pivotal role in this transformation, contributing to improved soil aggregation, reduced compaction, and enhanced water infiltration.

Bioturbation, the process by which living organisms, including tree roots, alter the structure of the soil, becomes a central theme in agroforestry. As tree roots penetrate the soil, they create channels and pores, improving aeration and water movement. This natural soil engineering enhances the porosity of the soil, allowing for better water retention and nutrient exchange.

The combination of deep-rooted trees and shallow-rooted crops in agroforestry systems creates a complementary network of root systems. This network stabilizes the soil, preventing erosion and reducing the risk of landslides. The improved soil structure also enhances the soil's capacity to hold water, mitigating the impacts of drought and promoting overall water use efficiency within agroforestry landscapes.

As agroforestry transforms the physical characteristics of the soil, it sets the stage for a resilient and adaptable agricultural ecosystem, capable of withstanding environmental challenges and promoting sustainable land use practices.

Reduced Erosion Risk

Erosion, a pervasive threat to agricultural landscapes, finds a formidable opponent in the design and principles of agroforestry. The intricate root systems of trees, combined with their protective canopy, act as natural shields against erosion, preserving the integrity of the topsoil and ensuring the long-term productivity of the land.

One of the primary mechanisms through which agroforestry mitigates erosion is through root reinforcement. The root systems of trees bind soil particles together, creating a stable structure that resists the erosive forces of water runoff. This physical binding not only reduces surface runoff but also prevents soil detachment, promoting soil stability on slopes and vulnerable areas.

Alley cropping, a common agroforestry practice where rows of trees are planted alongside rows of crops, further enhances erosion control. The arrangement creates a natural contour that reduces the speed and impact of water runoff. The result is a landscape where the risk of soil erosion is minimized, preserving the fertile topsoil and safeguarding the long-term sustainability of agricultural endeavors.

The protective canopy of trees in agroforestry systems adds another layer of defense against erosion. By intercepting rainfall, trees reduce the impact of water droplets on the soil surface, preventing soil compaction and splash erosion. This multifaceted approach to erosion control highlights the resilience and adaptability embedded in agroforestry practices.

Biodiversity Promotion

Agroforestry emerges as a biodiversity champion within agricultural settings, creating environments that support a rich tapestry of life both above and below the ground. The intentional integration of trees, crops, and understory vegetation contributes to diverse habitats that foster ecological balance and resilience (Fig 1).

Fig. 1: Enhancing Biodiversity through conservation

Above-ground biodiversity in agroforestry systems is evident in the varied vegetation that characterizes these landscapes. Trees provide habitat and food sources for a multitude of organisms, from insects and birds to mammals. The diverse canopy structure, with its varying heights and leaf types, attracts different bird species, contributing to natural pest control as these birds feed on insect pests.

The promotion of biodiversity extends below the ground, where the root systems of trees and associated vegetation create a thriving subterranean ecosystem. The diversity of root structures enhances soil structure and nutrient cycling. Mycorrhizal fungi, symbiotic organisms that form partnerships with plant roots, play a crucial role in nutrient exchange, benefiting both trees and crops. The intricate web of soil-dwelling organisms contributes to the breakdown of organic matter, further enriching the soil with nutrients.

Agroforestry systems act as corridors for wildlife movement, facilitating the dispersion of seeds and promoting genetic diversity within plant populations. The result is a balanced and resilient ecosystem where the interdependence of different species supports the overall health and functionality of the agroecosystem.

Moreover, the diversified landscape of agroforestry reduces the need for chemical pesticides by promoting natural pest control mechanisms. Beneficial insects, attracted by the diversity of flowering plants, act as predators, keeping pest populations in check. This biological control contributes to the ecological sustainability of agroforestry systems, aligning with principles of environmentally friendly and integrated pest management.

3. Agroforestry Systems for Soil Conservation

Agroforestry, with its diverse systems tailored for specific ecological contexts, emerges as a guardian of soil health and conservation. The intricate dance between trees, crops, and livestock in these systems orchestrates a symphony of ecological functions, mitigating erosion, enhancing soil structure, and fostering sustainable agricultural landscapes. Let's delve deeper into each of these agroforestry systems — Alley Cropping, Silvopasture, Windbreaks and Shelterbelts, and Riparian Buffer Strips — unraveling their nuanced contributions to soil conservation.

Alley Cropping

Alley cropping, a manifestation of agroforestry's ingenuity, intertwines the productivity of annual crops with the perennial benefits of trees, forming alleys of mutualistic coexistence. Beyond the aesthetic appeal of tree-lined rows, this system plays a pivotal role in soil conservation by design.

The alleys created in alley cropping function as natural contours, reducing the speed and impact of water runoff. This spatial arrangement mitigates soil erosion by breaking the flow of water and preventing the displacement of fertile topsoil. The canopy cover provided by trees acts as a shield against the erosive forces of rain, safeguarding the soil surface and preserving its integrity.

The dynamic nutrient exchange between trees and crops enriches the soil with organic matter and essential nutrients. Fallen leaves and tree litter become a natural mulch, reducing soil moisture evaporation and suppressing weed growth. This organic mulch further contributes to erosion control by creating a protective layer that shields the soil from direct exposure to raindrops.

Alley cropping, with its strategic interplay of trees and crops, exemplifies the art of balancing productivity with conservation. It transforms agricultural landscapes into resilient ecosystems where soil health is nurtured, and the cycles of erosion are harmoniously disrupted.

Silvopasture

Silvopasture, the marriage of trees, forage, and livestock, not only redefines livestock management but stands as a testament to sustainable soil conservation practices. In this agroforestry system, trees and forage provide shade and sustenance to livestock, creating a dynamic equilibrium that extends its benefits to the soil beneath.

The impact of silvopasture on soil structure is profound. Livestock, seeking shade and forage among the trees, distribute their grazing pressure more evenly. This prevents localized soil compaction, preserving the soil structure and allowing for better water infiltration. The root systems of trees, reaching deep into the soil, further enhance its structure, creating channels for water movement and nutrient cycling.

The litter from trees, a combination of leaves, twigs, and bark, becomes a natural mulch that covers the soil. This mulch regulates soil temperature, reduces evaporation, and suppresses weed growth. As it decomposes, it contributes to the organic matter content of the soil, enhancing its water-holding capacity and nutrient availability.

Silvopasture embodies a holistic approach where soil conservation intertwines with sustainable livestock practices. By harmonizing the needs of trees, forage, and livestock, this agroforestry system creates landscapes where ecological balance extends from the canopy to the soil, fostering resilience and sustainability.

Windbreaks and Shelterbelts

Windbreaks and shelterbelts, the unsung heroes of agricultural landscapes, are strategic arrangements of trees that stand as formidable shields against the erosive forces of wind. Designed to combat wind erosion, these agroforestry practices go beyond conservation; they sculpt a microenvironment that champions soil health and resilience.

The physical structure of windbreaks disrupts the wind flow, creating a barrier that reduces wind speed. This reduction in wind speed minimizes the detachment of soil particles, preventing wind erosion. The protective role of windbreaks is particularly crucial in arid and semi-arid regions, where the vulnerability of topsoil to wind erosion threatens agricultural productivity.

Beyond erosion control, windbreaks contribute to improved soil moisture retention. By reducing wind speed, these agroforestry features minimize water evaporation from the soil surface, creating a more favorable microclimate for plant growth. The shade provided by the trees moderates temperature extremes, further optimizing soil conditions for agricultural productivity.

Windbreaks and shelterbelts are a testament to the proactive role agroforestry can play in combating environmental challenges. These living barriers not only protect against wind erosion but create landscapes where soil conservation is intricately linked to the ecological functions of trees.

Riparian Buffer Strips

Riparian buffer strips, designed to protect watercourses and riparian zones, weave together the benefits of trees and vegetation in safeguarding both water quality and soil health. Alongside the banks of streams and rivers, these strips become vital ecological buffers that extend their influence beyond water bodies.

The root systems of trees in riparian buffer strips serve as anchors, stabilizing the soil along watercourses and preventing erosion. By holding the soil in place, these trees protect the adjacent agricultural land from the loss of fertile topsoil. The vegetation in buffer strips acts as a natural filter, trapping sediments and reducing nutrient runoff into water bodies.

Riparian buffer strips contribute to biodiversity by creating transitional zones between aquatic and terrestrial ecosystems. This diverse habitat supports a variety of species, including birds, insects, and amphibians, fostering ecological balance. The presence of diverse vegetation in these strips enhances nutrient cycling, promoting soil fertility and health.

The multifaceted role of riparian buffer strips showcases agroforestry's potential to address interconnected environmental challenges. By integrating trees into riparian zones, this agroforestry system exemplifies how soil conservation becomes intertwined with broader efforts to protect water resources and biodiversity.

4. Integration of Trees in Crop Production

The integration of trees into crop production is more than a mere adjustment to farming practices; it is a profound reimagining of agricultural systems. This holistic approach transcends traditional boundaries, offering a dynamic synergy between trees and crops that fosters ecological balance, resilience, and sustainable productivity. Within this integration, three key strategies — Mixed Cropping Systems, Agroecological Practices, and Tree Interplanting — stand as beacons of innovation, providing a blueprint for a regenerative agriculture that harmonizes with nature.

Mixed Cropping Systems

Mixed cropping, a time-honored agricultural practice, is elevated to new heights when trees are seamlessly integrated into the system. The essence of mixed cropping lies in the diversity of crops coexisting within the same landscape, each contributing to the overall health and productivity of the agroecosystem. When trees join this tapestry of diversity, their role becomes transformative.

The interplay of trees and crops in mixed cropping systems creates a multifaceted web of interactions. The canopy of trees provides shade for understory crops, fostering microclimates that enhance biodiversity. The fallen leaves from trees become a natural mulch, regulating soil temperature, reducing moisture evaporation, and suppressing weed growth. As these leaves decompose, they contribute to the organic matter content of the soil, enriching it with essential nutrients.

The root systems of trees and crops work in tandem, enhancing soil structure and preventing erosion. The deep roots of trees access nutrients from lower soil layers, while the shallow roots of crops stabilize the topsoil. This collaboration results in a resilient soil profile that supports sustained agricultural productivity.

Mixed cropping not only maximizes land use but also minimizes the risk of crop failure. The diversity inherent in this system acts as a natural defense mechanism against pests and diseases. While one crop may be susceptible, others in the mix provide a buffer, reducing the need for chemical inputs and fostering a more sustainable and agroecological approach.

Beyond its agronomic advantages, mixed cropping is a testament to the intricate dance between plants in a natural ecosystem. By integrating trees into this symphony of diversity, farmers embrace a model of agriculture that mirrors the resilience and adaptability found in nature itself.

Agroecological Practices

Agroecology, as both a science and a set of farming practices, places ecological principles at the forefront of agriculture. The integration of trees into crop production aligns seamlessly with the essence of agroecology, creating farming systems that prioritize biodiversity, ecological balance, and sustainable resource management.

Agroforestry, a prominent agroecological practice, embodies the integration of trees with crops or livestock to create dynamic and multifunctional landscapes. The principles of agroecology, applied within agroforestry, lead to enhanced nutrient cycling, improved water use efficiency, and a reduction in environmental impacts.

In agroecological practices, the role of trees extends beyond mere companionship with crops. Trees contribute to the overall health of the agroecosystem by providing habitat for beneficial insects and birds, contributing to natural pest control. The complex root systems of trees enhance soil structure, promoting water infiltration and preventing erosion.

Agroecology emphasizes the importance of understanding and working with the natural processes of ecosystems. Farmers practicing agroecological methods become stewards of the land, working in harmony with nature to create resilient and sustainable agricultural systems. This integration of trees becomes a cornerstone in the larger movement towards regenerative agriculture, where the farm is viewed as an interconnected and dynamic living organism.

The participatory nature of agroecological practices places farmers at the forefront of decision-making. Their intimate knowledge of the land and its intricacies combines with scientific insights, creating a dynamic and adaptive approach to agriculture. This collaboration between traditional wisdom and scientific understanding forms the bedrock of agroecological practices, with the integration of trees as a central theme in sustainable land management.

Tree Interplanting

Tree interplanting, a strategic arrangement where trees are intentionally planted within areas dedicated to annual or perennial crops, exemplifies the potential for harmonious coexistence between woody perennials and traditional crops. This method transcends the conventional boundaries of agriculture, optimizing land use and fostering synergistic relationships.

In tree interplanting systems, the choice of tree species becomes crucial. Fruit-bearing trees, such as apples or cherries, can be interplanted with crops like corn or beans to create a diversified and productive landscape. The spatial arrangement allows for efficient use of sunlight, with the taller trees providing shade for the shorter crops.

The benefits of tree interplanting extend beyond the agronomic realm. The aesthetic appeal of orchards or groves interspersed with crops transforms the landscape, creating visually pleasing and ecologically diverse agricultural systems. Beyond the visual appeal, the integration of trees contributes to the creation of microclimates that benefit crop growth and overall ecosystem health.

The root systems of trees and crops in interplanting systems interact in dynamic ways. The deep roots of trees access nutrients from lower soil layers, while the shallower roots of crops exploit the upper layers. This complementary utilization of soil resources enhances nutrient cycling and improves overall soil structure.

Beyond its ecological benefits, tree interplanting has the potential to diversify income sources for farmers. The fruit-bearing trees can serve as additional revenue streams, contributing to economic sustainability. This diversification aligns with the principles of resilient and regenerative agriculture, where the integration of trees becomes a strategic move towards sustainable land use.

5. Role of Trees in Nutrient Cycling

The role of trees in nutrient cycling is a crucial aspect of sustainable agriculture, illuminating the intricate ways in which woody perennials contribute to the enrichment of soil fertility. Understanding the dynamics of nutrient accumulation and release, the interactive processes within agroforestry systems, and the significant contribution of trees to organic matter, unveils a holistic perspective on how trees become active participants in the nutrient cycles of agricultural landscapes.

Nutrient Accumulation and Release

Trees, with their extensive root systems and perennial nature, play a pivotal role in the accumulation and subsequent release of nutrients within agroecosystems. The roots of trees delve deep into the soil, accessing nutrients from various soil layers that might be out of reach for annual crops. This unique ability of trees to tap into deeper nutrient reservoirs contributes to the overall efficiency of nutrient cycling.

As trees absorb nutrients from the soil, these elements are transported upwards through the tree's vascular system and incorporated into leaves, branches, and other biomass. Over time, as leaves fall and other organic matter accumulates, a process of nutrient release begins. Microbial activity in the soil breaks down this organic matter, releasing nutrients back into the soil where they become accessible to other plants, including companion crops.

This dual role of trees — as nutrient accumulators and providers — is particularly pronounced in agroforestry systems. The cyclic flow of nutrients between trees and crops creates a dynamic and self-renewing nutrient cycle. This not only reduces the dependence on external inputs but also fosters a more sustainable and resilient agricultural system.

Dynamic Interactions in Agroforestry Systems

Agroforestry systems exemplify the intricate dance of nutrient cycling, where trees and crops engage in dynamic interactions that enhance overall soil fertility. In these systems, the spatial arrangement of trees alongside crops creates microclimates and ecological niches that optimize nutrient utilization.

The roots of trees and crops in agroforestry systems engage in complementary interactions. While the deep-rooted trees access nutrients from lower soil layers, the shallower roots of crops explore the upper layers, creating a harmonious utilization of soil resources. This prevents nutrient competition and promotes a synergistic exchange where each component contributes to the other's nutritional needs.

Agroforestry systems also benefit from the presence of nitrogen-fixing trees. Certain tree species have specialized root nodules that house nitrogen-fixing bacteria. These bacteria convert atmospheric nitrogen into a form that plants can utilize, contributing to the nitrogen pool in the soil. This natural fertilization process enriches the soil with nitrogen, a vital element for plant growth.

Furthermore, the leaf litter and organic matter produced by trees in agroforestry systems serve as a continuous source of nutrients. As this organic matter decomposes, it releases essential elements such as nitrogen, phosphorus, and potassium into the soil. The gradual breakdown of organic matter contributes to the long-term sustainability of nutrient availability in the system.

The dynamic interactions within agroforestry systems underscore the potential for integrated and regenerative approaches to agriculture. The synergy between trees and crops not only optimizes nutrient cycling but also enhances water use efficiency, reduces pest pressures, and contributes to the overall resilience of the agroecosystem.

Organic Matter Contribution

One of the significant contributions of trees to nutrient cycling lies in their role as organic matter contributors. Organic matter, derived from plant and animal residues, plays a pivotal role in soil structure, water retention, and nutrient availability. Trees, through their continuous shedding of leaves, branches, and other biomass, become key players in the organic matter dynamics of agroecosystems.

The organic matter provided by trees serves as a substrate for microbial activity in the soil. Microorganisms, including bacteria and fungi, break down this organic matter, releasing nutrients in a form that is readily available to plants. The decomposition process is a fundamental component of nutrient cycling, facilitating the recycling of organic materials into essential elements for plant growth.

The diverse composition of tree litter contributes a variety of organic compounds to the soil. This diversity enhances the microbial community's functionality, promoting a balanced and resilient ecosystem. Different tree species may contribute specific compounds that influence soil pH, nutrient availability, and microbial diversity, creating a rich tapestry of interactions within the soil matrix.

Beyond nutrient cycling, organic matter improves soil structure by enhancing aggregation and water-holding capacity. The presence of organic matter creates pore spaces in the soil, allowing for better aeration and water infiltration. This, in turn, reduces the risk of soil erosion and enhances the soil's ability to retain moisture during dry periods.

6. Water Management in Agroforestry

Water management in agroforestry represents a sophisticated and ecologically sound approach to optimizing water use within agricultural landscapes. Through the intricate interplay of root systems, microclimate regulation, and the reduction of water runoff, agroforestry systems emerge as models of sustainable water management, mitigating water-related challenges while fostering ecological resilience.

Root Systems and Water Infiltration

The intricate relationship between tree root systems and water infiltration lies at the heart of effective water management in agroforestry systems. Trees, with their deep and extensive roots, exhibit a remarkable ability to access water from various soil layers. Unlike the shallow root systems of many annual crops, tree roots delve into the deeper reaches of the soil profile, tapping into hidden water reservoirs.

The deep-rooted nature of trees enhances soil structure, creating channels that facilitate water movement and infiltration. As trees absorb water, they contribute to the improvement of soil porosity and structure, allowing rainwater to penetrate more effectively. This reduces surface runoff, which is a key factor in preventing soil erosion and nutrient loss.

In agroforestry systems, the interaction between tree roots and mycorrhizal fungi adds another layer of sophistication to water absorption. Mycorrhizal associations, where fungi form symbiotic relationships with plant roots, extend the reach of the root system. This collaborative effort enhances nutrient and water uptake, contributing to the overall resilience and efficiency of water use in the agroforestry landscape.

Furthermore, the deep roots of trees play a role in maintaining soil moisture during dry periods. By accessing water from deeper layers, trees can withstand periods of water scarcity better than many annual crops. This resilience becomes particularly valuable in regions prone to drought, where the ability to access water from deeper soil layers can be a decisive factor for overall agricultural productivity.

Microclimate Regulation

Microclimate regulation within agroforestry systems is a dynamic process that influences not only temperature and humidity but also rainfall distribution. The canopy of trees in these systems acts as a natural shelter, providing shade and moderating temperature extremes. This shading effect plays a critical role in reducing evaporation, thus preserving soil moisture and creating a more favorable environment for plant growth.

The moderation of temperature fluctuations beneath the tree canopy contributes to water conservation. By preventing overheating during hot periods and reducing rapid cooling during cold spells, trees foster a stable microclimate that benefits both trees and companion crops. This stability in temperature also supports plants in managing water stress more effectively.

The canopy's role in intercepting rainfall is another aspect of microclimate regulation with profound implications for water management. The leaves and branches of trees intercept raindrops, slowing down their impact on the soil surface. This reduces soil compaction caused by the force of rain, promoting better water infiltration and minimizing runoff.

In regions with irregular rainfall patterns, the microclimate regulation provided by trees can influence the distribution of rainfall within the agroforestry system. The canopy intercepts and releases rainwater gradually, preventing rapid runoff and allowing for more effective absorption by the soil. This not only contributes to water conservation but also enhances the overall resilience of the agroecosystem.

Reducing Water Runoff

The reduction of water runoff is a hallmark feature of agroforestry systems, showcasing their ability to minimize soil erosion and conserve water resources. The complex structure of these systems, where trees are strategically interspersed among crops, creates a natural defense against runoff.

The root systems of trees, with their anchoring effect on the soil, significantly reduce surface runoff. In sloping landscapes, where the risk of erosion is heightened, trees act as stabilizers, preventing the displacement of soil during heavy rainfall. This not only safeguards the topsoil but also protects against nutrient loss, ensuring that valuable nutrients remain available for plant uptake.

Leaf litter and organic matter from trees contribute significantly to the reduction of water runoff. The absorbent nature of this organic material prevents rapid runoff during intense rainfall events. Instead, it allows water to infiltrate gradually, promoting sustained soil moisture levels and reducing the risk of flash floods.

Moreover, the buffering effect of agroforestry systems extends to downstream areas. By slowing down the flow of water, these systems protect against the detrimental effects of flash floods on adjacent landscapes. This not only benefits the agroforestry system itself but also contributes to broader watershed management, fostering a holistic and integrated approach to water conservation.

7. Agroforestry Practices for Sustainable Land Use

Agroforestry practices stand as beacons of sustainable land use, embodying a holistic approach that integrates trees into agricultural landscapes for ecological, economic, and societal benefits. The adoption of agroforestry is rooted in long-term planning and design, ensuring the resilience of agroecosystems. The economic viability of agroforestry practices not only sustains livelihoods but also fosters a harmonious relationship between agriculture and the environment. Moreover, the integration of social and environmental considerations into agroforestry practices emphasizes the role of these systems in promoting community well-being and ecological stewardship.

Long-Term Planning and Design

The success of agroforestry practices hinges on meticulous long-term planning and thoughtful design. This aspect distinguishes agroforestry from conventional farming methods, emphasizing sustainability, resilience, and the harmonious integration of trees into agricultural landscapes. Long-term planning involves a comprehensive understanding of the local ecosystem, climate conditions, and the specific requirements of selected crops and tree species.

In the planning phase, practitioners assess the ecological dynamics to create a system that maximizes synergies. Factors such as water availability, soil composition, and the growth habits of different tree species are taken into account. The aim is to develop a landscape where trees and crops complement each other, avoiding competition for resources and minimizing potential negative interactions.

Spatial arrangement is a crucial component of agroforestry design. Decisions about tree density, spacing, and positioning are made light exposure, nutrient cycling, and water utilization. Agroforestry systems may include alley cropping, where rows of trees are interspersed with rows of crops, or silvopasture, where trees are integrated into grazing lands. The careful layout ensures that the system is not only productive but also resilient to environmental changes.

The design process also considers potential challenges and incorporates solutions. For instance, shading effects from tree canopies on crops are mitigated by selecting tree species with open canopies or by adjusting the spatial arrangement. The overall aim is to create a dynamic, self-sustaining system that evolves with changing environmental conditions.

Furthermore, the principles of agroecology guide agroforestry planning. This holistic approach considers the farm as an interconnected and evolving ecosystem. Agroecological practices involve mimicking natural ecosystems to create resilient and sustainable agricultural systems. In the context of agroforestry, this means acknowledging the intricate relationships between trees, crops, soil, and biodiversity.

By adopting a long-term planning and design perspective, agroforestry practitioners contribute to the creation of landscapes that transcend short-term gains. The integration of perennial trees adds an enduring dimension to agriculture, fostering sustainability that extends beyond annual cropping cycles. This forward-thinking approach positions agroforestry as a model for resilient and adaptive land use in the face of climate change and evolving agricultural challenges.

Economic Viability

At the heart of the widespread adoption of agroforestry practices is their economic viability. Beyond ecological benefits, agroforestry is embraced for its potential to provide sustainable income streams, reduce costs, and enhance overall economic resilience for farmers.

The economic benefits of agroforestry stem from the diversification it introduces to traditional farming practices. In agroforestry systems, farmers cultivate a variety of tree species alongside traditional crops. This diversification translates into multiple sources of income throughout the year. Fruit-bearing trees, such as apples or nuts, yield marketable products at different times, ensuring a continuous flow of income and reducing vulnerability to seasonal fluctuations.

Timber-producing trees or those yielding non-timber forest products contribute significantly to economic viability. Timber harvested sustainably from agroforestry systems provides a valuable resource, and non-timber products like medicinal plants, resins, or nuts further diversify income sources. This diversity of products allows farmers to tap into various markets, increasing economic stability.

Moreover, agroforestry systems often lead to cost savings. The nutrient cycling facilitated by trees reduces the need for external inputs like synthetic fertilizers. The diverse ecological interactions within the system often provide natural pest control, minimizing the reliance on chemical pesticides. These efficiencies contribute not only to improved economic returns but also to the overall sustainability and resilience of farming operations.

The long-term economic benefits of agroforestry are especially evident in the sustained production of timber and non-timber products. Unlike traditional monoculture agriculture, where income is tied to a single crop, agroforestry's diversified income streams offer farmers more financial security over time.

Furthermore, agroforestry aligns with the principles of sustainable and regenerative agriculture. By cultivating a mix of crops and trees, farmers contribute to the long-term health of the land while deriving economic benefits. The integration of trees into agricultural landscapes becomes a strategic move toward not only economic prosperity but also environmental stewardship.

In summary, the economic viability of agroforestry practices stems from their ability to diversify income sources, reduce costs, and align with sustainable agricultural principles. As farmers navigate the complexities of modern agriculture, the adoption of agroforestry emerges as a pragmatic and economically sound solution that bridges the gap between traditional farming practices and a more sustainable and resilient future.

Social and Environmental Considerations

Agroforestry practices are inherently rooted in the principles of social and environmental sustainability. Beyond economic considerations, the well-being of communities and the health of the environment are integral components of agroforestry systems.

Socially, agroforestry contributes to community resilience by diversifying income sources and providing employment opportunities. The integration of trees into agricultural landscapes creates a visually appealing environment that fosters a sense of pride and connection to the land. Agroforestry practices often involve local communities in decision-making processes, empowering them to be stewards of their landscapes.

The social benefits extend to improved food security and nutrition. Agroforestry systems that incorporate fruit-bearing trees contribute to a more diverse and nutritious diet for communities. The availability of a variety of crops and tree products enhances food security by providing a buffer against crop failures or price fluctuations in the market.

Environmental considerations are embedded in the fabric of agroforestry practices. The presence of trees in agroforestry systems plays a pivotal role in mitigating the impacts of climate change. Trees sequester carbon dioxide, reducing greenhouse gas emissions and contributing to the global effort to combat climate change. The overall resilience of agroforestry landscapes to climate variability ensures the continued provision of ecosystem services.

Moreover, agroforestry systems act as biodiversity hubs. The diverse structure and composition of these systems provide habitat for a wide array of plant and animal species. This ecological richness contributes to the conservation of native flora and fauna, promoting a balanced and healthy ecosystem. The integration of trees into agricultural landscapes becomes a proactive measure in addressing environmental challenges, aligning with the broader goals of sustainable land management.

The reduction of soil erosion is another environmental benefit of agroforestry. The root systems of trees stabilize the soil, preventing erosion and safeguarding water resources. This not only preserves fertile topsoil but also protects downstream areas from the negative impacts of runoff.

Additionally, the aesthetic appeal of agroforestry landscapes enhances the quality of life for communities. Beyond economic and practical considerations, the presence of trees in agricultural areas contributes to a more pleasant and ecologically rich environment. This aesthetic value is not only culturally significant but also promotes a positive and sustainable connection between people and the land.

8. Challenges and Solutions in Agroforestry Implementation

Agroforestry, while offering numerous benefits, is not without its challenges. The successful implementation of agroforestry systems requires addressing various hurdles, ranging from land tenure issues to the intricate aspects of species selection and management. Additionally, knowledge transfer and training play a crucial role in ensuring that farmers and communities can effectively embrace and sustain agroforestry practices. This section delves into the challenges associated with agroforestry implementation and presents viable solutions to overcome these obstacles.

Land Tenure Issues

Challenges

Land tenure issues pose a significant hurdle to the widespread adoption of agroforestry. In many regions, unclear land ownership, insecure land tenure, and the potential for conflicts over land use can hinder farmers from making long-term investments in agroforestry practices. Uncertainty regarding land tenure may discourage farmers from planting trees, as they fear losing access to the land and the benefits derived from the trees in the future.

Lack of incentives for long-term investments

The lack of clear and secure land tenure may result in farmers prioritizing short-term gains over long-term investments in agroforestry. Without assurances of continued access to the land, farmers may be hesitant to commit to planting trees that take years to mature and provide economic benefits.

Solutions

1. **Community involvement and partnerships:** Engaging local communities and forming partnerships with local authorities can help address land tenure issues. Collaborative efforts can lead to the establishment of community-based agroforestry initiatives where land tenure is collectively managed, providing a sense of security for all involved.
2. **Policy support:** Advocating for policies that recognize and support agroforestry practices can contribute to resolving land tenure issues. Clear legal frameworks that protect the rights of farmers engaging in agroforestry and provide incentives for long-term investments can be instrumental in overcoming these challenges.
3. **Capacity building:** Educating farmers about their land rights and helping them navigate legal processes can empower them to secure tenure and make informed decisions about agroforestry. Training programs and workshops can provide valuable insights into the legal aspects of land tenure.

Species Selection and Management

Challenges

Choosing the right mix of tree and crop species and effectively managing them is crucial for the success of agroforestry systems. Inappropriate species selection or poor management practices can lead to competition for resources, reduced crop yields, and increased susceptibility to pests and diseases.

Balancing ecological and economic goals

Selecting species that align with both ecological and economic goals is a delicate balance. While certain trees may contribute to soil fertility and biodiversity, they might not provide immediate economic benefits. Striking the right balance between these goals requires careful consideration.

Ensuring compatibility between species

Incompatibility between tree and crop species can result in competition for sunlight, water, and nutrients. This can lead to reduced yields and compromise the overall productivity of the agroforestry system.

Solutions

1. **Site-specific planning:** Conducting thorough site assessments and planning agroforestry systems based on local conditions can help optimize species selection. Consideration of soil types, climate, and existing vegetation will guide the choice of trees and crops that are well-suited to the specific environment.
2. **Diversification:** Incorporating a diverse mix of tree and crop species in agroforestry systems can enhance resilience. Diversification not only mitigates risks associated with species-specific challenges but also provides a range of ecological and economic benefits.
3. **Complementary species:** Selecting tree and crop species that complement each other rather than compete is essential. For instance, nitrogen-fixing trees can enhance soil fertility for crops, while certain trees may provide shade that benefits shade-tolerant crops.
4. **Continuous monitoring and adaptive management:** Regular monitoring of agroforestry systems allows for the identification of issues promptly. Adaptive management strategies can then be employed to address challenges and optimize the performance of the system over time.

Knowledge Transfer and Training

Challenges

The successful implementation of agroforestry practices relies heavily on the knowledge and skills of farmers and communities. Limited access to information, insufficient training, and a lack of awareness about the benefits of agroforestry can impede its adoption.

Lack of awareness and education

In many regions, farmers may be unaware of the potential benefits of agroforestry or may lack the knowledge to implement these practices effectively. Traditional farming practices passed down through generations may hinder the acceptance of new, more sustainable approaches.

Barriers to knowledge transfer

The transfer of knowledge from researchers, extension services, or experienced farmers to those who are less informed can face barriers. These may include insufficient communication channels, language barriers, or the absence of effective extension services.

Solutions

1. **Extension services and farmer training:** Strengthening agricultural extension services and providing targeted training programs can bridge the knowledge gap. Extension officers can work directly with farmers, offering practical guidance on agroforestry practices, species selection, and management techniques.
2. **Demonstration plots:** Establishing demonstration plots showcasing successful agroforestry systems can be an effective educational tool. Farmers can witness firsthand the benefits of these practices and gain confidence in adopting them on their farms.
3. **Local knowledge sharing:** Encouraging the sharing of local knowledge and experiences among farmers fosters a sense of community and mutual learning. Farmer-to-farmer knowledge transfer can be facilitated through workshops, field days, and community gatherings.
4. **Customized communication strategies:** Tailoring communication strategies to local contexts and using accessible formats, such as visual aids or local languages, can enhance the effectiveness of knowledge transfer. This ensures that information is conveyed in a way that resonates with the target audience.
5. **Incentives and recognition:** Providing incentives, such as financial support or recognition, for farmers who successfully implement agroforestry practices can motivate others to follow suit. This creates a positive feedback loop, encouraging the dissemination of knowledge and the adoption of sustainable land management practices.

Conclusion

In the intricate canvas of agroforestry and its role in sustainable land use for soil management and conservation, the threads of environmental harmony, resilience, and responsible stewardship are woven together, creating a narrative that goes beyond traditional agricultural practices. Agroforestry, as a holistic approach, extends its roots into the very fabric of ecological balance and sustainability.

At its essence, agroforestry is a commitment to environmental harmony. The deliberate integration of trees into agricultural landscapes goes beyond the pragmatic aspects of land utilization; it is a conscious effort to emulate and coexist with natural ecosystems. By adopting the layered structure of forests, agroforestry creates microenvironments that foster biodiversity. The presence of various plant and animal species within these systems contributes not only to the health of the land but also to the intricate dance of ecological relationships.

One of the distinctive contributions of agroforestry is its role in carbon sequestration. In the global pursuit of mitigating climate change, agroforestry emerges as a frontline ally. Trees, with their inherent capacity to sequester carbon dioxide, play a pivotal role in reducing greenhouse gas emissions. In this way, agroforestry aligns itself with sustainable land use practices that prioritize not only agricultural productivity but also the broader health of the planet.

Resilience is a hallmark of agroforestry, providing a robust response to environmental challenges. The dynamic interactions within agroforestry systems create a resilient tapestry capable of withstanding various stressors. Climate variability, unpredictable weather patterns, and the evolving landscapes of the future are met with adaptability within agroforestry. The diversity of species, encompassing both crops and trees, serves as a natural buffer, offering a safeguard against the uncertainties imposed by a changing climate.

The ability of agroforestry to reduce soil erosion stands out as a testament to its adaptive land management qualities. The intricate root systems of trees stabilize the soil, preventing erosion and maintaining the integrity of fertile topsoil. This not only protects agricultural productivity but also contributes to the broader goal of sustainable land use. Water retention is another key feature, with agroforestry systems effectively harnessing and utilizing water resources. This dual benefit of erosion control and water management showcases agroforestry's multifaceted contribution to sustainable soil practices.

Beyond the practical advantages, agroforestry embodies a philosophy of responsible stewardship. It encourages a shift from exploitative agricultural practices to a more nurturing and symbiotic relationship with the land. The cyclical nature of nutrient cycling within agroforestry systems, where trees contribute organic matter through leaf litter and root turnover, mirrors the circular and sustainable ethos of responsible land management.

In conclusion, agroforestry's role in sustainable land use for soil management and conservation transcends the boundaries of conventional agriculture. Its environmental harmony, resilience, and responsible stewardship distinguish it as a beacon of sustainable practices. As we navigate an era marked by environmental challenges, agroforestry stands as a model that not only sustains agricultural productivity but also contributes to the broader imperative of planetary health. It is a living testament to the possibility of cultivating the land in a way that enriches both the soil and the intricate web of life it supports.

8

Irrigation Management for Water Efficiency and Conservation

[1]Niranjan B.N., [2]Kh. Chandrakumar Singh, [3]Savita Jangde [4]Wajid Hasan and [5]Sheetanshu Gupta

[1]Dakshina Kannada Milk Union (KMF), Mangalore, Karnataka
[2]VCSGU University of Horticulture and Forestry, Bharsar, Uttarakhand
[3]Department of Plant Physiology, Institute of Agricultural Sciences B.H.U. Varanasi, U.P.
[4]KVK Jehanabad, Bihar Agricultural University Sabour, Bihar, India
[5]Department of Biotechnology, Institute of Technology and Management BKT, Lucknow, U.P.

1. Introduction

1.1 Background

The historical backdrop of irrigation is rich and diverse, reflecting the ingenuity of civilizations in adapting to their environments to meet the basic human need for sustenance. Ancient cultures such as the Sumerians, Egyptians, and Chinese were pioneers in developing rudimentary irrigation systems, channeling water from rivers, and constructing canals to nourish their crops. These early practices laid the foundation for the evolution of irrigation, marking a significant transition from relying solely on rainfall to actively controlling water distribution.

Over the centuries, irrigation methods continued to evolve, influenced by technological advancements and the expansion of agricultural practices. From the rudimentary techniques of diversion channels, societies progressed to more sophisticated systems, such as the qanats in Persia and the extensive canal networks of the Indus Valley civilization. The advent of the Industrial Revolution further accelerated the mechanization of irrigation, introducing steam and later electric power to pump water more efficiently (Fig 1).

Fig. 1: Irrigation through Water management

The twentieth century witnessed a paradigm shift with the widespread adoption of modern irrigation technologies, including sprinkler systems and drip irrigation. These innovations revolutionized water delivery, enabling precise control over the amount and timing of irrigation. The intersection of technology and agriculture continues to shape contemporary irrigation practices, with smart irrigation systems and sensor-based monitoring becoming integral components of modern farming.

1.2 Importance of Irrigation Management

The importance of irrigation management in the present context cannot be overstated, given the complex challenges facing agriculture. With a global population surpassing 7 billion and projections indicating further growth, the demand for food has never been more pronounced. Irrigation, as a key component of agricultural systems, holds the potential to bridge the gap between food supply and demand.

However, the unchecked expansion of irrigation without proper management can lead to detrimental consequences. Unoptimized water use can result in water wastage, soil degradation, and adverse environmental impacts. Moreover, climate change has introduced new uncertainties, affecting precipitation patterns and increasing the frequency of extreme weather events. In this context, effective irrigation management emerges as a strategic imperative for sustainable agriculture.

Water scarcity, exacerbated by climate change and population growth, poses a significant threat to global food security. In many regions, water resources are already stretched to their limits, and unsustainable irrigation practices contribute to the depletion of aquifers and the deterioration of water quality. Irrigation management becomes a linchpin in addressing these challenges by promoting judicious water use, reducing wastage, and optimizing irrigation systems to enhance water efficiency.

The economic viability of agriculture is intricately linked to irrigation management. Well-managed irrigation systems contribute to increased crop yields, which, in turn, support the livelihoods of farmers and contribute to rural development. Conversely, inadequate or inefficient irrigation practices can lead to reduced yields, economic losses, and, in extreme cases, agricultural abandonment.

Beyond the economic realm, irrigation management plays a vital role in environmental sustainability. The excessive withdrawal of water from rivers and aquifers can disrupt ecosystems, threaten biodiversity, and contribute to the degradation of natural habitats. By implementing sustainable irrigation practices, we can strike a balance between meeting agricultural needs and safeguarding the health of ecosystems.

In the context of a rapidly changing climate, the adaptability and resilience of agriculture hinge on effective irrigation management. Variable weather patterns, including prolonged droughts and intense rainfall, require dynamic and responsive approaches to water allocation. Irrigation management practices that integrate climate data, weather forecasting, and adaptive strategies are essential for ensuring agricultural productivity in the face of climate uncertainty.

2. Basics of Irrigation

2.1 Types of Irrigation Systems

The cultivation of crops is intricately linked with the availability and efficient use of water, making the selection of appropriate irrigation systems a cornerstone of agricultural success. The diverse array of irrigation methods has evolved over centuries, adapting to the unique requirements of different regions and crops. A deeper exploration of these methods provides valuable insights into the technological, environmental, and economic considerations that shape modern agricultural practices (Table 1).

Table 1: Types of Irrigation Systems

S.No	Irrigation System	Description
1	Surface Irrigation	- Water is applied directly to the soil surface and flows by gravity across the field. Methods include furrow, basin, border, and flood irrigation.
2	Drip Irrigation	- Water is delivered directly to the base of each plant through a network of tubes, pipes, and emitters. Drip irrigation is efficient and minimizes water wastage.
3	Sprinkler Irrigation	- Water is sprayed over the crops in the form of droplets through a system of pipes and pumps. Common types include center pivot and lateral move systems.
4	Subsurface Drip Irrigation	- Similar to drip irrigation, but water is delivered below the soil surface through buried tubes. It reduces evaporation losses and is suitable for row crops and orchards.

S.No	Irrigation System	Description
5	Center Pivot Irrigation	- A type of sprinkler irrigation where a rotating sprinkler system pivots around a central point. It is commonly used for large, circular fields.
6	Lateral Move Irrigation	- Similar to center pivot irrigation, but the equipment moves laterally along straight paths. It is suitable for rectangular fields.
7	Subsurface Irrigation	- Water is applied below the soil surface through a network of buried pipes or tubes. This method reduces evaporation and minimizes soil surface wetting.
8	Furrow Irrigation	- Water is directed down small channels (furrows) between crop rows. It is a form of surface irrigation and is common for row crops.
9	Basin Irrigation	- Fields are divided into leveled basins, and water is applied to each basin. It is suitable for crops with good water retention capabilities.
10	Localized Irrigation	- Water is applied directly to individual plants or groups of plants. Drip irrigation and soaker hoses are examples of localized irrigation.
11	Manual Irrigation	- Watering is done by hand using hoses, watering cans, or other manual methods. It is suitable for small gardens, nurseries, and specific areas.
12	Gravity Irrigation	- Relies on the force of gravity to move water through the irrigation system. Methods include furrow, basin, and flood irrigation.

Surface Irrigation Systems

2.1.1 Flood Irrigation: Flood irrigation, with its roots in ancient agricultural practices, involves the flooding of fields with water. This method is especially suited for level terrains, allowing water to cover the entire field evenly. While simple and cost-effective, flood irrigation may lead to water wastage due to runoff, necessitating careful land leveling and contouring to optimize water distribution.

2.1.2 Furrow Irrigation: Furrow irrigation introduces a more controlled approach by directing water into small channels or furrows between rows of crops. This method reduces water contact with foliage, minimizing the risk of diseases. However, challenges arise in sloping terrains where water distribution may become uneven. Proper land management practices, including contouring, can enhance the effectiveness of furrow irrigation.

2.1.3 Basin Irrigation: Basin irrigation involves creating basins or depressions around individual plants or groups of plants, allowing water to accumulate around the root zones. This method is particularly advantageous in orchards and vineyards, concentrating water where it is needed most. Similar to furrow irrigation, basin systems require careful design and management to ensure uniform water distribution across the entire field.

Subsurface Irrigation Systems

2.1.4 Drip Irrigation: Drip irrigation marks a significant advancement in water efficiency by delivering water directly to the root zone through a network of tubes or pipes with emitters. This precise method minimizes water wastage, promotes consistent soil moisture, and enables controlled nutrient application. Drip systems are versatile, adapting to various crops and terrains, making them an integral part of sustainable agriculture practices.

2.1.5 Subsurface Drip Irrigation: Subsurface drip irrigation takes the principles of drip irrigation below the soil surface, further reducing evaporation losses and surface runoff. Particularly beneficial in arid and semi-arid regions, this method optimizes water use efficiency by directly targeting the root zone. The reduced contact with foliage also mitigates disease risks, enhancing overall crop health.

2.1.6 Sprinkler Irrigation: Sprinkler irrigation simulates natural rainfall by dispersing water through a system of pipes and pumps. This versatile method is suitable for a wide range of crops and terrains. Challenges, such as energy consumption and potential water wastage due to evaporation, necessitate precise calibration and maintenance. Sprinkler systems play a crucial role in modern agriculture, offering adaptability and efficiency.

Localized Irrigation Systems

2.1.7 Micro-Irrigation: Micro-irrigation, comprising both drip and sprinkler systems on a smaller scale, addresses the specific needs of horticultural crops and gardens. The precision in water delivery minimizes weed growth, allows for targeted nutrient application, and conserves water. Micro-irrigation systems are especially valuable in regions where water resources are limited, emphasizing the importance of water conservation.

2.1.8 Trickle Irrigation: Trickle irrigation involves the slow, continuous release of water directly onto the soil surface or root zone through tubes or pipes. With its low flow rate, this method excels in water efficiency, minimizing runoff and soil erosion. Commonly used in orchards and vineyards, trickle irrigation fosters optimal conditions for plant growth while conserving precious water resources.

Specialized Irrigation Systems

2.1.9 Greenhouse Irrigation: Greenhouse cultivation demands precise control over environmental variables, including water management. Capillary matting, ebb and flow systems, and nutrient film techniques are commonly used in greenhouses. These systems enable growers to fine-tune water delivery, nutrient concentrations, and environmental conditions, showcasing the integration of technology with traditional agricultural practices.

2.1.10 Vertical Farming Irrigation: Vertical farming represents a frontier in agriculture, challenging conventional spatial constraints. In hydroponic and aeroponic systems, crops are grown without soil, and nutrient-rich water is circulated directly to plant roots. This closed-loop approach maximizes resource efficiency, making vertical farming a sustainable solution for urban agriculture and addressing issues of space constraints.

2.2 Components of Irrigation Systems

Irrigation systems are intricate networks that integrate various components, working synergistically to deliver water precisely to crops. A deeper understanding of these components is pivotal for ensuring optimal performance, resource efficiency, and sustainability in agricultural practices.

2.2.1 Water Source: The selection of an appropriate water source is a foundational decision in designing an irrigation system. Natural sources such as rivers, lakes, and underground aquifers, as well as artificial reservoirs, provide the necessary water supply. Factors influencing the choice include water availability, quality, and proximity to the agricultural area. Sustainable water management involves assessing the potential impact on the ecosystem and ensuring the long-term viability of the water source.

2.2.2 Conveyance System: Conveyance systems play a critical role in transporting water from its source to the fields. These systems include canals, pipelines, and channels. Proper design and maintenance are essential to minimize water losses during transportation. Canals and channels may require periodic desilting to prevent blockages, while pipelines demand regular inspections for leaks or damages. The efficiency of water conveyance directly affects the overall effectiveness of the irrigation system.

2.2.3 Control Structures: Control structures, such as gates, valves, and regulators, are strategically positioned along the conveyance system to manage the flow of water. These components allow for precise control over water distribution, ensuring it reaches the fields in the required quantity and at the right time. Automated control structures, integrated with monitoring systems, enable real-time adjustments based on factors like soil moisture levels and weather conditions, optimizing water use.

2.2.4 Storage Reservoirs: Storage reservoirs serve as critical components, particularly in regions with seasonal variations in water availability. These reservoirs store water during periods of excess and release it during times of scarcity. Well-designed storage reservoirs contribute to the reliability and resilience of irrigation systems, providing a buffer against unpredictable weather patterns or prolonged dry spells.

2.2.5 Pumping Stations: Pumping stations are essential for elevating water from lower to higher elevations, overcoming topographical challenges. Pumps are employed to maintain the flow of water, ensuring it reaches areas with higher elevation, such as terraced fields. Energy-efficient pumps, powered by renewable sources whenever possible, contribute to the sustainability of irrigation systems.

2.2.6 Filtration Systems: Filtration systems are integral to preventing impurities and debris from compromising the efficiency of the irrigation system. Filters remove particles that could clog emitters in drip irrigation systems or nozzles in sprinkler systems. Regular maintenance of filtration systems is crucial to sustaining water quality and avoiding disruptions in water distribution.

2.2.7 Distribution Network: The distribution network comprises pipelines, tubes, or hoses responsible for delivering water from the conveyance system to the fields. This network must be carefully designed to minimize pressure losses and ensure uniform water distribution across the entire cultivated area. Proper sizing, material selection, and layout planning contribute to the efficiency of the distribution network.

2.2.8 Emitters (Drip Irrigation) or Sprinklers: Emitters, in the case of drip irrigation, or sprinklers, in sprinkler irrigation, are the devices responsible for releasing water onto the crops. The selection and placement of emitters or nozzles are critical factors in achieving uniform coverage and preventing water wastage. Innovations in emitter technology, such as pressure-compensating emitters, contribute to enhanced uniformity and efficiency.

2.2.9 Monitoring and Control Systems: Modern irrigation systems leverage monitoring and control systems for data-driven decision-making. Sensors for soil moisture, weather conditions, and water usage, combined with automated controllers, enable real-time adjustments. These systems optimize irrigation schedules, contributing to water-use efficiency and overall sustainability. Integration with precision agriculture technologies further enhances the adaptive capacity of irrigation systems.

2.3 Water Distribution in Irrigation

Efficient water distribution is a fundamental aspect of irrigation management, directly impacting crop health, resource conservation, and environmental sustainability. A closer examination of water distribution principles reveals the complexities involved in achieving optimal outcomes.

2.3.1 Uniformity of Water Distribution: Uniform water distribution is a measure of the consistency with which water reaches every part of the cultivated area. High uniformity ensures that each plant receives the required amount of water, fostering equal crop growth. Achieving uniformity involves careful consideration of factors such as pipe sizing, emitter spacing, and maintenance practices. Regular checks and adjustments are necessary to address any deviations and optimize the overall performance of the irrigation system.

2.3.2 Application Rate: The application rate, indicating the volume of water delivered per unit of time and area, is a crucial parameter in irrigation management. Understanding the application rate guides decisions on the duration of irrigation sessions, preventing overwatering or underwatering. It varies based on the specific requirements of different crops and the type of irrigation system employed. Fine-tuning the application rate contributes to water-use efficiency and resource conservation.

2.3.3 Water Penetration and Infiltration: The ability of water to penetrate the soil and infiltrate to the root zone is critical for effective irrigation. Factors influencing penetration and infiltration include soil type, compaction, and the rate of water application. Proper soil moisture management ensures that water reaches the root zones of plants, promoting optimal growth. Techniques such as furrow diking and contour ploughing contribute to enhanced water penetration.

2.3.4 Water Losses and Conservation: Minimizing water losses during distribution is a key consideration in sustainable irrigation practices. Techniques such as lining canals to prevent seepage, prompt repair of leaks, and adopting water-saving technologies contribute to efficient water conservation. Tailwater recovery systems, designed to collect and reuse excess water, further minimize overall water consumption. A holistic approach to water conservation acknowledges the interconnectedness of environmental, economic, and agricultural considerations.

2.3.5 Soil Moisture Management: Understanding soil moisture dynamics is fundamental for optimizing water distribution. Soil moisture sensors provide real-time data, allowing for precise irrigation scheduling. Evapotranspiration models, considering factors such as temperature and humidity, aid in determining when and how much water is needed. Fine-tuned soil moisture management prevents over-irrigation, reduces water wastage, and promotes healthy crop development.

2.3.6 Tailwater Recovery: Efficient irrigation systems incorporate tailwater recovery mechanisms to capture and reuse excess water at the end of a field. Tailwater recovery not only minimizes water wastage but also contributes to overall resource efficiency. Properly designed recovery systems prevent nutrient runoff, reducing the environmental impact and supporting sustainable agricultural practices.

2.3.7 Environmental Considerations: Balancing the water needs of crops with environmental sustainability is an essential aspect of responsible irrigation management. Excessive water extraction can harm ecosystems and aquatic habitats, emphasizing the importance of considering the ecological impact. Conservation measures, coupled with technological innovations, contribute to environmentally conscious irrigation practices that align with broader ecological goals.

3. Water Efficiency in Irrigation

3.1 Definition and Importance

3.1.1 Definition: Water efficiency in irrigation represents a comprehensive approach to water management in agriculture, encompassing practices and technologies that optimize water use. At its core, it involves the judicious allocation and application of water resources to ensure that crops receive the required amount while minimizing losses.

3.1.2 Importance: The importance of water efficiency in irrigation cannot be overstated, especially in the face of escalating challenges such as water scarcity, population growth, and climate change. With agriculture being a major consumer of water globally, adopting water-efficient practices is paramount for sustaining food production, conserving water resources, and mitigating the environmental impact of irrigation. Beyond addressing immediate agricultural needs, water efficiency contributes to the long-term resilience of ecosystems and the equitable distribution of water for various societal needs.

3.2 Factors Affecting Water Efficiency

3.2.1 Irrigation System Design: The design of an irrigation system plays a pivotal role in determining its efficiency. Factors such as topography, soil type, and crop water requirements influence the choice of irrigation method and system layout. Precision irrigation technologies, including drip and sprinkler systems, contribute to the optimization of water distribution. Advanced design considerations also involve slope management to prevent runoff and the integration of smart controllers for adaptive irrigation.

3.2.2 Soil Management: Effective soil management is integral to water efficiency. Certain soil management practices enhance water retention and infiltration. Mulching, for instance, reduces evaporation, retains soil moisture, and mitigates erosion. Cover cropping improves soil structure, promoting water infiltration and reducing surface runoff. Contour ploughing helps retain water on the field, preventing downhill movement and promoting even distribution.

3.2.3 Crop Selection and Rotation: Water-efficient agriculture begins with thoughtful crop selection and rotation. Different crops have varying water requirements and sensitivities to water stress. Selecting crops suited to the local climate and seasonal variations optimizes water use. Crop rotation helps break pest cycles and allows for more efficient use of nutrients, contributing to overall water efficiency in the agricultural system.

3.2.4 Irrigation Scheduling: Optimizing irrigation schedules aligns water applications with the dynamic needs of crops. Sensor-based technologies, weather forecasts, and soil moisture monitoring enable precise scheduling. Adaptive scheduling systems can dynamically adjust irrigation based on real-time conditions, preventing over-irrigation and conserving water resources. Implementing these technologies ensures that water is applied when and where it is most needed.

3.2.5 Water Quality and Filtration: Water quality is crucial for both crop health and the efficiency of irrigation systems. Poor water quality can lead to clogging of emitters and nozzles, reducing uniform water distribution. Filtration systems, ranging from screen filters to more advanced technologies like disc filters, remove particles and debris, ensuring that irrigation infrastructure functions optimally. Regular maintenance of filtration systems is essential for sustained water quality and system efficiency.

3.2.6 Technology and Automation: The integration of technology and automation is a game-changer for water efficiency. Sensor-based monitoring systems provide real-time data on soil moisture levels, weather conditions, and water usage. Automated controllers adjust irrigation schedules based on these inputs, optimizing water delivery. Additionally, the use of remote sensing and data analytics allows for data-driven decision-making, enhancing the adaptability and responsiveness of irrigation systems.

3.3 Benefits of Water-efficient Irrigation

3.3.1 Conservation of Water Resources: Water-efficient irrigation practices contribute significantly to the conservation of water resources. By minimizing losses through evaporation, runoff, and deep percolation, farmers can stretch available water supplies to meet the increasing demands of agriculture. This conservation aspect is crucial in regions facing water scarcity, ensuring sustainable water use for future generations.

3.3.2 Cost Savings: Water efficiency translates directly into cost savings for farmers. Reduced water consumption leads to lower pumping and energy costs. While there may be initial investments in adopting water-efficient technologies, the long-term financial benefits, including decreased operational costs and increased crop yields, outweigh these initial expenses.

3.3.3 Increased Crop Yields: One of the most tangible benefits of water-efficient irrigation is the direct correlation with increased crop yields. By ensuring that crops receive the right amount of water at the right time, farmers can promote healthy growth and maximize productivity. This not only contributes to food security but also bolsters economic stability for farming communities.

3.3.4 Mitigation of Environmental Impact: Water-efficient irrigation practices play a vital role in mitigating the environmental impact associated with excessive water use. Preventing over-extraction of water from rivers and aquifers safeguards ecosystems, preserves biodiversity, and maintains the health of aquatic habitats. The reduction in nutrient runoff due to optimized water use further minimizes the environmental footprint of agriculture.

3.3.5 Adaptation to Climate Change: As climate change introduces increased variability in precipitation patterns and temperature, water-efficient irrigation practices become indispensable for adaptation. Precision irrigation technologies and responsive scheduling enable agriculture to withstand the challenges posed by a changing climate. Adaptive strategies contribute to the resilience of farming systems in the face of uncertain weather conditions.

3.3.6 Regulatory Compliance and Social Responsibility: Water-efficient practices align with regulatory requirements focused on sustainable water management. By adhering to these practices, farmers not only comply with environmental regulations but also demonstrate social responsibility. The adoption of water-efficient irrigation reflects a commitment to environmental stewardship, contributing to broader community and global water conservation efforts.

4. Conservation Practices

4.1 Overview of Water Conservation

4.1.1 The Need for Water Conservation: The imperative for water conservation in agriculture arises from the intensifying challenges posed by water scarcity, population growth, and the unpredictable impacts of climate change. As global demands for food production escalate, the judicious use of water resources becomes paramount. Water conservation practices seek to strike a balance between fulfilling the growing food requirements and ensuring the sustainability of water supplies. This proactive approach aims to mitigate the environmental impact of irrigation and foster resilience in agricultural systems.

4.1.2 Balancing Agricultural Productivity and Sustainability: Water conservation involves navigating the delicate equilibrium between maximizing agricultural productivity and maintaining the sustainability of water resources. As the global population continues to expand, agricultural practices must evolve to meet escalating food demands. Water conservation practices play a pivotal role in optimizing water use efficiency, reducing waste, and fostering sustainable land and water management. This balance is crucial for ensuring the long-term viability of agricultural systems and safeguarding water availability for future generations.

4.1.3 Integrating Conservation into Agricultural Practices: Modern agricultural practices are increasingly characterized by the integration of water conservation measures. Farmers are adopting innovative technologies, precision irrigation methods, and sustainable land management techniques to minimize water losses and enhance overall efficiency. Beyond the immediate benefits for farmers, the incorporation of water conservation practices aligns with broader societal goals of environmental preservation and responsible resource stewardship. It reflects a commitment to sustainable agriculture that considers the interplay between food production, water use, and ecosystem health.

4.2 Techniques for Water Conservation in Irrigation

4.2.1 Drip Irrigation: Drip irrigation stands out as a flagship technique for water conservation in agriculture. This method involves the precise delivery of water directly to the root zones of plants through a network of tubes or pipes with emitters. The controlled release of water minimizes surface evaporation and reduces contact with foliage, resulting in optimal water use. Drip irrigation not only conserves water but also enables targeted nutrient application, contributing to overall resource efficiency and crop health.

4.2.2 Soil Moisture Sensors: Soil moisture sensors play a pivotal role in water conservation efforts by providing real-time data on soil moisture levels. This information allows farmers to tailor irrigation schedules to the specific needs of crops, avoiding over-irrigation and ensuring that water is applied precisely where and when it is needed. By facilitating informed decision-making, soil moisture sensors contribute to water-use efficiency and resource conservation.

4.2.3 Rainwater Harvesting: Rainwater harvesting is a sustainable practice that involves collecting and storing rainwater for later use in irrigation. This technique reduces reliance on external water sources, especially during the rainy season. By capturing and utilizing rainwater, farmers can optimize water availability, reduce dependency on conventional water sources, and decrease the environmental impact associated with excessive groundwater extraction.

4.2.4 Cover Cropping: Cover cropping is a strategic technique where non-harvested or specific cover crops are planted during periods when the main crop is not actively growing. This practice protects the soil from erosion, improves soil structure, and enhances water retention. By preventing runoff and promoting water infiltration, cover cropping contributes to water conservation and overall soil health. Additionally, cover crops offer additional benefits, including weed suppression and nutrient cycling.

4.2.5 Mulching: Mulching, a simple yet effective technique, involves applying a layer of organic or inorganic material on the soil surface. This protective layer reduces evaporation, suppresses weed growth, and helps maintain soil moisture. Organic mulches, such as straw or wood chips, break down over time, contributing to soil fertility. Mulching is particularly valuable in arid and semi-arid regions, where water conservation is critical for sustaining agriculture.

4.2.6 Conservation Tillage: Conservation tillage practices involve minimizing soil disturbance during planting. Reduced tillage or no-till methods help maintain soil structure, prevent water runoff, and enhance water infiltration. By preserving soil moisture and structure, conservation tillage contributes to water-use efficiency and mitigates the risk of soil erosion. This practice is especially relevant in regions prone to water scarcity and soil degradation.

4.2.7 Smart Irrigation Controllers: Smart irrigation controllers leverage technology to optimize water use in irrigation. These controllers integrate data from various sources, including weather forecasts, soil moisture sensors, and evapotranspiration models, to dynamically adjust irrigation schedules. By responding to real-time conditions, smart controllers prevent over-irrigation, reduce water wastage, and enhance overall water-use efficiency. The intelligent application of water resources contributes to both economic savings and environmental sustainability.

4.2.8 Terracing and Contouring: Terracing and contouring are land management practices designed to reduce water runoff on sloping terrain. By creating level areas along the contours of the land, these practices slow down water movement, allowing it to infiltrate the soil. Terracing and contouring help prevent soil erosion, improve water retention, and promote sustainable land use in hilly or mountainous regions. These techniques are valuable for conserving water and preserving soil structure in areas with varied topography.

4.2.9 Variable Rate Irrigation (VRI): Variable Rate Irrigation involves adjusting the rate of water application based on spatial variability within a field. This technology enables farmers to tailor irrigation levels to the specific needs of different areas, considering factors such as soil type, topography, and crop requirements. VRI promotes precision irrigation, optimizing water use across the entire field. By customizing water applications, farmers can conserve water resources while ensuring optimal crop growth.

4.2.10 Laser Land Leveling: Laser land leveling is a precision technique that ensures a uniform and level field surface. By eliminating high and low spots, water is distributed more evenly during irrigation, reducing the risk of water pooling and runoff. Laser land leveling enhances water-use efficiency, particularly in areas with undulating topography. The leveled surface allows for better control over water application, preventing water wastage and optimizing the overall irrigation process.

Water conservation techniques are not only critical for sustaining agricultural productivity but also for mitigating the environmental impact of irrigation. By implementing these practices, farmers contribute to responsible water management, safeguarding water resources for future generations.

5. Irrigation Scheduling

Efficient irrigation scheduling is a foundational element in the sustainable and responsible management of water resources in agriculture. This section delves into the significance of scheduling, various methodologies employed for irrigation scheduling, and best practices for effective implementation.

5.1 Importance of Scheduling

5.1.1 Resource Optimization: Irrigation scheduling serves as a linchpin for the optimal utilization of water resources in agriculture. Aligning irrigation practices with the actual water requirements of crops is paramount in preventing over-irrigation. By doing so, farmers can curtail water wastage, mitigating the environmental impact associated with excessive water use. This optimization becomes particularly critical in regions grappling with water scarcity, where prudent water management is vital for sustainable agricultural practices.

5.1.2 Crop Health and Yield: Accurate irrigation scheduling ensures that crops receive the precise amount of water they need at specific growth stages. This targeted approach promotes healthy crop development, optimizing yields. Under-irrigation can lead to water stress, impacting crop growth and reducing productivity, while over-irrigation may result in waterlogged soils, affecting root health. Proper scheduling seeks to strike a harmonious balance, enhancing both crop health and overall agricultural output.

5.1.3 Energy Efficiency: Irrigation processes often involve energy-intensive components, such as pumping water from sources to fields. Efficient scheduling plays a role in reducing the overall energy consumption associated with irrigation. By delivering water precisely when and where it is needed, farmers can optimize energy use, leading to cost savings and a more sustainable agricultural operation.

5.1.4 Environmental Stewardship: Implementing precise irrigation schedules contributes to environmental stewardship. Avoiding unnecessary water applications minimizes the risk of nutrient runoff and soil erosion, which are environmental concerns associated with indiscriminate water use. This proactive approach aligns with the broader objectives of sustainable agriculture, emphasizing responsible resource utilization and the preservation of ecosystems.

5.2 Methods of Irrigation Scheduling

5.2.1 Soil Moisture-Based Scheduling: Soil moisture-based scheduling relies on real-time monitoring of moisture levels in the crop root zone. Employing soil moisture sensors strategically placed in the field, this method provides valuable data for determining when and how much water to apply. By avoiding over-irrigation and ensuring crops have access to adequate moisture, soil moisture-based scheduling enhances water-use efficiency and minimizes wastage.

5.2.2 Evapotranspiration (ET) Methods: Evapotranspiration methods estimate crop water needs by considering both soil surface evaporation and plant transpiration. Reference evapotranspiration (ET_o), calculated using meteorological data, is adjusted with crop-specific coefficients to determine crop evapotranspiration (ETc). This comprehensive approach offers insights into actual water requirements, aiding the development of precise irrigation schedules.

5.2.3 Weather-Based Scheduling: Weather-based scheduling integrates meteorological data—temperature, humidity, wind speed, and solar radiation—to determine crop water needs. This approach adapts irrigation schedules based on current weather conditions, enhancing accuracy. By responding to variations in climate, weather-based scheduling contributes to the precision of water applications.

5.2.4 Plant-Based Scheduling: Plant-based scheduling involves assessing the water status of plants to determine irrigation timing. Techniques such as leaf-potential measurements or infrared thermometry gauge plant water stress directly. By evaluating the physiological responses of plants to water availability, this method provides insights into the actual water needs of crops, facilitating precise irrigation management.

5.2.5 Time-Based Scheduling: Time-based scheduling relies on predefined irrigation schedules based on historical or typical crop water requirements. Although less precise than real-time methods, time-based scheduling can be effective when supported by accurate knowledge of crop water needs. This method is often used in combination with other scheduling approaches to provide a practical and straightforward irrigation strategy.

5.3 Implementing Effective Scheduling Practices

5.3.1 Data Collection and Monitoring: Effective irrigation scheduling commences with meticulous data collection and monitoring. Soil moisture sensors, weather stations, and relevant monitoring tools provide essential information for making informed decisions. Regular updates ensure that schedules remain adaptive to changing conditions and the evolving growth stages of crops.

5.3.2 Crop-Specific Knowledge: A comprehensive understanding of the water needs of specific crops at different growth stages is indispensable for precise scheduling. Crop coefficients, reflecting the relationship between reference evapotranspiration and crop evapotranspiration, guide tailored schedules. This crop-specific knowledge empowers farmers to adapt schedules to the unique requirements of each cultivated crop.

5.3.3 Technology Integration: The integration of technology amplifies the effectiveness of irrigation scheduling. Automated systems, such as smart controllers and remote sensing technologies, facilitate real-time data analysis and adaptive scheduling. These technologies streamline the scheduling process, allowing for swift adjustments based on changing environmental conditions.

5.3.4 Irrigation Infrastructure Optimization: Efficient irrigation scheduling is complemented by well-designed and properly maintained irrigation infrastructure. Regular maintenance of pipes, pumps, and emitters minimizes water losses during distribution. Ensuring that the irrigation system is in good condition contributes to the overall effectiveness of irrigation scheduling.

5.3.5 Training and Education: Educating farmers in effective irrigation scheduling practices is essential for successful implementation. Providing insights into various scheduling methods, data interpretation, and the significance of precise irrigation enhances the capacity of farmers to make informed decisions. Extension services and agricultural training programs play a pivotal role in disseminating knowledge and fostering best practices.

5.3.6 Adaptive Management: Irrigation scheduling should be viewed as an adaptive process. Regular reviews and adjustments based on feedback from the field, weather forecasts, and crop responses ensure continuous improvement. This adaptive management approach guarantees that schedules remain responsive to dynamic agricultural and environmental conditions.

Efficient irrigation scheduling is a dynamic and multidimensional process that combines scientific principles with practical insights. By recognizing the importance of scheduling, employing diverse scheduling methods, and incorporating technological advancements, farmers can achieve optimal water-use efficiency

6. Technologies for Water Management

In the pursuit of sustainable agriculture, the integration of advanced technologies is proving instrumental in revolutionizing water management practices. This section explores cutting-edge technologies that enhance water efficiency, encompassing advanced irrigation technologies, remote sensing, GIS applications, and automation in irrigation systems.

6.1 Advanced Irrigation Technologies

6.1.1 Drip Irrigation Systems: Drip irrigation stands at the forefront of advanced irrigation technologies, offering a precision-driven approach to water distribution. This system delivers water directly to the root zones of plants through a network of tubes or pipes with emitters. By minimizing surface evaporation and ensuring targeted water delivery, drip irrigation conserves water and optimizes resource utilization. Additionally, this technology facilitates controlled nutrient application, contributing to overall resource efficiency and crop health.

6.1.2 Sprinkler Irrigation Systems: Advanced sprinkler irrigation systems, including center pivot and lateral move systems, utilize a network of pipes with strategically placed nozzles to disperse water over crops. These systems are equipped with sensors and control mechanisms that enable precise water distribution. Center pivot systems, for example, rotate around a central pivot point, ensuring even water coverage. Lateral move systems follow a similar principle but move along a straight path. These technologies enhance water distribution efficiency and allow for tailored irrigation schedules based on factors such as soil moisture and weather conditions.

6.1.3 Precision Irrigation Technologies: Precision irrigation technologies represent a paradigm shift in water management. Variable Rate Irrigation (VRI) systems, for instance, dynamically adjust water application rates based on field variability. This technology takes into account factors such as soil type, topography, and crop requirements, optimizing water use across the entire field. Precision mobile irrigation systems utilize GPS technology for accurate water distribution, ensuring that water is applied precisely where and when it is needed. These technologies not only conserve water but also contribute to the sustainability of agricultural practices.

6.1.4 Subsurface Drip Irrigation (SDI): Subsurface Drip Irrigation (SDI) takes drip irrigation a step further by placing the drip lines below the soil surface. This technique minimizes water evaporation and reduces surface wetting, resulting in efficient water use. SDI is particularly beneficial for row crops and orchards, as it provides targeted water delivery directly to the root zone while minimizing water contact with leaves. The underground placement of drip lines also protects them from physical damage and reduces weed growth.

6.2 Remote Sensing and GIS Applications

6.2.1 Satellite Imagery: Satellite-based remote sensing offers a comprehensive view of agricultural landscapes. Advanced imaging technologies capture data on vegetation health, soil moisture levels, and crop conditions. This data is invaluable for assessing irrigation needs, detecting water stress, and optimizing water management strategies. High-resolution satellite imagery provides a bird's-eye perspective, allowing farmers to monitor large agricultural areas and make informed decisions based on real-time information.

6.2.2 Unmanned Aerial Vehicles (UAVs): Unmanned Aerial Vehicles (UAVs), commonly known as drones, have emerged as powerful tools for precision agriculture. Equipped with multispectral and thermal sensors, drones can capture high-resolution aerial imagery. This data enables farmers to monitor crop health, assess water distribution, and identify irrigation inefficiencies. UAVs offer the advantage of agility and flexibility, allowing for timely and targeted data collection in various agricultural settings.

6.2.3 Geographic Information System (GIS): Geographic Information System (GIS) technology integrates spatial data to analyze and visualize patterns in agricultural landscapes. GIS applications in water management include mapping soil types, topography, and water availability. By overlaying various layers of information, GIS aids in decision-making related to irrigation scheduling, resource allocation, and land-use planning. GIS is a powerful tool for creating spatial models that enhance the understanding of the complex interactions within agricultural ecosystems.

6.2.4 Soil Moisture Monitoring: Advanced sensors for soil moisture monitoring provide real-time data on the water content in the soil. These sensors can be integrated with remote sensing technologies to offer a comprehensive understanding of soil conditions. Accurate soil moisture information is crucial for optimizing irrigation schedules and preventing both under- and over-irrigation. Continuous monitoring allows farmers to adapt their irrigation strategies based on dynamic soil moisture levels, ensuring efficient water use.

6.3 Automation in Irrigation Systems

6.3.1 Smart Irrigation Controllers: Smart irrigation controllers represent a significant advancement in automated water management. These controllers leverage data from various sources, including weather forecasts, soil moisture sensors, and evapotranspiration models, to dynamically adjust irrigation schedules. The integration of smart controllers enables automated and adaptive irrigation, optimizing water delivery based on real-time conditions. This technology not only enhances water-use efficiency but also contributes to resource conservation and cost savings for farmers.

6.3.2 Automated Drip Irrigation Systems: Automated drip irrigation systems take the efficiency of drip technology to the next level by incorporating sensors and controllers for complete automation. These systems can be programmed to deliver precise amounts of water at specific intervals. The integration of sensors ensures that irrigation is triggered based on actual soil moisture levels, preventing over-irrigation. Automated drip irrigation systems contribute to water conservation and efficient resource utilization while minimizing the need for manual intervention.

6.3.3 Remote Control and Monitoring: Automation in irrigation extends to remote control and monitoring systems, providing farmers with unprecedented control over their irrigation infrastructure. Through mobile applications or computer interfaces, farmers can remotely monitor and control irrigation systems. Real-time data on soil moisture, weather conditions, and system performance allow for quick decision-making. This technology enhances flexibility and convenience in managing irrigation operations, particularly in large agricultural setups.

6.3.4 Internet of Things (IoT) in Agriculture: The Internet of Things (IoT) has found extensive applications in agriculture, with IoT devices revolutionizing water management. Soil moisture sensors, weather stations, and automated valves are interconnected to form a comprehensive irrigation management system. These devices communicate data in real time, allowing for centralized control and monitoring. The integration of IoT in agriculture enhances the efficiency and responsiveness of irrigation systems, contributing to sustainable and data-driven farming practices.

7. Challenges and Barriers

7.1 Common Challenges in Irrigation Management

7.1.1 Water Scarcity: Water scarcity is a pervasive challenge that directly impacts irrigation management. As global water resources come under increasing stress due to population growth and climate change, the availability of water for irrigation becomes limited. Competing demands from various sectors further exacerbate the challenge, necessitating a careful balancing act to ensure sustainable water use in agriculture.

7.1.2 Inefficient Water Use: Inefficient water use in irrigation is a significant challenge, often stemming from outdated or poorly maintained irrigation systems. Water losses due to evaporation, runoff, and inefficient distribution can lead to suboptimal crop yields and contribute to water wastage. Addressing inefficiencies in irrigation practices is crucial for maximizing water use efficiency.

7.1.3 Lack of Infrastructure: In some regions, the lack of proper irrigation infrastructure poses a challenge to effective water management. Outdated or poorly maintained irrigation systems can result in uneven water distribution and inadequate coverage. Investing in the development and maintenance of irrigation infrastructure is essential to overcome this challenge and enhance overall water-use efficiency.

7.1.4 Financial Constraints: Farmers, particularly smallholders, often face financial constraints that limit their ability to invest in modern irrigation technologies. The initial costs associated with upgrading or adopting advanced irrigation systems can be prohibitive. Overcoming financial barriers requires innovative financing mechanisms, subsidies, and support systems to make water-efficient technologies accessible to a broader spectrum of farmers.

7.1.5 Lack of Technical Knowledge: A lack of technical knowledge among farmers about modern irrigation practices and technologies is a prevalent challenge. Many farmers may not be aware of water-efficient irrigation methods or lack the necessary skills to implement them effectively. Providing education, training, and extension services is crucial for disseminating knowledge and empowering farmers to adopt water-efficient practices.

7.2 Barriers to Implementing Water Efficiency Measures

7.2.1 Resistance to Change: Resistance to change, whether due to traditional practices or a fear of the unknown, can be a significant barrier to implementing water efficiency measures. Farmers and communities accustomed to existing irrigation practices may be hesitant to adopt new technologies or modify their established routines. Overcoming this barrier requires effective communication, education, and demonstrating the tangible benefits of water-efficient measures.

7.2.2 Policy and Regulatory Barriers: Inadequate or restrictive policies and regulations can impede the adoption of water efficiency measures in irrigation. Complex permitting processes, unclear water rights, and conflicting regulations across jurisdictions can create obstacles for farmers seeking to implement efficient irrigation practices. Addressing policy and regulatory barriers involves collaborative efforts between stakeholders, policymakers, and advocacy groups to streamline regulations and provide incentives for water-efficient practices.

7.2.3 Limited Access to Information: Limited access to information about water-efficient technologies and practices is a significant barrier for many farmers. Lack of awareness about available solutions, their benefits, and how to implement them can hinder adoption. Improving access to information through extension services, demonstration projects, and knowledge-sharing platforms is essential for breaking down this barrier.

7.2.4 Short-Term Economic Pressures: Farmers facing immediate economic pressures may prioritize short-term gains over long-term investments in water efficiency. The financial challenges of implementing new irrigation technologies or practices, especially during periods of economic uncertainty, can deter farmers from making necessary changes. Addressing short-term economic pressures requires a combination of financial support, incentives, and risk mitigation strategies.

7.2.5 Fragmented Stakeholder Engagement: Fragmented engagement among stakeholders, including farmers, policymakers, researchers, and industry players, can hinder the coordinated implementation of water efficiency measures. Collaboration and coordination are essential for developing holistic strategies that address the diverse challenges associated with irrigation management. Facilitating stakeholder engagement through partnerships and forums can help overcome this barrier.

7.3 Overcoming Challenges

7.3.1 Integrated Water Management Approaches: Adopting integrated water management approaches that consider the entire water cycle is crucial. This includes watershed management, water harvesting, and recycling practices. By addressing water management at a broader scale, it becomes possible to optimize water use, improve water quality, and enhance overall sustainability.

7.3.2 Investment in Research and Development: Investing in research and development of water-efficient technologies is essential for overcoming technical challenges. This includes the development of advanced irrigation systems, sensor technologies, and precision agriculture tools. Research efforts should focus on making these technologies affordable, user-friendly, and tailored to the specific needs of diverse farming communities.

7.3.3 Capacity Building and Education: Capacity building and education initiatives play a pivotal role in overcoming knowledge barriers. Providing farmers with the necessary skills and information about water-efficient practices empowers them to make informed decisions. Extension services, training programs, and knowledge-sharing platforms are effective tools for disseminating information and building technical capacity.

7.3.4 Policy Reforms and Incentives: Addressing policy and regulatory barriers requires proactive policy reforms and the implementation of incentives. Policymakers should strive to create an enabling environment that promotes water-efficient practices. This may include streamlining permitting processes, clarifying water rights, and introducing financial incentives for adopting water-efficient technologies.

7.3.5 Financial Support Mechanisms: Overcoming financial constraints necessitates the development of innovative financial support mechanisms. This may involve government subsidies, low-interest loans, or public-private partnerships that make water-efficient technologies financially accessible to farmers. Collaborative efforts between financial institutions, government agencies, and the private sector are crucial for designing and implementing these support mechanisms.

Conclusion

The overarching importance of efficient irrigation practices in the context of global water scarcity is a central theme. Agriculture, being a primary consumer of water, plays a pivotal role in shaping the responsible and sustainable use of this finite resource. Recognizing the challenges and opportunities associated with irrigation management sets the stage for a comprehensive understanding of its role in contemporary agriculture.

Fundamental to this understanding is the exploration of various irrigation systems, ranging from traditional to modern. Tailoring irrigation practices to specific crop needs, local climate conditions, and water availability emerges as a crucial consideration. This foundational knowledge underscores the need for precision and adaptability in irrigation management.

Examining the components of irrigation systems further reveals the intricate infrastructure that facilitates effective water distribution. From pumps and pipes to emitters and controllers, each component plays a vital role in optimizing water use. The interplay of these elements emphasizes the complexity of irrigation management, requiring not only technological sophistication but also diligent maintenance and strategic design.

Water efficiency, a core theme in the chapter, emerges as a linchpin for sustainable agriculture. Defining water efficiency and exploring its importance highlights the need for judicious water use beyond mere conservation. The multifactorial nature of water efficiency, influenced by variables like soil moisture, weather conditions, and crop requirements, underscores the dynamic nature of irrigation management.

Conservation practices in irrigation underscore the practical applications of sustainable water use. Techniques such as soil moisture management and cover cropping showcase how farmers can minimize water usage while optimizing crop health and reducing environmental impact. These practices reflect the potential for achieving a delicate balance between agricultural productivity and resource conservation.

Irrigation scheduling emerges as a strategic tool in the pursuit of water efficiency. The various methods explored, from soil moisture-based approaches to plant and weather-based strategies, underscore the need for adaptability and precision. Implementing effective scheduling practices aligns water applications with the ever-changing needs of crops and environmental conditions, contributing to enhanced efficiency.

The integration of advanced technologies in water management represents a transformative shift in modern agriculture. From precision irrigation systems to remote sensing and automation, these technologies empower farmers to make data-driven decisions, optimize resource use, and mitigate environmental impact. The continuous evolution of these technologies holds the key to creating sustainable, resilient, and water-efficient agricultural systems for the future.

9

Soil Water Plant Relationship in Climate Change

[1]Savita Jangde, [2]Santosh Marahatta, [3]Jeetendra Kumar [4]Sheetanshu Gupta and [5]Adarsh Pandey

[1]Department of Plant Physiology, Institute of Agricultural Sciences B.H.U. Varanasi, U.P.
[2]Department of Agronomy, Faculty of Agriculture. Agriculture and Forestry University (AFU), Rampur, Chitwan, Nepal
[3]Krishi Vigyan Kendra, Jehanabad, Bihar Agricultural University, Bihar
[4]Department of Biotechnology, Institute of Technology and Management BKT, Lucknow, U.P.
[5]Department of Botany. Swami Shukdevanand College Shahjahanpur, Uttar Pradesh

Plant Relationship with Soil in Climate Change

1. Gravimetric Moisture

Significance of Gravimetric Analysis

Gravimetric moisture analysis, being the foundation of soil water assessment, allows for a meticulous examination of the water content in the soil. The process involves obtaining a representative soil sample, subjecting it to controlled drying conditions, and determining the weight loss after moisture removal. This method not only quantifies the total water content but also helps in understanding the soil's water-holding capacity and drainage characteristics. Gravimetric analysis is indispensable in precision agriculture, aiding farmers in optimizing irrigation schedules and preventing overwatering or underhydration.

Impact on Plant Health

Gravimetric moisture content is directly linked to plant health and growth. Plants rely on the water present in the soil for various physiological processes, including nutrient uptake, photosynthesis, and transpiration. The gravimetric analysis provides insights into the water availability for plants, enabling informed decisions on irrigation management. Understanding the dynamics of gravimetric moisture content is crucial in regions facing water scarcity, allowing for sustainable water use practices to ensure optimal crop yields without compromising environmental integrity.

Challenges and Advances in Gravimetric Analysis

While gravimetric analysis is a reliable method, it is not without challenges. Accurate representation of soil moisture requires careful sampling and consideration of soil heterogeneity. Advances in technology, such as the use of precision instruments and remote sensing, enhance the accuracy of gravimetric measurements. Integrating these technological innovations with traditional gravimetric methods provides a comprehensive understanding of soil water dynamics, facilitating more effective water resource management.

Capillary Moisture

Role of Capillary Action in Soil Water Movement: Capillary moisture, driven by capillary action, plays a pivotal role in facilitating the upward movement of water through soil. Capillarity is the result of the cohesive and adhesive forces between water molecules and soil particles. This upward movement defies the force of gravity, allowing water to reach the root zones of plants. Understanding capillary dynamics is essential in designing irrigation systems that harness natural capillary processes, ensuring efficient water distribution in agriculture and landscaping.

Soil Structure and Capillary Water Retention: The capillary water held within soil contributes significantly to its structure. The retention of water in capillaries influences soil texture, porosity, and overall fertility. Fine-textured soils, with smaller capillaries, tend to retain water more effectively than coarse-textured soils. The balance between water retention and drainage is critical for preventing waterlogging and optimizing nutrient availability. Agriculture practices benefit from tailoring soil management techniques based on the capillary characteristics of different soil types.

Capillary Moisture and Climate Change: Climate change introduces new challenges to capillary moisture dynamics. Altered precipitation patterns, increased evaporation rates, and shifts in temperature can disrupt traditional capillary processes. Understanding these changes is vital for adapting agricultural practices to ensure water sustainability. Additionally, advances in capillary sensor technologies provide real-time data on soil moisture dynamics, aiding in the development of adaptive strategies to mitigate the impact of climate change on capillary water availability.

Hygroscopic Moisture

The Intricacies of Hygroscopic Water: Hygroscopic moisture, characterized by its affinity to soil particles, forms a thin film around each grain. This water is tightly bound and is not readily available to plants. The hygroscopic nature of this moisture is crucial for maintaining soil structure and stability. The water film influences soil aggregation, preventing erosion and promoting aeration. Researchers delve into the molecular interactions between soil particles and hygroscopic water to unlock insights into soil physics and the broader ecological impact.

Influence on Nutrient Availability: Hygroscopic water, while not directly accessible to plants, plays a crucial role in nutrient availability. It acts as a reservoir for dissolved nutrients, facilitating their movement within the soil matrix. The hygroscopic film around soil particles creates a microenvironment where ions can interact, promoting the exchange of essential nutrients. Soil fertility is intricately linked to hygroscopic moisture, and understanding this relationship is paramount for sustainable agriculture practices.

Challenges in Hygroscopic Moisture Measurement: Measuring hygroscopic moisture poses challenges due to its tightly bound nature. Traditional methods involve subjecting soil samples to extremely dry conditions to induce water absorption from the air. However, this process is intricate and requires precise control of environmental conditions. Advances in analytical techniques, such as nuclear magnetic resonance (NMR) and neutron scattering, offer non-destructive ways to study hygroscopic moisture, shedding light on its distribution and behavior within the soil profile.

2. Retention and Drying Characteristics of Soil Moisture

Soil Moisture Retention

Forces at Play: The retention of moisture in soil is a dynamic process influenced by a multitude of factors. Primary among these are capillary forces, gravitational forces, and the cohesive and adhesive properties of water molecules. Capillary forces draw water upwards against gravity, while gravitational forces pull water downward. The balance between these forces, alongside soil texture and structure, determines how much water the soil can retain. Understanding these intricate relationships is essential for tailoring irrigation strategies and preventing issues like waterlogging or drought stress.

Soil Texture's Role: Soil texture, characterized by the proportions of sand, silt, and clay particles, significantly influences moisture retention. Sandy soils, with larger particles, generally have lower water retention capacities but allow for rapid drainage. Conversely, clayey soils, with smaller particles, possess higher water retention capacities but may experience drainage challenges. The delicate balance between these soil textures impacts the availability of water for plant roots and necessitates thoughtful soil management practices.

Organic Matter's Impact: The presence of organic matter in soil acts as a game-changer in moisture retention dynamics. Organic matter enhances soil structure, fostering the formation of aggregates that create pore spaces. These pores serve as reservoirs for water retention, ensuring a steady supply of moisture to plant roots. Additionally, organic matter's water-holding capacity plays a vital role in buffering against extreme weather conditions, offering resilience to both drought and heavy rainfall.

Soil Moisture Drying Characteristics

Evaporation and Transpiration: Once water is absorbed by the soil, the drying process begins, primarily driven by evaporation and transpiration. Evaporation involves the conversion of liquid water into vapor from the soil surface, influenced by factors such as temperature, wind, and humidity. Transpiration, the release of water vapor through plant leaves, further contributes to soil moisture loss. The intricate coupling of these processes dictates the rate at which soil moisture dries, impacting plant water availability and overall ecosystem health.

Soil Structure and Drying Rates: Soil structure plays a crucial role in determining how rapidly or slowly soil moisture dries. Well-aggregated soils with sufficient pore spaces allow for efficient drainage and aeration, promoting faster drying rates. In contrast, compacted soils hinder drainage and impede air movement, leading to slower drying rates and an increased risk of waterlogged conditions. Agricultural practices that prioritize soil structure maintenance are vital for optimizing drying characteristics and preventing water-related challenges.

Climate Change and Drying Dynamics: As climate change alters precipitation patterns and increases temperatures, the drying characteristics of soil moisture undergo significant transformations. Prolonged droughts intensify drying rates, posing challenges for agriculture and natural ecosystems. Conversely, extreme rainfall events can lead to rapid saturation and subsequent runoff. Understanding the evolving dynamics of soil moisture drying in the context of climate change is crucial for developing adaptive strategies that ensure sustainable land use and water resource management.

Advanced Techniques in Studying Retention and Drying

Remote Sensing and Modeling: Technological advancements have revolutionized the study of soil moisture retention and drying characteristics. Remote sensing techniques, such as satellite-based sensors, provide real-time data on soil moisture content over large areas, enabling precise monitoring and analysis. Additionally, sophisticated modeling approaches, including numerical simulations and machine learning algorithms, contribute to a more comprehensive understanding of the complex interplay of factors influencing soil moisture dynamics.

Soil Moisture Sensors: In-situ soil moisture sensors offer a ground-level perspective, providing accurate and real-time data on soil moisture levels. These sensors, strategically placed in the root zone, allow for precise irrigation management, minimizing water wastage and optimizing plant health. Continuous advancements in sensor technology enhance their reliability and accessibility, making them indispensable tools for farmers and researchers alike.

3. Potential Energy in Soil Water

Gravitational Potential Energy: The gravitational potential energy in soil water is a fundamental force shaping the movement and distribution of water within the soil profile. As precipitation infiltrates the soil, gravitational potential energy is harnessed, initiating a cascade of interactions. The water descends through the soil layers, driven by the force of gravity. The potential energy stored in elevated water positions is gradually transformed into kinetic energy, influencing the rate and direction of water movement. Understanding this energy conversion is critical for predicting how water percolates through the soil, impacting groundwater recharge, and influencing the overall water balance in ecosystems.

Matric Potential Energy: Matric potential energy, often termed soil water potential, represents the energy state associated with the adhesion of water to soil particles. This intricate force is a key determinant of soil moisture retention. The forces at play in matric potential are influenced by soil texture, structure, and moisture content. For instance, finer-textured soils with smaller particles exhibit higher matric potential, enhancing their water-holding capacity. Understanding matric potential is essential for elucidating how water is retained in the soil, contributing to sustained plant growth and providing resilience against periods of water scarcity.

Kinetic Energy in Soil Water

Capillary Rise and Capillary Fringe: Kinetic energy manifests prominently in capillary rise, an intriguing phenomenon driven by capillary action. Water, propelled by cohesive and adhesive forces, ascends through small soil pores against the force of gravity. The kinetic energy associated with this upward movement is instrumental in understanding water availability to plant roots in both surface and subsurface soils. The capillary fringe, situated at the interface between saturated and unsaturated zones, becomes a dynamic space where kinetic energy shapes water retention and facilitates nutrient transport. Investigating capillary dynamics is crucial for optimizing irrigation practices and fostering sustainable agricultural systems.

Transpiration and Kinetic Energy: The kinetic energy involved in soil water dynamics extends to the process of transpiration, where plants release water vapor through their leaves. As water is drawn from the soil through plant roots and transported upwards, kinetic energy governs this upward flow. The delicate equilibrium between transpiration rates and soil water availability is pivotal for maintaining plant health and ecosystem functioning. Recognizing the role of kinetic energy in transpiration offers insights into the intricate balance between water demand and supply in plant physiology.

Soil Water Energy and Plant Uptake

Osmotic Potential and Plant Root Uptake: Osmotic potential, a critical component of soil water potential, dictates plant water uptake. This energy state represents the force required to move water across cell membranes and into plant roots. The osmotic potential gradient between soil water and plant roots influences the rate and efficiency of water absorption. Understanding this interplay is essential for optimizing irrigation practices, particularly in regions where water resources are limited. The osmotic potential concept provides a lens through which we can enhance water use efficiency in agriculture, minimizing water wastage while sustaining crop productivity.

Water Movement in the Rhizosphere: The rhizosphere, a dynamic zone influenced by plant roots, serves as a nexus where soil water energy influences nutrient and water uptake. Root exudates, compounds released by plant roots, alter the physical and chemical properties of the soil. This dynamic interaction impacts water availability and nutrient cycling in the rhizosphere. Investigating the energy dynamics in this zone provides valuable insights into the complex relationships that drive soil-plant interactions. Understanding how soil water energy influences the rhizosphere is crucial for optimizing agricultural practices, fostering nutrient uptake, and supporting plant health.

Implications for Soil and Ecosystem Health

Soil Structure and Stability: The implications of soil water energy on soil structure and stability are profound. Matric potential, a key player in soil water energy, influences soil aggregation and particle arrangement. This, in turn, impacts soil porosity, permeability, and water-holding capacity. A deeper understanding of these implications allows for targeted soil management practices. Techniques such as cover cropping, organic matter addition, and conservation tillage become essential strategies for maintaining soil structure, preventing erosion, and promoting sustainable land use.

Climate Change and Energy Dynamics: As climate change continues to manifest, alterations in precipitation patterns, temperature regimes, and extreme weather events significantly impact the energy dynamics of soil water. Changes in energy states influence water availability, soil moisture retention, and the resilience of ecosystems. In-depth investigations into these dynamics are essential for predicting and mitigating the effects of climate change on soil health, water resources, and agricultural productivity. Research efforts focus on developing adaptive strategies that account for evolving energy dynamics in the soil-water-plant continuum, ensuring sustainable land management practices in the face of a changing climate.

Advanced Techniques in Studying Soil Water Energy

Soil Water Potential Sensors: The advent of advanced soil water potential sensors marks a significant leap in our ability to monitor and understand the energy states of soil water. These sensors, employing principles of tensiometry or psychrometry, offer real-time data on matric potential variations in the soil. The integration of these technologies into precision agriculture provides a means to optimize irrigation practices, minimize water wastage, and enhance water use efficiency. As these sensors become more accessible, they empower farmers and researchers with critical information for informed decision-making.

Modeling and Simulation: Numerical modeling and simulation techniques have emerged as powerful tools for comprehensively understanding soil water energy dynamics. These models integrate a plethora of factors, including soil properties, climate conditions, and vegetation cover, to simulate water movement and energy transformations within the soil-plant-atmosphere continuum. By leveraging these modeling approaches, researchers gain insights into the complex interplay of forces influencing soil water dynamics. Simulations provide a platform for predicting the impact of various land management practices and climate change scenarios on soil water energy states, facilitating the development of adaptive strategies.

4. Soil Water Potential

Thermodynamic Foundation: Soil water potential serves as a cornerstone in the understanding of soil-water-plant interactions. At its core, it embodies a thermodynamic perspective, representing the potential energy per unit quantity of water in the soil. This thermodynamic foundation provides a theoretical framework for comprehending the various forces influencing water within the soil matrix.

Each component of soil water potential contributes to the overall energy state, orchestrating a dynamic interplay of physical and chemical forces. Recognizing soil water potential as an energy metric enables scientists, agronomists, and environmental researchers to interpret water movement, availability, and the delicate equilibrium within ecosystems.

Components of Soil Water Potential

1. **Gravitational Potential (Ψg):** Gravitational potential energy originates from the force of gravity acting on water within the soil profile. As precipitation or irrigation infiltrates the soil, water descends under the influence of gravity. The gravitational potential decreases as water moves

downwards, transforming into kinetic energy. This transformation is fundamental to the initiation of water movement in the soil.

2. **Matric Potential (Ψm):** Matric potential arises from the capillary and adsorptive forces between water and soil particles. In the unsaturated zone, where soils contain both air and water, matric potential influences water retention. Fine-textured soils with smaller pores exhibit higher matric potential, enhancing water retention capacity. The understanding of matric potential is crucial for predicting how water is held in the soil and the subsequent implications for plant water uptake.

3. **Osmotic Potential (Ψo):** Osmotic potential is a reflection of the concentration of solutes in the soil solution. It represents the energy required to move water across semipermeable membranes, such as those found in plant root cells. Osmotic potential is vital in the context of plant water uptake, as roots absorb water from the soil solution. Understanding osmotic potential provides insights into how soil salinity and solute concentrations impact plant water relations.

4. **Pressure Potential (Ψp):** Pressure potential accounts for the physical pressure exerted on water within the soil. Positive pressure potential results from factors such as soil water saturation, root water uptake, or irrigation practices. Negative pressure potential indicates tension within the soil matrix, often associated with water deficit conditions. The interplay between pressure potential and other components influences soil water movement and availability.

Significance of Soil Water Potential

Plant Water Uptake: The significance of soil water potential becomes evident in its role in plant water uptake. Plants have evolved sophisticated mechanisms to regulate water movement based on gradients of matric and osmotic potentials. Roots absorb water from the soil, and the difference in potential energy between the soil water and the root governs the movement of water into the plant. This finely tuned system ensures that plants receive adequate water for growth and development, preventing both water stress and oversaturation.

Soil Moisture Retention: Soil water potential is a key determinant of soil moisture retention. Matric potential, in particular, plays a pivotal role in the unsaturated zone by influencing the ability of soil to retain water against the force of gravity. This retention capacity, influenced by soil texture, organic matter content, and other factors, contributes to soil structure, porosity, and nutrient transport. By understanding these dynamics, researchers and land managers can optimize soil conditions for water availability to plants and sustainable ecosystem functioning.

Irrigation Management: In the realm of agriculture, the significance of soil water potential is underscored by its role in guiding irrigation practices. Precision irrigation, tailored to the water needs of specific crops, hinges on a thorough understanding of soil water potential. By monitoring and assessing soil water potential, farmers can fine-tune irrigation schedules to match the water requirements of crops. This not only conserves water resources but also enhances crop yield, mitigates water stress, and contributes to sustainable agricultural practices.

Evaluation of Total and Component Potentials

Measurement Techniques: Accurate measurement of soil water potential and its components is crucial for meaningful assessments. Various techniques are employed, each catering to specific ranges and conditions. Tensiometers, based on capillary action, measure soil matric potential in the field. Psychrometers assess the potential based on the water vapor pressure in soil, providing insights into atmospheric interactions. Pressure chambers gauge the pressure potential, offering a snapshot of the physical pressure exerted on soil water. Laboratory methods, such as osmometers and pressure plates, complement these field measurements, providing a comprehensive evaluation.

Total Potential Assessment:

The total soil water potential (Ψt) is the sum of its components:

$$\Psi t=\Psi g+\Psi m+\Psi o+\Psi p$$

This summation captures the complete energy state of soil water, integrating gravitational, matric, osmotic, and pressure potentials. Evaluating the total potential allows for a holistic understanding of the factors influencing water movement within the soil-plant-atmosphere continuum. This comprehensive assessment is instrumental for predicting water dynamics, identifying limiting factors, and optimizing land management practices.

Component Potential Interactions: Understanding the interactions among individual components of soil water potential adds a layer of complexity to our comprehension of soil-water dynamics. For instance, during periods of water scarcity, matric potential may become the dominant force, limiting water availability to plants. Conversely, in saline soils, high osmotic potential can influence both plant water uptake and soil moisture retention. Evaluating these interactions is crucial for devising strategies to mitigate water stress, enhance irrigation efficiency, and promote sustainable land use practices.

5. Soil Water Retention

Gravimetric Method: The gravimetric method provides a detailed insight into soil water retention by assessing the relationship between soil moisture content and soil water potential. This method involves saturating soil samples to their full water-holding capacity and then allowing them to drain. The remaining water content represents the water retained against gravity. By plotting moisture content against soil water potential, researchers can construct a soil water retention curve. This curve characterizes the soil's ability to hold water at various matric potentials, providing valuable information for predicting water movement and availability.

Understanding the gravimetric method's outcomes aids in developing irrigation strategies tailored to a specific soil's water retention characteristics. It allows for the identification of soil types that may be prone to waterlogging or drought stress, facilitating precise water management in agriculture and natural ecosystems.

Soil Texture Influence

Soil texture significantly influences water retention, and understanding its impact is crucial for effective land management. Sandy soils, characterized by larger particles, typically have lower water retention capacities due to larger pore spaces that facilitate rapid drainage. Conversely, clayey soils, with finer particles, exhibit higher water retention capabilities as their smaller pore spaces retain water more effectively.

The balance between sand, silt, and clay in a soil's composition determines its texture and subsequently influences its water retention dynamics. This knowledge is indispensable for tailoring agricultural practices to the specific needs of different soils, optimizing water use, and preventing issues such as water stress or waterlogged conditions.

Field Capacity

Measurement and Determination

Field capacity is determined through field measurements or laboratory experiments. In the field, a soil sample is saturated with water, left to drain, and then the remaining water content is measured. This process mimics natural rainfall or irrigation and provides insights into the soil's drainage characteristics. In laboratory settings, controlled conditions are applied to soil samples, and the moisture content is monitored until drainage ceases, allowing for the calculation of field capacity.

Field capacity is essential for optimizing irrigation schedules in agriculture. By knowing the soil's maximum water-holding capacity, farmers can schedule irrigations to coincide with the point when the soil is nearing its field capacity. This ensures that plants receive adequate moisture without leading to waterlogged conditions, contributing to sustainable and efficient water use.

Wilting Point

Plant Wilting and Water Availability

The wilting point holds significant implications for plant growth and survival during periods of water scarcity. At this point, the soil retains water so tightly that plant roots can no longer extract it effectively, leading to wilting. Recognizing the onset of wilting is critical for predicting drought stress in plants and implementing timely irrigation practices.

The wilting point is a direct indicator of a soil's ability to sustain plant life under challenging water conditions. This knowledge is instrumental for farmers and land managers to develop strategies for mitigating water stress, selecting drought-resistant crops, and implementing water-conserving measures.

Laboratory Determination

Determining the wilting point involves laboratory experiments where soil samples are subjected to controlled drying conditions. The moisture content is measured at intervals until plants exhibit wilting. The point at which wilting occurs is then identified as the wilting point. This laboratory determination provides a quantitative measure of a soil's water availability and aids in making informed decisions about irrigation needs and water conservation practices.

Water-Holding Capacity

Factors Influencing Water-Holding Capacity

Water-holding capacity is a comprehensive metric that considers both field capacity and the wilting point. Several factors influence water-holding capacity, with soil texture, organic matter content, and soil structure playing pivotal roles. Soils rich in organic matter exhibit higher water-holding capacities due to increased aggregation, which creates pore spaces that retain water more effectively.

Understanding the factors influencing water-holding capacity provides valuable insights into soil management practices. Incorporating organic amendments, cover cropping, and other techniques to improve soil structure and increase organic matter content can enhance a soil's water-holding capacity, promoting sustainable agriculture and ecosystem health.

Practical Implications

The practical implications of water-holding capacity are significant for farmers seeking to optimize crop yields and conserve water resources. Soils with higher water-holding capacities can sustain crops for longer periods between irrigation events, providing a crucial buffer against water scarcity. Conversely, soils with lower water-holding capacities may necessitate more frequent irrigation to meet the water demands of plants.

Available Water

Calculating Available Water

Available water is the range of soil moisture content lying between field capacity and the wilting point, representing the portion of water accessible to plants. Calculating available water involves subtracting the wilting point from the field capacity:

Available Water=Field Capacity−Wilting PointAvailable Water=Field Capacity−Wilting Point

This calculation provides a quantitative measure of the water available for plant use and is essential for precision agriculture and water resource management.

Precision Agriculture

In precision agriculture, the concept of available water is instrumental in optimizing irrigation practices. Soil moisture sensors and advanced technologies are often deployed to monitor real-time soil moisture conditions. This enables farmers to make data-driven decisions, applying irrigation precisely when needed and conserving water resources. The information on available water guides farmers in scheduling irrigation events based on the soil's water storage capacity and the water requirements of specific crops.

6. Soil Water Storage

Infiltration and Saturation

The process of soil water storage initiates with the infiltration of water into the soil. This crucial mechanism is influenced by various factors, such as soil texture, structure, and land cover. Infiltration occurs when precipitation or irrigation water enters the soil, filling the pore spaces. The rate and depth of infiltration depend on soil characteristics, with well-structured soils and those rich in organic matter exhibiting higher infiltration rates.

Understanding infiltration dynamics is essential for predicting water availability in the root zone. Infiltrated water eventually reaches different layers of the soil profile, leading to saturation. Saturation refers to the state where soil pores are filled with water. The capacity of the soil to store water during saturation is determined by its texture, structure, and porosity. Grasping these dynamics provides valuable insights into the initial state of soil water storage, which is foundational for managing water resources effectively.

Capillary Rise and Root Zone Storage

Capillary rise is a phenomenon where water moves upward through soil pores due to capillary forces. This process contributes significantly to soil water storage in the root zone, making water accessible for plant uptake. The extent of capillary rise varies with soil texture, influencing the height to which water can ascend. Finer-textured soils facilitate more prominent capillary rise, resulting in a deeper root zone storage capacity.

The storage capacity in the root zone is pivotal for sustaining plant growth and minimizing water stress. It is influenced by factors such as root density, plant species, and soil properties. A comprehensive understanding of capillary rise and root zone storage enables land managers to optimize irrigation practices, fostering the resilience of crops to variable water conditions. By recognizing the interplay between capillary rise and root zone storage, practitioners can tailor irrigation strategies to enhance water use efficiency and promote sustainable agriculture.

Factors Influencing Soil Water Depletion

Evapotranspiration

Evapotranspiration, a combined process of soil surface evaporation and plant transpiration, is a major driver of soil water depletion. The sun's energy heats the soil, leading to water evaporation from the surface, while plants simultaneously absorb water through their roots, releasing it through transpiration. The rate of evapotranspiration is influenced by climatic factors such as temperature, humidity, wind speed, and solar radiation. The type and density of vegetation also play a significant role in regulating this process.

Understanding the dynamics of evapotranspiration is crucial for predicting soil water depletion rates, especially in regions prone to water scarcity. By comprehending how climatic and vegetative factors impact evapotranspiration, land managers can implement measures to mitigate excessive soil water depletion, such as adjusting planting schedules or selecting drought-resistant plant varieties.

Soil Texture and Structure

Soil texture and structure profoundly impact the rate at which water is depleted from the soil. Sandy soils, characterized by larger particles, generally have faster drainage rates, leading to quicker soil water depletion. On the other hand, clayey soils, with smaller particles, have slower drainage but may retain water for longer periods. The arrangement of soil particles in the soil structure influences water movement and retention, significantly affecting the overall soil water depletion pattern.

Understanding the interplay between soil texture, structure, and water depletion is essential for implementing effective land management practices. Practitioners can adapt soil conservation techniques, such as cover cropping or contour ploughing, to modify soil structure and optimize water retention. By considering soil characteristics, land managers can tailor their approach to mitigate soil water depletion, fostering sustainable water use in agriculture and preserving soil health.

Land Use and Management Practices

Land use decisions and management practices have a direct impact on soil water depletion. Agricultural practices, including tillage, cover cropping, and irrigation methods, influence the soil's water-holding capacity and drainage characteristics. Deforestation and urbanization can alter the natural hydrological cycle, affecting groundwater recharge and increasing the risk of soil water depletion.

Land managers must carefully assess the consequences of land use decisions on soil water dynamics. Adopting sustainable land management practices, such as agroforestry or water conservation measures, can contribute to maintaining soil water availability and mitigating depletion. Integrating practices that promote soil health and water retention helps create resilient ecosystems capable of withstanding changing environmental conditions.

Monitoring and Managing Soil Water Storage

Soil Moisture Monitoring

Effective soil water management necessitates continuous monitoring of soil moisture levels. Various tools, including soil moisture sensors, tensiometers, and other monitoring devices, provide real-time data on the soil's water content and potential. This information empowers land managers to make informed decisions regarding irrigation schedules, optimizing water use efficiency, and minimizing soil water depletion.

Soil moisture monitoring is particularly crucial in precision agriculture, where data-driven decisions can significantly impact crop yield and resource conservation. Integrating advanced technologies for soil moisture assessment ensures timely responses to changing soil water conditions, promoting sustainable water use practices.

Irrigation Strategies

Adopting efficient irrigation strategies is fundamental for sustainable soil water management. Precision irrigation techniques, such as drip or sprinkler irrigation, deliver water directly to the root zone, minimizing surface evaporation and optimizing water use. Additionally, scheduling irrigations based on soil moisture data and crop water requirements helps prevent both water stress and unnecessary soil water depletion.

Understanding the water requirements of specific crops, the characteristics of the soil, and the prevailing climatic conditions aids in developing tailored irrigation strategies. This approach contributes to efficient water use, reduces the risk of soil water depletion, and enhances overall agricultural productivity. Implementing best practices in irrigation management is crucial for the sustainability of water resources and the resilience of agricultural systems.

Soil Conservation Practices

Implementing soil conservation practices is instrumental in maintaining soil water storage and minimizing depletion. Techniques such as cover cropping, mulching, and contour ploughing contribute to soil health, reduce surface runoff, and enhance water infiltration. These practices foster a more resilient soil environment, better equipped to withstand fluctuations in water availability.

Cover cropping, for instance, not only protects the soil from erosion but also enhances organic matter content, improving soil structure and water-holding capacity. Mulching conserves soil moisture by reducing evaporation, particularly in arid or semi-arid regions. Contour ploughing minimizes water runoff, promoting water infiltration and reducing soil erosion.

7. Mechanism of Soil Moisture Absorption by Plants

Understanding how plants absorb moisture from the soil involves a detailed exploration of the complex mechanisms at the root-soil interface. This process, known as plant water uptake, is a critical physiological function that ensures the survival and growth of terrestrial plants.

Root Structure and Function:

Roots are the primary organs responsible for water absorption in plants. The root system is a highly organized structure that facilitates the uptake of water and essential nutrients from the soil. The root hairs, fine extensions found near the root tips, significantly increase the surface area available for water absorption. These microscopic structures play a crucial role in enhancing the efficiency of water uptake.

The absorption of water is an active process that requires energy. Root cells actively pump ions, creating a concentration gradient that facilitates the movement of water into the roots. This process is driven by the plant's energy currency, adenosine triphosphate (ATP). The root cell membranes act as selective barriers, allowing water molecules to pass through while excluding other solutes.

Osmosis

Osmosis is a fundamental mechanism governing the movement of water across plant cell membranes. This process occurs through a semi-permeable membrane, such as the cell membranes of root cells. Water molecules move from areas of low solute concentration (in the soil) to areas of high solute concentration (inside the root cells). This movement is crucial for maintaining turgor pressure in plant cells, providing structural support and facilitating nutrient transport.

Osmosis is the driving force behind the absorption of water from the soil into the roots. As water moves into the root cells, it creates a positive pressure, allowing it to ascend through the plant's vascular system, ultimately reaching the leaves and contributing to various physiological processes.

Role of Mycorrhizal Associations

Mycorrhizal associations are symbiotic relationships between plant roots and fungi. These relationships significantly enhance the plant's ability to absorb water and nutrients from the soil. Mycorrhizal fungi form an extensive network of hyphae that extends beyond the reach of the plant's root system, effectively increasing the volume of soil explored for resources.

In return for providing fungi with sugars produced through photosynthesis, the plant receives improved access to water and nutrients, including phosphorus and nitrogen. This mutualistic association not only benefits individual plants but also plays a crucial role in nutrient cycling and ecosystem health.

Transpiration Pull

Transpiration, the loss of water vapor from plant leaves, creates a negative pressure or tension in the plant's xylem vessels. This negative pressure, often referred to as the transpiration pull, is a cohesive and adhesive force that sustains a continuous column of water within the plant's vascular system.

As water molecules evaporate from the stomata on the leaf surface, they create a negative pressure that pulls water from the roots, through the stem, and into the leaves. The cohesive nature of water molecules and the adhesive forces between water and cell walls contribute to this upward movement. The transpiration pull is essential for the transport of water and nutrients from the roots to the leaves, maintaining the plant's hydration and facilitating various physiological processes.

Soil Moisture Determination

Accurate determination of soil moisture content is fundamental for optimizing irrigation practices, assessing water availability to plants, and managing water resources effectively. Various methods, each with its advantages and limitations, are employed to measure soil moisture content:

Gravimetric Method

The gravimetric method is a traditional and accurate technique for determining soil moisture content. It involves physically extracting soil samples, weighing them, drying them, and re-weighing to calculate the moisture content. This method provides precise results but is labor-intensive and time-consuming. Despite its limitations in terms of speed and practicality, the gravimetric method is often considered the standard for calibration and validation of other techniques.

Tensiometers

Tensiometers are devices that measure soil water potential, indicating the soil's ability to retain water against gravity. These instruments consist of a porous ceramic cup connected to a water-filled tube. The tension in the tube corresponds to the soil water potential, offering real-time insights into soil moisture status. Tensiometers are particularly useful for monitoring water availability in the root zone and guiding irrigation decisions.

While tensiometers provide valuable information about soil water potential, they are limited to the specific depth at which they are installed. Multiple sensors at different soil depths may be required to obtain a comprehensive understanding of soil moisture distribution.

Time-Domain Reflectometry (TDR)

Time-domain reflectometry (TDR) is a modern and efficient method for measuring soil moisture content. This technique involves sending electromagnetic pulses through the soil and analyzing the time it takes for the pulses to return. The soil's dielectric constant, influenced by its moisture content, determines the travel time. TDR is a rapid and non-destructive method for assessing soil moisture, offering advantages in terms of speed and ease of use.

TDR is particularly effective for continuous monitoring of soil moisture profiles and is suitable for a wide range of soil types. It allows for real-time data collection, enabling prompt adjustments to irrigation schedules and water management practices.

Neutron Moisture Meters

Neutron moisture meters measure soil moisture by detecting neutrons scattered by hydrogen atoms in water. These instruments emit neutrons into the soil, and the scattered neutrons are measured to estimate soil moisture content. Neutron moisture meters are especially effective for large-scale and deep soil moisture assessments. They provide valuable insights into soil moisture distribution in the root zone and beyond.

While neutron moisture meters offer advantages in terms of depth penetration, they require specialized training for operation and should be used with caution due to radiation concerns. Additionally, their use may be restricted in certain environments or regulated by local authorities.

Soil Moisture Stress and Plant Growth

Soil moisture stress, resulting from inadequate water availability in the soil, has profound implications for plant growth and overall ecosystem health. The relationship between soil moisture stress and plant growth is complex and involves various physiological and biochemical responses.

Effects of Soil Moisture Stress

1. **Stomatal Closure:** In response to water scarcity, plants often close their stomata to reduce water loss through transpiration. While this adaptive response conserves water, it also limits carbon dioxide uptake for photosynthesis. Stomatal closure is a crucial mechanism to maintain water balance but comes at the cost of reduced photosynthetic activity.

2. **Reduced Cell Expansion:** Water is essential for cell expansion and maintaining turgor pressure. Soil moisture stress leads to decreased cell expansion, resulting in stunted growth and compromised structural integrity. The reduction in cell expansion is particularly evident in growing tissues, impacting both root and shoot development.

3. **Impaired Nutrient Uptake:** Water is crucial for the transport of nutrients to plant roots. In water-stressed conditions, the flow of water through the soil is restricted, leading to impaired nutrient uptake. Nutrient deficiencies can further hinder various metabolic processes and growth-related activities.

4. **Decreased Photosynthesis:** Water stress negatively affects photosynthesis, the process by which plants convert sunlight into energy. Reduced water availability limits the plant's ability to produce sugars and other essential compounds. This, in turn, affects energy allocation for growth and reproduction.

5. **Impact on Reproductive Development:** Water stress during critical growth stages can adversely affect reproductive development. Plants may experience reduced flower formation, pollen viability, and seed or fruit production. This can have long-term consequences for plant reproduction and overall ecosystem dynamics.

Plant Responses to Soil Moisture Stress

1. **Root Growth and Exploration:** Plants can respond to soil moisture stress by altering their root architecture to explore a larger volume of soil for water. This adaptive response enhances the plant's ability to access available water in the soil. Increased root branching and elongation are common strategies employed by plants to cope with water scarcity.

2. **Osmotic Adjustment:** Some plants can undergo osmotic adjustment, a process where they modify the concentration of solutes in their cells to maintain water uptake under water-stressed conditions. By adjusting the osmotic potential of their cells, plants can sustain turgor pressure and maintain essential physiological processes even in challenging water conditions.

3. **Dormancy and Senescence:** Under severe water stress, plants may enter dormancy or undergo senescence. Dormancy is a state of reduced metabolic activity, allowing the plant to conserve energy and water during periods of extreme stress. Senescence involves the shedding of leaves or older plant tissues, redirecting resources to support vital functions.
4. **Symbiotic Relationships:** Plants in water-stressed environments may form symbiotic relationships with microorganisms to enhance their resilience. For example, nitrogen-fixing bacteria or mycorrhizal fungi can improve nutrient acquisition, supporting plant growth even in nutrient-poor soils. These symbiotic relationships contribute to the overall adaptability of plants to challenging environmental conditions.

Understanding the effects of soil moisture stress on plants and their adaptive responses is crucial for devising effective water management strategies in agriculture and natural ecosystems. The ability of plants to adjust their physiological processes and morphology in response to changing water conditions underscores the resilience of the plant kingdom in the face of environmental challenges.

8. Estimation of Water Requirement by Empirical Formulae

Estimating water requirements is a fundamental aspect of agricultural planning and water resource management. Empirical formulae, derived from observed relationships between climatic parameters and evapotranspiration, provide valuable tools for making these estimations. Three widely used empirical formulae for estimating water requirements are the Thornthwaite equation, Penman's equation, and the Blaney-Criddle equation.

Thornthwaite Equation:

The Thornthwaite equation, developed by climatologist C.W. Thornthwaite, is an empirical method specifically designed for estimating potential evapotranspiration (PET). Potential evapotranspiration represents the evaporative demand of the atmosphere under idealized conditions. The Thornthwaite equation is particularly suitable for regions with a temperate climate, where it has found widespread applicability.

$PET=16\times L\times \text{days in month}\times(10T/I)^{a}/(10T/I)^{b}+1$

In this equation

- PET is the potential evapotranspiration in inches,
- L is the day length in hours,
- T is the average monthly temperature in Fahrenheit,
- I is a heat index calculated as the sum of 12 monthly values of $(T/5)^{1.514}$,
- a and b are coefficients.

The Thornthwaite equation provides an estimate of potential evapotranspiration, a critical parameter for irrigation planning and water resource management. It accounts for the influence of temperature and day length on the atmospheric demand for water, making it a valuable tool for understanding climatic impacts on water availability.

Penman's Equation

Penman's equation is a comprehensive method for estimating actual evapotranspiration, taking into account both energy and aerodynamic components. Developed by Howard Penman, this equation provides a more accurate representation of actual evapotranspiration and is suitable for a wide range of climates.

$$ET = 0.0023 \times (T+17)^2 \times \sqrt{} U \times (Smax - Smin)/(T+17) + 0.36 \times (T+17) \times U$$

Here:

- ET is the evapotranspiration rate in inches per day,
- T is the average daily temperature in Celsius,
- U is the wind speed at 2 meters above the ground in meters per second,
- Smax and Smin are the saturation vapor pressure and actual vapor pressure, respectively.

Penman's equation incorporates temperature, wind speed, and vapor pressure to provide a more accurate estimation of actual evapotranspiration. Its versatility makes it suitable for various climates and applications, from agricultural irrigation planning to hydrological modeling.

Blaney-Criddle Equations

The Blaney-Criddle equations are simpler, widely used empirical methods for estimating monthly and annual crop water requirements. These equations are based on the relationship between potential evapotranspiration, temperature, and day length. While less complex than some other methods, they offer practical utility for quick assessments in different climates.

Monthly Equation

$$ET_m = 0.5 \times (Tmax + Tmin) \times P_m / \text{crop coefficient}$$

In this equation

- ETm is the monthly evapotranspiration in inches,
- Tmax and Tmin are the maximum and minimum monthly temperatures in Fahrenheit,

- Pm is the monthly percentage of day length,
- The crop coefficient accounts for the specific water requirements of the crop.

Annual Equation

$ET_{annual}=\sum^{12}_{m=1} ET_{m}$

The Blaney-Criddle equations provide a simplified approach for estimating crop water requirements. While they may lack the complexity of more sophisticated models, their ease of use and reliance on readily available data make them valuable for quick assessments in diverse agricultural settings.

In summary, empirical formulae such as the Thornthwaite equation, Penman's equation, and the Blaney-Criddle equations play pivotal roles in estimating water requirements for agricultural and water resource management. The choice of formula depends on factors such as data availability, the desired level of accuracy, and the specific climatic conditions of the region under consideration. These formulae aid in optimizing irrigation schedules, ensuring sustainable water use in agriculture, and managing water resources effectively in the face of varying climatic conditions.

Empirical methods have proven their utility in diverse applications, contributing valuable insights into the complex relationship between climatic variables and evapotranspiration. As technological advancements continue, these empirical formulae provide foundational tools for water resource professionals, researchers, and policymakers working towards sustainable water management practices.

9. Irrigation Water Quality about Soil and Plant Water

Irrigation water quality is a pivotal factor that intricately influences soil health and plant growth, significantly impacting agricultural productivity. The complex interplay between the composition of irrigation water, soil structure, and plant water dynamics underscores the importance of understanding these relationships for sustainable agricultural practices. An in-depth exploration of key aspects highlights the significance of irrigation water quality and its profound effects on soil and plant water dynamics.

1. Water Composition and Soil Structure

The composition of irrigation water plays a multifaceted role in shaping soil structure, a fundamental aspect of a healthy and productive soil environment. Water with elevated salt levels, particularly sodium, can trigger soil salinization, a process where salts accumulate in the soil, adversely affecting its physical properties. One of the primary consequences is the dispersion of soil particles, leading to the formation of compacted layers with reduced water infiltration rates.

This compaction hinders root penetration, restricting the ability of plants to access water and nutrients in the deeper soil layers.

Balancing the mineral content of irrigation water is crucial to maintaining favorable soil structure. Waters with excessive salinity can be detrimental, emphasizing the importance of periodic water testing and the implementation of soil management practices to mitigate the adverse effects of saline irrigation water on soil structure.

2. Impact on Soil pH

The pH of irrigation water can significantly influence the pH of the soil, a parameter critical for nutrient availability. Alkaline irrigation water can contribute to an increase in soil pH, potentially leading to nutrient imbalances and reducing the availability of certain essential elements to plants. Conversely, acidic irrigation water may contribute to soil acidification, affecting the solubility of various minerals.

Monitoring and managing the pH of both irrigation water and soil are essential for creating an optimal environment for plant growth. Regular pH testing allows for adjustments in management practices, such as the application of soil amendments, to ensure that nutrient availability aligns with the needs of the crops being cultivated.

3. Nutrient Content in Irrigation Water

The nutrient content of irrigation water directly impacts soil fertility, influencing plant growth and development. High-quality irrigation water, containing essential nutrients such as nitrogen, phosphorus, and potassium, can contribute positively to soil fertility. However, excessive levels of certain nutrients in irrigation water can have adverse effects on both soil and plants.

For instance, elevated levels of chloride or sodium in irrigation water can disrupt the delicate balance of cations in the soil, potentially leading to soil structure deterioration. Boron toxicity from irrigation water can adversely affect plant growth. Thus, a comprehensive understanding of the nutrient content of irrigation water is crucial for informed nutrient management practices, ensuring optimal soil fertility and plant nutrition.

4. Microbial Contamination

Microbial contamination in irrigation water poses risks to both soil and plant health. Waterborne pathogens, including bacteria and fungi, can cause soil-borne diseases, particularly affecting plant roots. Contaminated water can introduce harmful organisms into the soil, disrupting the delicate balance of beneficial microbial communities that play a vital role in nutrient cycling and soil health.

Regular testing and treatment of irrigation water for microbial contaminants are imperative to prevent the spread of diseases and maintain a healthy soil environment. Implementing preventive measures, such as using clean water sources or employing water treatment technologies, is essential to safeguard soil and plant health.

5. Effect on Plant Physiology

Irrigation water quality directly influences plant physiology, impacting processes such as photosynthesis, transpiration, and nutrient uptake. Waterborne contaminants can enter plant tissues through the roots, potentially disrupting essential physiological functions. High salinity in irrigation water can induce osmotic stress in plants, leading to reduced water uptake and overall stunted growth.

Understanding the specific water quality requirements of different crops is essential for optimizing plant performance. It involves selecting appropriate irrigation water sources and implementing strategies to mitigate the physiological stress caused by poor water quality.

6. Long-Term Soil Health

Consistent use of poor-quality irrigation water can have enduring effects on soil health. The gradual accumulation of salts, changes in soil structure, and nutrient imbalances can lead to reduced fertility and productivity over time. Sustainable irrigation practices necessitate periodic assessments of water quality and adjustments to management practices to preserve long-term soil health.

Implementing strategies such as leaching to flush excess salts, using soil amendments to improve structure, and selecting crops that are more tolerant to specific water quality conditions contribute to the maintenance of soil health over the long term.

7. Mitigation Strategies

Effectively managing the impact of irrigation water quality on soil and plants requires the implementation of proactive mitigation strategies. Some of these strategies include:

- **Water Treatment:** Employing water treatment techniques, such as filtration, reverse osmosis, or adding amendments to adjust water chemistry, can improve water quality before irrigation.
- **Leaching:** Implementing leaching practices helps flush excess salts from the root zone, preventing soil salinization. Regular monitoring of soil salinity levels and adjusting leaching practices accordingly are essential for maintaining soil health.

- **Soil Amendments:** Adding organic matter or specific soil amendments can help improve soil structure and nutrient retention. Amendments such as gypsum can also aid in mitigating the adverse effects of sodic irrigation water.
- **Crop Selection:** Choosing crop varieties that are tolerant to specific water quality conditions can help optimize yield in challenging environments. Crop rotation and diversification strategies can further contribute to soil health management.

10. Measurement of Soil Water: Lysimeter, Tensiometer, Moisture Meter, Pressure Plate, and More

Accurate measurement of soil water content is paramount for effective water management in agriculture, environmental science, and various research fields. The intricate relationship between soil water dynamics and plant health underscores the importance of employing diverse instruments and methods for comprehensive and precise soil water measurements.

1. Lysimeter

Advancements in Lysimetry: Lysimeters, traditionally used for measuring water flux through the soil, have evolved with technological advancements. Modern weighing lysimeters incorporate electronic sensors and data loggers for real-time monitoring of water changes in the soil profile. Automated lysimeter systems enable continuous data collection, providing insights into daily and seasonal variations in soil water dynamics. Additionally, lysimeters equipped with environmental sensors can capture additional parameters such as temperature and humidity, enhancing the understanding of the soil-plant-atmosphere continuum.

Applications in Precision Agriculture: Lysimeters play a crucial role in precision agriculture, where precise water management is essential. They aid in optimizing irrigation schedules, understanding the impact of different irrigation practices on water use efficiency, and assessing the leaching potential of nutrients. Furthermore, lysimeters contribute to the development of crop-specific water-use models, guiding farmers in making informed decisions for sustainable crop production.

2. Tensiometer

Advancements in Tensiometry: Tensiometers, measuring soil water tension or suction, have seen advancements in design and materials. Modern tensiometers incorporate improved ceramics for better durability and increased accuracy. Additionally, wireless tensiometer systems have emerged, allowing for remote monitoring of soil water tension. These systems transmit real-time data, enabling quick response to changes in soil moisture conditions. Integration with mobile applications and cloud-based platforms facilitates data visualization and analysis, making tensiometers valuable tools in precision agriculture.

Role in Precision Irrigation: Tensiometers play a pivotal role in precision irrigation by providing direct insights into the soil's water status. Farmers can use this information to tailor irrigation schedules to the specific needs of crops, preventing overwatering and conserving water resources. The ability to monitor soil moisture at different depths enhances precision irrigation strategies, ensuring optimal water availability for different root zones.

3. Moisture Meter

Technological Innovations in Moisture Meters: Moisture meters, leveraging various technologies such as capacitance, TDR, or frequency domain, have undergone significant technological innovations. Portable handheld moisture meters now feature digital displays, user-friendly interfaces, and the capability to store and analyze data. Some advanced models offer wireless connectivity, allowing seamless integration with other monitoring systems.

Applications in Agriculture and Research: Moisture meters find widespread applications in agriculture and research. Farmers use them for on-site assessments of soil moisture levels, aiding in timely irrigation decisions. Researchers leverage moisture meters for spatial and temporal mapping of soil moisture variations, contributing to a deeper understanding of hydrological processes. The versatility of moisture meters makes them essential tools for both large-scale agricultural operations and scientific investigations.

4. Pressure Plate Apparatus

Advancements in Laboratory Techniques: The pressure plate apparatus, a laboratory instrument for measuring soil water retention, has seen advancements in automation and precision. Automated pressure plate apparatus systems allow for high-throughput measurements, reducing the time required for analyzing soil water retention curves. These systems often integrate with computer software for streamlined data collection and analysis, enhancing efficiency in soil physics research.

Importance in Soil Characterization: The pressure plate apparatus remains a fundamental tool for characterizing soil moisture properties. Researchers use it to determine critical soil parameters such as field capacity and wilting point, essential for understanding plant-available water. The detailed insights obtained from pressure plate measurements contribute to improved soil management practices, especially in the context of irrigation and water conservation.

5. Neutron Moisture Meter

Innovations in Neutron Moisture Measurement: Neutron moisture meters, relying on the scattering of neutrons by hydrogen atoms in water, have witnessed innovations in design and safety. Modern neutron probes feature improved shielding materials, reducing exposure risks for operators. Additionally, advancements in electronics and neutron detection technologies have enhanced the precision and reliability of neutron moisture measurements.

Deep Soil Moisture Monitoring: Neutron moisture meters excel in deep soil moisture monitoring, making them valuable for understanding water dynamics beyond the root zone. Researchers use neutron probes to assess water availability in deeper soil layers, providing insights into groundwater recharge, especially in regions where deep-rooted crops or trees are prevalent. The ability to quantify soil moisture at different depths contributes to a more comprehensive understanding of the entire soil profile.

6. Gravimetric Method

Integration with Modern Analysis Techniques: While the gravimetric method remains a traditional technique for soil moisture measurement, it continues to be a cornerstone for calibration and validation of other methods. Modern gravimetric methods leverage advanced analytical techniques such as precision balances and moisture content analyzers, enhancing accuracy and reproducibility. Researchers combine gravimetric measurements with other methods to establish robust soil moisture calibration curves.

Role in Soil Health Assessments: The gravimetric method plays a crucial role in soil health assessments. By providing direct measurements of soil moisture content, it contributes to understanding the moisture-holding capacity of soils. This information is vital for evaluating soil fertility, nutrient availability, and microbial activity. Gravimetric measurements, coupled with soil texture analysis, offer a comprehensive picture of soil water dynamics and support sustainable soil management practices.

7. Time-Domain Reflectometry (TDR)

Advancements in TDR Technology: Time-Domain Reflectometry (TDR), a modern and efficient method for measuring soil moisture, has experienced advancements in sensor design and data processing. TDR probes now feature improved waveguide materials, enhancing durability and longevity in various soil conditions. Advanced TDR systems integrate with software platforms that enable real-time data visualization, analysis, and interpretation.

Applications in Precision Agriculture and Research: TDR technology finds extensive applications in precision agriculture and research. Its ability to provide continuous and non-destructive soil moisture measurements makes it valuable for monitoring dynamic changes in soil water content. TDR probes can be installed at multiple depths, offering a three-dimensional view of soil moisture profiles. This spatial and temporal resolution is crucial for understanding the heterogeneity of soil water distribution and optimizing irrigation practices.

8. Capacitance Sensors

Wireless Connectivity and Precision: Capacitance sensors, measuring soil moisture based on changes in dielectric properties, have embraced wireless connectivity and precision technology. Advanced capacitance sensors can transmit data wirelessly to central control systems, allowing for real-time monitoring across large agricultural fields. High-precision capacitance sensors offer accurate measurements, even in soils with varying salinity levels, making them versatile tools for precision agriculture.

Role in Precision Irrigation Management: Capacitance sensors play a pivotal role in precision irrigation management. By providing continuous and accurate soil moisture data, these sensors enable farmers to optimize irrigation schedules based on the specific water needs of crops. The ability to integrate capacitance sensors with automated irrigation systems enhances water use efficiency, conserving resources and minimizing environmental impact.

9. Infrared Thermometry

Advancements in Remote Sensing: Infrared thermometry, which measures surface temperatures of leaves or soil, has advanced with the integration of remote sensing technologies. Satellite and aerial platforms equipped with infrared sensors offer large-scale and high-resolution thermal imagery. These advancements facilitate remote monitoring of surface temperatures over extensive agricultural landscapes, providing valuable insights into variations in soil moisture conditions.

Applications in Remote Sensing: Infrared thermometry plays a pivotal role in remote sensing applications for agriculture and environmental monitoring. Remote sensing platforms equipped with infrared sensors can detect temperature differentials associated with soil moisture variations. This information aids in generating maps of soil moisture distribution, helping farmers and researchers make informed decisions about irrigation, drought monitoring, and land management.

10. Dielectric Probes

Integration with IoT and Precision Agriculture: Dielectric probes, often used in conjunction with capacitance sensors, have evolved with the integration of the Internet of Things (IoT) in precision agriculture. Modern dielectric probes can be part of sensor networks that communicate seamlessly with central control systems. This integration allows for real-time data collection and analysis, enabling timely decisions for irrigation and soil water management.

Role in Continuous Monitoring: Dielectric probes excel in continuous monitoring of soil moisture. Their ability to provide real-time data at multiple depths contributes to a dynamic understanding of soil water dynamics. Farmers and researchers can use this information to adapt irrigation strategies based on changing environmental conditions, ensuring optimal water availability for crops.

10

Rainwater Harvesting, Agricultural Drainage and Water Conservation

[1]Niranjan B.N., [2]Kh. Chandrakumar Singh, [3]Savita Jangde [4]Wajid Hasan and [5]Er. Jeetendra Kumar

[1]Dakshina Kannada Milk Union (KMF), Mangalore, Karnataka
[2]VCSGU University of Horticulture and Forestry, Bharsar, Uttarakhand
[3]Department of Plant Physiology, Institute of Agricultural Sciences B.H.U. Varanasi, U.P
[4]KVK Jehanabad, Bihar Agricultural University Sabour, Bihar
[5]Krishi Vigyan Kendra, Jehanabad, Bihar Agricultural University Bihar

1. Monsoon: Types and Behavior in India

The monsoon is a dynamic and vital climatic phenomenon in India, playing a pivotal role in shaping the country's weather patterns, agricultural practices, and overall water resource management. This seasonal shift in wind patterns brings about distinct wet and dry periods, impacting various regions differently. Understanding the types and behavior of the monsoon is essential for effective planning and adaptation to this critical aspect of India's climate (fig 1).

Fig. 1: Monsoon in India

Types of Monsoon in India

1. Southwest Monsoon (June to September)

The southwest monsoon is the lifeline of India's agriculture, bringing the much-needed rainfall that supports a majority of the country's crops. The onset of the southwest monsoon is a highly anticipated event, typically occurring around the first week of June in the southern state of Kerala. From there, it progresses northward, covering the entire subcontinent.

This monsoon season, which lasts from June to September, is characterized by the influx of moist winds from the Indian Ocean. These winds carry heavy rainfall, especially to the western coast and northeastern regions. The southwest monsoon's influence is so profound that it contributes to approximately 70% of India's annual precipitation.

The agricultural significance of the southwest monsoon cannot be overstated. Crops like rice, sugarcane, and cotton are heavily dependent on the timely and adequate rainfall brought by this monsoon. The season also plays a crucial role in replenishing water reservoirs, ensuring water availability for the dry months ahead.

2. Northeast Monsoon (October to December)

Following the southwest monsoon, the northeast monsoon, also known as the retreating monsoon, occurs from October to December. This monsoon is characterized by winds moving from the northeast towards the Bay of Bengal. While it doesn't bring as much rainfall as its counterpart, it has particular importance for the southeastern coastal regions, including Tamil Nadu, Andhra Pradesh, and parts of Karnataka.

The northeast monsoon is associated with the retreating southwest monsoon, and its onset signifies the end of the rainy season in most parts of India. It brings some rainfall to the southeastern coastal areas, contributing to agricultural activities in these regions. However, its overall impact is not as significant as the southwest monsoon, and its duration is shorter.

Behavior of the Monsoon in India

1. Onset and Progression

The onset of the monsoon, particularly the southwest monsoon, is a critical event for India. The India Meteorological Department (IMD) closely monitors and predicts the onset, usually declaring it around June 1st. The onset begins in Kerala and gradually progresses northward, covering the entire country by early July.

The progression of the monsoon is closely tracked to anticipate the timing and distribution of rainfall. The IMD issues regular updates on the monsoon's advance, helping farmers plan their agricultural activities. The timing of the monsoon's onset and progression has a direct impact on cropping patterns and water management strategies.

2. Rainfall Distribution

The distribution of rainfall during the monsoon season is not uniform across India. Some regions experience heavy rainfall, while others may face water scarcity. The monsoon trough, an elongated low-pressure area, plays a crucial role in determining rainfall distribution. It moves northward and influences the movement of monsoon winds, impacting the areas that receive significant rainfall (fig 2).

Fig. 2: Rainfall Disturbances causing unexpected precipitation

Regions like the western coast and northeastern states receive abundant rainfall, contributing to their lush green landscapes. In contrast, parts of the northwest and the Deccan plateau may experience lower rainfall, leading to concerns about drought and water scarcity.

3. Withdrawal

The withdrawal of the monsoon marks the conclusion of the rainy season and the transition to drier conditions. The retreat usually begins in northwest India around September, and the IMD monitors this process closely. The withdrawal of the southwest monsoon is followed by the onset of the northeast monsoon in the southeastern coastal regions.

The withdrawal has implications for agricultural practices and water management. Farmers need to adjust their cropping schedules based on the changing weather conditions, and water reservoirs must be managed efficiently to meet the demands of the dry season.

4. Impact on Agriculture

The monsoon profoundly influences agricultural practices in India. The timing and adequacy of rainfall during the monsoon season directly impact crop yields and overall agricultural productivity.

Kharif crops, which include rice, millets, and pulses, are sown during the monsoon season. These crops are highly dependent on the southwest monsoon for their growth and development. Adequate rainfall during this period ensures a good harvest and contributes to food security.

Rabi crops, such as wheat and barley, depend on post-monsoon rainfall. The moisture retained in the soil from the monsoon season facilitates the cultivation of these crops during the drier months.

However, the variability in monsoon rainfall poses challenges for agriculture. Deficient or excess rainfall can lead to droughts or floods, adversely affecting crop yields. Farmers and policymakers need to implement adaptive strategies to mitigate the impact of unpredictable monsoon patterns on agriculture.

5. Monsoon and Climate Change

The monsoon in India is not immune to the effects of climate change. Changes in global climate patterns, including rising temperatures and alterations in sea surface temperatures, can influence the behavior of the monsoon.

Scientists and meteorologists are studying these changes to understand their implications for India's monsoon. There are concerns about shifts in rainfall patterns, increased frequency of extreme weather events, and potential impacts on water resources.

Adaptation strategies are crucial to mitigate the impact of climate change on the monsoon. These may include improved water management practices, crop diversification, and the development of resilient agricultural systems.

6. Adaptation Strategies

Adapting to the dynamic behavior of the monsoon is essential for sustainable development in India. Several adaptation strategies can help mitigate the impact of monsoon variability:

- **Water Harvesting and Conservation:** Implementing rainwater harvesting systems can help capture and store rainwater during the monsoon for later use during dry periods. Conservation measures, such as efficient irrigation practices, also play a vital role in water management.
- **Drought-Resistant Crops:** Research and promotion of drought-resistant crop varieties can enhance agricultural resilience to erratic monsoon patterns. These crops can withstand periods of water scarcity and contribute to food security.

- **Early Warning Systems:** Developing and enhancing early warning systems for extreme weather events, such as floods or prolonged droughts, can help communities prepare and respond effectively. Timely information is crucial for farmers to make informed decisions.
- **Infrastructure Development:** Investing in resilient infrastructure, including water storage facilities, irrigation systems, and flood control measures, can minimize the impact of extreme monsoon events on communities and agriculture.
- **Climate-Smart Agriculture:** Promoting climate-smart agricultural practices, such as precision farming and organic farming, can enhance the resilience of the agriculture sector. These practices consider climate variability and aim for sustainable production.

7. Monsoon and Society

The monsoon is deeply embedded in the cultural and social fabric of Indian society. Festivals like Teej and Raksha Bandhan are celebrated during the monsoon season, reflecting the significance of rain in people's lives.

At the same time, the unpredictability of the monsoon poses challenges for vulnerable communities. Farmers, in particular, are exposed to the risks associated with fluctuating rainfall patterns. Droughts and floods can have severe consequences for livelihoods and food security.

Government initiatives, community-based adaptation programs, and awareness campaigns are crucial for building resilience in vulnerable communities. The involvement of local communities in water management and agricultural practices is essential for sustainable development.

8. Future Trends and Innovations

As India faces the challenges of a changing climate, ongoing research and innovations are essential for understanding and adapting to future monsoon trends. Some key trends and innovations include:

- **Climate Modeling:** Advancements in climate modeling can provide more accurate predictions of monsoon patterns. Improved forecasting can help communities and policymakers prepare for and respond to changing weather conditions.
- **Remote Sensing Technologies:** Satellite-based remote sensing technologies contribute to monitoring and assessing vegetation, soil moisture, and rainfall patterns. These tools enhance our understanding of the monsoon's behavior and support informed decision-making.

- **Hydrological Modeling:** Modeling the behavior of rivers, reservoirs, and groundwater systems can improve water resource management. Predictive models help plan for water availability during and after the monsoon season.
- **Smart Agriculture Practices:** The integration of technology in agriculture, such as precision farming, sensor-based irrigation, and data analytics, can enhance productivity while optimizing water use. Smart agriculture practices can contribute to sustainable food production in the face of monsoon variability.
- **Community-Driven Initiatives:** Empowering local communities to take ownership of water management and agricultural practices fosters resilience. Participatory approaches, combined with traditional knowledge, can contribute to sustainable development.

2. Rainfall in India: Characteristics, Distribution, Onset, and Withdrawal of Effective Rains

Rainfall is a dynamic and multifaceted climatic element that plays a pivotal role in shaping India's diverse ecosystems, agricultural practices, and overall socio-economic landscape. Examining the intricate characteristics and distribution of rainfall, as well as understanding the onset and withdrawal of effective rains, is crucial for sustainable water resource management and societal well-being.

Characteristics of Rainfall in India

1. Seasonality

India experiences a pronounced seasonal cycle of wet and dry periods, primarily governed by the monsoon winds. The southwest monsoon, a key player in this cycle, brings the bulk of the annual rainfall during its onset from June to September. The seasonality of rainfall significantly influences agricultural planning, as crops are strategically aligned with the monsoon season.

The dry periods, occurring outside the monsoon season, prompt the need for effective water management strategies. Water storage, conservation, and distribution systems become essential components of mitigating the impact of dry spells on agriculture and water availability for various purposes.

2. Intensity

Rainfall intensity varies widely across India, contributing to the country's diverse landscapes and climates. Coastal regions, particularly the western coast and northeastern states, often experience heavy and intense rainfall. This intensity is a result of the orographic lifting effect, where moist air is forced to ascend over mountain ranges, leading to significant precipitation.

In contrast, interior and rain-shadow regions may witness lighter and sporadic rainfall. Understanding these variations in intensity is crucial for planning infrastructure, agriculture, and water management systems tailored to the specific needs of each region.

3. Spatial Distribution

The spatial distribution of rainfall in India is intricately linked to its diverse topography and geographical features. The western coast, facing the Arabian Sea, and the northeastern states receive substantial rainfall due to their proximity to moisture sources and orographic effects. Coastal regions are marked by lush greenery and abundant water resources.

On the other hand, regions such as the Deccan Plateau and northwestern areas, including parts of Rajasthan, experience moderate to low rainfall. The influence of mountain ranges, like the Western Ghats and the Aravalli Range, contributes to spatial variations in rainfall distribution. This spatial diversity underscores the need for region-specific approaches in water resource management and agriculture.

4. Inter-annual Variability

India's rainfall patterns exhibit inter-annual variability influenced by global climatic phenomena. Events like El Niño, characterized by warmer-than-average sea surface temperatures in the Pacific Ocean, can lead to reduced rainfall and drought conditions. Conversely, La Niña, marked by cooler-than-average sea surface temperatures, may enhance rainfall and contribute to floods.

Understanding and predicting these inter-annual variations are essential for implementing adaptive strategies. Early warning systems and climate modeling play crucial roles in preparing for and mitigating the impacts of such variability on agriculture, water resources, and disaster management.

Distribution of Rainfall in India

1. Western Coast

The western coast of India, including states like Kerala, Karnataka, Goa, and Maharashtra, receives abundant rainfall during the monsoon season. This region is influenced by the Western Ghats, which acts as a barrier to the moist southwest monsoon winds. The orographic lifting of air masses results in heavy rainfall, supporting rich biodiversity, agriculture, and hydropower generation.

The spatial distribution of rainfall along the western coast highlights the need for sustainable land use practices, considering the ecological sensitivity of the Western Ghats and the importance of water resources for both urban and rural communities.

2. Northeastern States

The northeastern states of India, encompassing Assam, Meghalaya, and Arunachal Pradesh, experience some of the highest rainfall in the country. The Eastern Himalayas contribute to orographic lifting, enhancing precipitation. Additionally, the Bay of Bengal provides moisture, making this region one of the wettest in the world.

The distribution of rainfall in the northeast is vital for its unique ecosystems, including diverse flora and fauna. Conservation efforts and sustainable development practices are crucial to maintaining the delicate balance between the region's rich biodiversity and the needs of its human population.

3. Gangetic Plain

The Gangetic plain, comprising states like Uttar Pradesh and Bihar, receives moderate to high rainfall. The flat topography facilitates the even distribution of rainfall across the plains. This region is agriculturally significant, contributing substantially to the country's food production.

Balancing agricultural practices with water conservation measures is essential for sustainable development in the Gangetic plain. Efficient irrigation systems, crop diversification, and soil conservation efforts contribute to optimizing water use in this fertile region.

4. Deccan Plateau

The Deccan Plateau, covering parts of Maharashtra, Madhya Pradesh, and Karnataka, experiences moderate rainfall. The distribution of rainfall is influenced by the Western Ghats, which leads to leeward areas facing drier conditions. The plateau's topography results in variations in rainfall across different zones.

In this region, water management becomes crucial, necessitating the development of efficient irrigation systems and sustainable agricultural practices. Recognizing the variability in rainfall distribution within the Deccan Plateau is vital for implementing targeted interventions for water resource management.

5. Northwestern Regions

The northwestern regions of India, including Rajasthan and parts of Gujarat, are characterized by arid to semi-arid conditions. These areas receive lower rainfall, and the Thar Desert in Rajasthan experiences scanty precipitation. The Aravalli Range acts as a barrier, limiting the moisture-carrying winds from reaching these regions.

The distribution of rainfall in the northwest presents challenges for agriculture and water availability. Sustainable water management practices, including rainwater harvesting and watershed development, are essential for mitigating the impacts of arid conditions.

Onset and Withdrawal of Effective Rains

1. Onset of Monsoon

The onset of the monsoon, particularly the southwest monsoon, is a pivotal event for India. The India Meteorological Department (IMD) closely monitors and predicts the onset, typically declared around June 1st. The monsoon onset begins in the southern state of Kerala and progresses northward, covering the entire country by early July.

The onset of effective rains is marked by a shift in wind patterns, with moist winds from the southwest bringing the first significant rainfall. This event triggers the planting of kharif crops, such as rice and millets, across the country. The timely onset of the monsoon is essential for agricultural planning and water resource management.

2. Withdrawal of Monsoon

The withdrawal of the monsoon is as crucial as its onset, influencing the transition to drier conditions. The retreat usually begins in northwest India around September. The withdrawal process is closely monitored by the IMD, and its progression has implications for agriculture and water reservoir management.

The withdrawal is followed by the onset of the northeast monsoon, affecting the southeastern coastal regions. The conclusion of the monsoon season is marked by a reduction in rainfall, indicating the shift towards the post-monsoon and winter seasons.

3. Effective Rains

Effective rains refer to rainfall that contributes significantly to soil moisture and supports agricultural activities. The timing, duration, and distribution of effective rains are critical for crop growth and overall water availability. Adequate effective rains ensure a successful crop season, while deficiencies can lead to drought conditions and impact food production.

Efficient water storage and distribution systems, combined with meteorological forecasting technologies, play a crucial role in optimizing the benefits of effective rains. The agricultural calendar is intricately linked to the availability of effective rains, highlighting the importance of synchronized planning and adaptive strategies.

4. Impact on Agriculture

The onset and withdrawal of effective rains have direct implications for agriculture, particularly the cultivation of kharif and rabi crops. Kharif crops, which include rice, millets, and pulses, are sown during the monsoon season and rely on effective rains for germination and growth.

The withdrawal of the monsoon signals the beginning of the post-monsoon season, crucial for planting rabi crops like wheat and barley. Timely effective rains during this period are essential for the development of these crops, contributing to overall food production and agricultural sustainability.

Agricultural practices must align with the variability in rainfall patterns, emphasizing the need for adaptive measures. Crop diversification, water-efficient irrigation techniques, and resilient seed varieties are key components of ensuring agricultural productivity in the face of changing rainfall dynamics.

5. Challenges and Adaptation

Challenges associated with the onset and withdrawal of effective rains include the variability in monsoon patterns, unpredictable shifts in weather conditions, and the impact of climate change. Farmers and policymakers need to adapt to these challenges through:

- **Water Management:** Implementing efficient water management practices, including rainwater harvesting and irrigation, helps mitigate the impact of irregular rainfall patterns.
- **Crop Diversification:** Diversifying crops based on the suitability of rainfall patterns can enhance resilience. Choosing drought-resistant varieties and adjusting planting schedules contribute to adaptive agriculture.
- **Weather Forecasting:** Improved weather forecasting technologies and early warning systems enable farmers to make informed decisions regarding planting, harvesting, and water resource management.
- **Community Engagement:** Involving local communities in decision-making processes related to agriculture and water management fosters resilience. Traditional knowledge and community-based adaptation strategies play a crucial role in addressing challenges.

Adaptive strategies must also account for the interconnected nature of socio-economic and environmental factors. Collaborative efforts between communities, researchers, and policymakers are essential for developing and implementing effective adaptation measures that are locally relevant and sustainable.

3. Dry Spells and Wet Spells: Impacts on Water Loss from Soil

Dry spells and wet spells are crucial components of a region's climate that significantly influence the water dynamics within the soil. The variability in precipitation patterns during these spells plays a pivotal role in shaping ecosystems, agricultural practices, and water resource management. A comprehensive understanding of the impacts of dry spells, identification of critical periods, and assessment of water loss from the soil during such intervals are imperative for sustainable water use and planning.

Dry Spells

Definition and Characteristics

Dry spells refer to periods of below-average precipitation, resulting in a deficiency in soil moisture. These intervals can range from a few days to several weeks or months, contingent upon the region and its climatic patterns. The intensity of dry spells can vary, with some causing mild water stress and others leading to severe drought conditions.

Impacts on Soil Moisture

1. **Reduced Soil Moisture Content:** The most immediate consequence of dry spells is a reduction in soil moisture content. Limited rainfall leads to desiccation of the soil, affecting its ability to support plant growth and other biological processes.
2. **Increased Evaporation and Transpiration:** During dry spells, the lack of moisture in the soil amplifies evaporation and transpiration rates. This further depletes soil moisture, creating a feedback loop that exacerbates water stress for vegetation.
3. **Impact on Agriculture:** Dry spells pose significant challenges for agriculture. Crop yields can be adversely affected, resulting in decreased productivity and potential economic losses for farmers. Adequate irrigation becomes crucial during prolonged dry spells to mitigate these impacts.
4. **Groundwater Depletion:** Extended dry spells can contribute to the depletion of groundwater reserves. When surface water sources are insufficient, reliance on groundwater increases, putting additional pressure on aquifers.
5. **Ecosystem Stress:** Natural ecosystems, including forests and grasslands, are highly sensitive to dry spells. Reduced soil moisture can lead to vegetation stress, affecting biodiversity and ecosystem health.

Wet Spells

Definition and Characteristics

Wet spells, in contrast, denote periods of above-average precipitation. These can manifest as heavy rainfall or prolonged periods of consistent rain. Wet spells are vital for replenishing soil moisture, supporting plant growth, and maintaining water balance in ecosystems.

Impacts on Soil Moisture

1. **Soil Rehydration:** Wet spells contribute to the rehydration of the soil. Adequate rainfall allows the soil to regain moisture, promoting optimal conditions for plant growth and microbial activity.

2. **Runoff and Erosion:** Intense rainfall during wet spells can lead to runoff and soil erosion. This can impact soil structure and nutrient content, affecting both agricultural and natural ecosystems.

3. **Groundwater Recharge:** Excess rainfall during wet spells facilitates groundwater recharge. Aquifers are replenished, contributing to sustained water availability for wells and springs.

4. **Positive Impact on Agriculture:** Wet spells are generally beneficial for agriculture, supporting crop growth and ensuring adequate water supply. However, excessive rainfall can also lead to waterlogging, posing challenges for certain crops.

5. **Ecological Benefits:** Natural ecosystems thrive during wet spells, with increased vegetation growth, improved habitat conditions, and enhanced biodiversity. Wet spells contribute to the overall health and resilience of ecosystems.

Critical Dry Spells

Definition and Significance

Critical dry spells refer to periods of prolonged and severe moisture deficiency in the soil, surpassing the typical duration and intensity of regular dry spells. These periods have heightened significance due to their potential to cause substantial ecological, agricultural, and societal impacts.

Impacts on Soil Moisture

1. **Severe Agricultural Impact:** Critical dry spells can have severe consequences for agriculture, leading to crop failure, reduced yields, and economic losses for farmers. Water scarcity during critical periods can be challenging to address through irrigation alone.

2. **Ecological Stress:** Natural ecosystems face heightened stress during critical dry spells. This stress can result in the decline of vegetation, increased susceptibility to wildfires, and disruptions in the natural balance of ecosystems.

3. **Water Supply Challenges:** Critical dry spells can strain water supply systems, affecting both surface water sources and groundwater availability. This can lead to water shortages for domestic, industrial, and agricultural use.

4. **Increased Vulnerability to Drought:** Critical dry spells are often precursors to drought conditions. The cumulative impact of prolonged moisture deficiency can escalate into drought, with widespread implications for water resources, food security, and livelihoods.

Water Loss from the Soil

Processes and Factors

1. **Evaporation:** Water loss from the soil occurs through the process of evaporation, where water at the soil surface is converted into vapor and released into the atmosphere. Dry spells intensify evaporation, leading to increased water loss.

2. **Transpiration:** Transpiration is the release of water vapor from plant leaves. During dry spells, plants may undergo water stress, reducing transpiration. However, in the absence of sufficient soil moisture, this process can contribute to further water loss.

3. **Infiltration and Runoff:** Inadequate soil moisture during dry spells reduces the infiltration capacity—the process by which water penetrates the soil. Instead, water may run off the surface, contributing to erosion and preventing effective moisture retention.

4. **Soil Moisture Depletion:** The overall soil moisture content decreases during dry spells due to the combined effects of evaporation, transpiration, and inadequate replenishment through precipitation. This depletion can negatively impact vegetation and ecosystems.

Mitigation Strategies

1. **Water Conservation Practices:** Implementing water conservation practices, such as rainwater harvesting, can enhance soil moisture retention during both wet and dry spells.

2. **Drought-Resistant Crops:** Cultivating drought-resistant crop varieties helps mitigate the impact of dry spells on agriculture. These crops are adapted to withstand moisture deficiency and maintain productivity.

3. **Irrigation Management:** Efficient irrigation practices, including precision irrigation and soil moisture monitoring, can optimize water use and minimize losses during dry spells.

4. **Land Use Planning:** Sustainable land use planning, including afforestation and soil conservation measures, contributes to maintaining soil moisture and reducing the risk of water loss.

5. **Early Warning Systems:** Establishing early warning systems for drought and dry spell conditions allows for timely intervention, helping communities, farmers, and policymakers prepare for and mitigate potential impacts.

6. **Community Engagement:** Involving local communities in water management and conservation efforts enhances the effectiveness of mitigation strategies. Traditional knowledge and community-based approaches contribute to sustainable practices.

4. Measurement and Factors Influencing the Hydrological Cycle

The hydrological cycle, often referred to as the water cycle, is a continuous and intricate process that governs the movement of water within the Earth's atmosphere, oceans, and land. Understanding the dynamics of this cycle is essential for effective water resource management, climate prediction, and environmental conservation. This comprehensive overview explores the measurement techniques employed for different components of the hydrological cycle and delves into the diverse factors that influence this vital Earth system.

Measurement of Hydrological Cycle Components

1. Precipitation

- **Measurement Techniques:** Precipitation, in its various forms such as rain, snow, sleet, and hail, is a crucial component of the hydrological cycle. Measurement techniques include rain gauges for liquid precipitation and snow gauges for snowfall. Advanced technologies like weather radars provide a broader perspective, offering insights into precipitation patterns over larger areas.

- **Factors Influencing Measurement:** The accuracy of precipitation measurements is influenced by several factors. Wind, evaporation, and the placement of gauges can impact results. Regular calibration, maintenance, and the use of complementary measurement methods enhance the reliability of precipitation data.

2. Evaporation

- **Measurement Techniques:** Evaporation, the process by which water changes from liquid to vapor, is challenging to measure directly. Indirect methods include pan evaporation measurements using Class A evaporation pans, lysimeters that measure water loss from soil, and remote sensing technologies.

- **Factors Influencing Measurement:** The rate of evaporation is influenced by factors such as wind speed, humidity, temperature, and the type of surface (water or land). Proper siting of instruments and calibration are critical for accurate and representative evaporation measurements.

3. Transpiration

- **Measurement Techniques:** Transpiration, the release of water vapor from plant leaves, is often measured indirectly. Techniques include sap flow meters, porometers that measure gas exchange in leaves, and lysimeters to estimate the water loss from plants.

- **Factors Influencing Measurement:** Transpiration measurements are influenced by factors such as plant species, physiological condition, and environmental factors like light intensity and atmospheric humidity. Careful consideration of these factors is essential for accurate transpiration assessments.

4. Runoff

- **Measurement Techniques:** Runoff, the flow of water over the land surface into rivers and streams, is measured using stream gauging stations. These stations record the flow of water, providing valuable information on the quantity and timing of runoff.
- **Factors Influencing Measurement:** Changes in land use, urbanization, and alterations in vegetation cover can impact runoff patterns. Proper calibration of gauging stations and consideration of local factors enhance the accuracy of runoff measurements.

5. Groundwater Flow

- **Measurement Techniques:** Groundwater flow is measured using wells equipped with piezometers and other instruments. These wells help monitor changes in groundwater levels and assess the direction and velocity of groundwater movement.
- **Factors Influencing Measurement:** The permeability of the aquifer, geological formations, and human activities such as pumping wells can influence groundwater flow. Regular monitoring and assessment of these factors are essential for accurate groundwater flow measurements.

Factors Influencing the Hydrological Cycle

1. Solar Energy

- **Influence on Hydrological Cycle:** Solar energy is the driving force behind the hydrological cycle. The uneven heating of the Earth's surface by solar radiation results in the evaporation of water from oceans, lakes, and rivers, initiating the cycle.

2. Topography

- **Influence on Hydrological Cycle:** The topography of a region plays a significant role in influencing precipitation patterns. Mountains, for example, can lead to orographic precipitation, where moist air is lifted, cooled, and condensed, resulting in increased rainfall on windward slopes and rain shadows on the leeward side.

3. Land Use and Land Cover Changes

- **Influence on Hydrological Cycle:** Human-induced changes in land use and land cover, such as deforestation, urbanization, and agriculture, can alter the natural flow of water. Changes in land cover impact evaporation rates, runoff patterns, and the overall water balance of a region.

4. Climate Change

- **Influence on Hydrological Cycle:** Climate change, driven by factors like increased greenhouse gas emissions, alters temperature and precipitation patterns. Changes in climate can disrupt the balance of the hydrological cycle, leading to shifts in precipitation and evaporation patterns.

5. Human Activities

- **Influence on Hydrological Cycle:** Human activities, including dam construction, water extraction for agriculture and industry, and urbanization, can modify the natural flow of rivers and impact groundwater recharge. Sustainable water management practices are crucial to mitigate these impacts.

6. Vegetation and Soil

- **Influence on Hydrological Cycle:** Vegetation influences the hydrological cycle through processes like transpiration. Changes in vegetation cover and soil properties can impact local hydrological processes, including evaporation rates and the retention of water.

7. Atmospheric Circulation

- **Influence on Hydrological Cycle:** Atmospheric circulation, influenced by the rotation of the Earth and solar heating, plays a vital role in the movement of air masses and the distribution of precipitation. High and low-pressure systems influence weather patterns.

8. Ocean Currents

- **Influence on Hydrological Cycle:** Ocean currents redistribute heat globally, influencing regional climate patterns. The exchange of heat between the oceans and the atmosphere contributes to the moisture available for precipitation.

9. Ice and Snow

- **Influence on the Hydrological Cycle:** Ice and snow, particularly in polar regions and mountainous areas, play a crucial role in the hydrological cycle. Changes in snow cover and glacier melt influence downstream runoff and water availability.

10. Urban Heat Island Effect

- **Influence on Hydrological Cycle:** Urban areas, with impervious surfaces and altered heat absorption and release characteristics, contribute to the urban heat island effect. This phenomenon can influence local precipitation patterns and runoff in urbanized regions.

5. Rainwater Harvesting: Methods, Classes, Benefits, and Approaches

Rainwater harvesting is a sustainable and increasingly vital practice that involves the collection and storage of rainwater for various purposes. This method harnesses the abundant and free resource of rainwater, providing numerous environmental, economic, and social benefits. In this comprehensive overview, we will explore various rainwater harvesting methods, classify them based on their applications, delve into the benefits, and discuss approaches to implement this eco-friendly practice.

Rainwater Harvesting Methods (Table 1)

Table 1: Methods of Rainwater Harvesting

S.No	Rainwater Harvesting Method	Description
1	Surface Runoff Harvesting	Capturing rainwater as it flows over the ground surface and directing it to storage systems, such as ponds, tanks, or infiltration basins. Common in urban areas with paved surfaces.
2	Rooftop Rainwater Harvesting	Collecting rainwater that falls on building rooftops. Gutters and downspouts direct the water to storage tanks or cisterns for later use. A common method for domestic water supply.
3	Groundwater Recharge	Allowing rainwater to percolate into the ground, recharging underground aquifers. Methods include recharge pits, trenches, and injection wells to replenish depleted groundwater resources.
4	Check Dam Rainwater Harvesting	Constructing check dams across small streams or rivers to slow down water flow, encouraging sediment deposition and facilitating groundwater recharge. Often used in hilly terrains.
5	Contour Bunding	Building low embankments (bunds) along the contour lines of the land to capture and retain rainwater. This method helps reduce soil erosion and promotes water infiltration.
6	Percolation Pits	Excavating pits in the ground to collect and allow rainwater to percolate into the soil. Percolation pits enhance groundwater recharge and are commonly used in urban and rural areas.
7	Sub-Surface Dykes	Constructing underground dykes or barriers to slow down and capture rainwater, promoting infiltration and preventing soil erosion. Suitable for areas with gentle slopes.

8	Check Basin Irrigation	Creating small basins around plants or trees to capture and retain rainwater for localized irrigation. Check basin irrigation is common in agriculture, especially for fruit orchards.
9	Fog Harvesting	Collecting water from fog using mesh or nets that capture water droplets from the air. This method is particularly useful in arid and coastal areas where fog is a regular occurrence.
10	Ferrocement Tanks	Building water storage tanks using ferrocement technology, which involves a combination of cement mortar and metal mesh. Ferrocement tanks are durable and suitable for rainwater storage.
11	Swales and Contour Trenches	Creating shallow, broad channels (swales) or trenches along contour lines to capture and slow down rainwater, promoting infiltration and reducing surface runoff.
12	Permeable Pavements	Using permeable surfaces for roads, driveways, or walkways that allow rainwater to pass through and be absorbed into the ground, reducing surface runoff and enhancing groundwater recharge.

1. **Surface Runoff Harvesting:** Surface runoff harvesting is a method that captures rainwater as it flows over the ground surface. This approach is particularly effective in areas where traditional water sources may be scarce or unreliable. By directing surface runoff into storage tanks, ponds, or other reservoirs, communities can ensure a more reliable and consistent water supply. Implementation of this method often involves the installation of gutters and downspouts on rooftops to guide rainwater efficiently into storage systems. In urban areas, where impervious surfaces can exacerbate runoff, surface runoff harvesting offers a pragmatic solution to mitigate water scarcity.

2. **Rooftop Rainwater Harvesting:** Rooftop rainwater harvesting is a widely adopted method, especially in urban settings where space constraints necessitate innovative solutions. In this approach, rainwater falling on rooftops is collected and conveyed through gutters and downspouts into storage tanks. The simplicity and efficiency of this method make it accessible for both residential and commercial buildings. Urban areas with limited open space can significantly benefit from rooftop rainwater harvesting as it reduces dependence on centralized water supply systems, eases the burden on stormwater drainage, and provides an on-site water source for various non-potable uses.

3. **Green Roof Rainwater Harvesting:** Green roofs, covered with vegetation, not only contribute to environmental sustainability but also serve as effective rainwater harvesting systems. These roofs absorb and retain rainwater, reducing runoff and providing insulation benefits. Green roof systems

consist of layers that allow for the gradual release of rainwater, preventing sudden and excessive runoff. Beyond rainwater harvesting, green roofs contribute to biodiversity, mitigate urban heat island effects, and enhance the aesthetic value of urban spaces. This method exemplifies a harmonious integration of ecological principles into urban infrastructure.

4. **In-Ground Rainwater Harvesting:** In-ground rainwater harvesting involves directing rainwater underground into permeable layers or storage systems. This method is particularly suitable for areas with high water tables, where traditional above-ground storage may be challenging. Percolation pits, trenches, or underground reservoirs are utilized to capture and store rainwater, preventing surface runoff and promoting groundwater recharge. By adopting in-ground rainwater harvesting, communities can tap into a decentralized and sustainable water source that complements existing water supply systems.

5. **Fog Net Systems:** Fog net systems represent an innovative approach to rainwater harvesting, especially in arid and fog-prone coastal areas. These systems capture moisture from fog, converting it into liquid water for collection. Mesh nets are strategically placed to capture tiny water droplets from fog, channeling them into storage containers. While not as widely applicable as other methods, fog net systems showcase the adaptability and creativity involved in rainwater harvesting. In regions where traditional rainfall is scarce, tapping into alternative water sources like fog can make a substantial difference in water availability.

Classes of Rainwater Harvesting

1. **Domestic Rainwater Harvesting:** Domestic rainwater harvesting involves collecting rainwater from rooftops for household purposes such as drinking, cooking, washing, and gardening. This class of rainwater harvesting empowers individuals and households to become more self-reliant in terms of water supply. In areas facing water scarcity or unreliable municipal water sources, domestic rainwater harvesting can significantly contribute to water security. The water collected is typically filtered and stored in tanks, ensuring a clean and accessible water source for various daily activities.

2. **Agricultural Rainwater Harvesting:** Agricultural rainwater harvesting is a class that specifically targets the needs of the agricultural sector. Rainwater collected through various methods, such as surface runoff harvesting and in-ground systems, is utilized for crop irrigation. This application addresses the challenges posed by unpredictable rainfall patterns and water scarcity in many agricultural regions. By capturing and storing rainwater during periods of abundance, farmers can mitigate the impact of droughts, reduce dependence on conventional irrigation sources, and enhance the overall efficiency of water use in agriculture.

3. **Urban Rainwater Harvesting:** Urban rainwater harvesting is tailored to the specific needs of urban environments, where large populations and increased impervious surfaces contribute to water stress. This class involves collecting rainwater for non-potable uses such as landscaping, industrial processes, and cooling systems. Urban rainwater harvesting reduces the strain on municipal water supplies, especially during peak demand periods. By incorporating rainwater into urban water management strategies, cities can alleviate pressure on centralized water systems, reduce costs, and enhance overall resilience to water-related challenges.

4. **Industrial Rainwater Harvesting:** Industrial rainwater harvesting is designed to meet the water needs of industrial processes and facilities. Industries can capture and store rainwater for various purposes, including cooling systems, equipment maintenance, and non-drinking uses. This class of rainwater harvesting aligns with sustainable business practices, reducing the environmental impact of industrial water consumption. By integrating rainwater harvesting into industrial operations, businesses can contribute to water conservation, lower operational costs, and enhance their corporate sustainability profiles.

Benefits of Rainwater Harvesting

1. **Water Conservation:** One of the primary benefits of rainwater harvesting is water conservation. By capturing rainwater, especially during periods of abundant rainfall, communities can reduce their dependence on traditional water sources. This conservation-oriented approach contributes to the sustainable management of water resources, ensuring that water is available for current and future generations.

2. **Mitigation of Flooding:** Rainwater harvesting helps mitigate the impact of flooding in urban and peri-urban areas. By capturing and storing rainwater, especially from impervious surfaces like rooftops, the volume of water entering stormwater drainage systems is reduced. This proactive flood mitigation approach enhances urban resilience to extreme weather events and minimizes the risks associated with flooding.

3. **Groundwater Recharge:** In regions where groundwater is a significant source of drinking water, rainwater harvesting plays a vital role in groundwater recharge. Methods such as in-ground rainwater harvesting facilitate the percolation of rainwater into the ground, replenishing aquifers and maintaining sustainable groundwater levels. This benefits both rural and urban communities that rely on groundwater for their water supply.

4. **Cost Savings:** Utilizing rainwater for various purposes can result in significant cost savings for individuals, communities, and businesses. The

expenses associated with treating and distributing water from centralized systems are reduced when rainwater is harnessed locally. Additionally, industries can lower their water procurement costs by incorporating rainwater into their processes.

5. **Erosion Control:** Rainwater harvesting contributes to erosion control by capturing and directing runoff. In areas prone to soil erosion, this method helps retain the topsoil and prevents the degradation of agricultural land. By minimizing soil erosion, rainwater harvesting supports sustainable land use practices and protects the natural productivity of landscapes.
6. **Improved Water Quality:** Stored rainwater is generally free from the pollutants commonly found in surface water sources. The absence of contaminants makes harvested rainwater suitable for various non-potable uses with minimal treatment. This improvement in water quality enhances the overall environmental sustainability of rainwater harvesting practices.
7. **Energy Efficiency:** Decentralized rainwater harvesting systems contribute to energy efficiency in water supply. Unlike centralized water treatment and distribution systems, which require significant energy inputs, local rainwater harvesting systems operate with lower energy demands. This energy efficiency aligns with broader sustainability goals and reduces the environmental footprint associated with water supply infrastructure.

Approaches to Rainwater Harvesting

1. **Community Engagement:** Community engagement is a cornerstone of successful rainwater harvesting projects. Building awareness, fostering understanding, and encouraging active participation within communities create a sense of ownership and responsibility. Engaging community members in the planning, implementation, and maintenance of rainwater harvesting systems enhances the long-term sustainability and success of these initiatives.
2. **Legislation and Policies:** Governments play a pivotal role in promoting and regulating rainwater harvesting through legislation and policies. Enacting laws that encourage rainwater harvesting, providing incentives for its adoption, and integrating it into building codes are essential steps. Clear policies create an enabling environment for individuals, businesses, and communities to embrace rainwater harvesting practices.
3. **Technological Innovation:** Advancements in rainwater harvesting technologies contribute to the efficiency and effectiveness of these systems. Ongoing research and development efforts focus on improving gutter systems, filtration devices, storage solutions, and overall system performance. Embracing technological innovations ensures that rainwater

harvesting remains a viable and attractive solution for diverse settings and conditions.

4. **Educational Initiatives:** Educational programs at various levels, from schools to community workshops, play a crucial role in promoting the benefits of rainwater harvesting. These initiatives raise awareness about the importance of water conservation, the practicalities of rainwater harvesting, and the positive impacts it can have on local water resources. Educational efforts contribute to a culture of sustainability and responsible water use.
5. **Integration with Land Use Planning:** Urban and rural planning should integrate rainwater harvesting as an integral component. Zoning regulations can incentivize the incorporation of rainwater harvesting systems in new constructions. Integrating rainwater harvesting into land use planning ensures that it becomes an integral part of sustainable urban and rural development, contributing to water resilience in the face of changing climate conditions.
6. **Research and Development:** Ongoing research is essential to advance rainwater harvesting techniques, address challenges associated with specific climates and geographies, and explore innovative solutions. Research and development efforts contribute to the continuous improvement of rainwater harvesting systems, making them more adaptable, efficient, and accessible to diverse communities around the world.
7. **Financial Support:** Providing financial support in the form of incentives, subsidies, or low-interest loans encourages individuals, communities, and businesses to invest in rainwater harvesting infrastructure. Financial support reduces the upfront costs associated with installing rainwater harvesting systems, making them more accessible and attractive to a broader range of stakeholders.
6. **Water-Saving Technologies, Rainwater Harvesting, and Drought Mitigation:** Water scarcity is a global challenge exacerbated by population growth, climate change, and inefficient water management. Innovative solutions, including water-saving technologies, rainwater harvesting, and drought mitigation strategies, play a pivotal role in ensuring sustainable water use and resilience in the face of evolving environmental conditions.

Water-Saving Technologies

1. **Smart Irrigation Systems:** Smart irrigation systems leverage technology to revolutionize traditional watering practices. Incorporating real-time data from weather forecasts and soil moisture sensors, these systems optimize irrigation schedules. By delivering precise amounts of water when and where it is needed, smart irrigation minimizes water wastage. The

integration of mobile apps allows users to remotely monitor and control irrigation systems, further enhancing water-use efficiency. This technology is especially beneficial in agriculture, where it can lead to substantial water savings and improved crop yields.

2. **Low-Flow Plumbing Fixtures:** Low-flow plumbing fixtures are instrumental in curbing domestic water consumption. These fixtures, including toilets, faucets, and showerheads, are designed to deliver adequate functionality with reduced water flow. Dual-flush toilets, for example, offer different flushing options for liquid and solid waste, optimizing water use. The widespread adoption of low-flow plumbing fixtures in residential and commercial buildings contributes significantly to water conservation without compromising comfort or hygiene.

3. **Drip Irrigation:** Drip irrigation represents a precision irrigation method that delivers water directly to the root zone of plants. This targeted approach minimizes water loss through evaporation and runoff, making it highly efficient in agriculture and landscaping. The system employs a network of tubes and emitters, ensuring a slow, consistent flow of water to individual plants. Drip irrigation not only conserves water but also promotes healthy plant growth by preventing waterlogging and minimizing weed growth.

4. **Greywater Recycling Systems:** Greywater recycling systems address the challenge of non-potable water demand in households. These systems capture and treat wastewater generated from activities such as washing and bathing. After treatment, the recycled greywater can be used for purposes like landscape irrigation and toilet flushing. Implementing greywater recycling reduces reliance on freshwater for non-drinking uses, contributing to sustainable water management and minimizing the environmental impact of household water consumption.

5. **Soil Moisture Sensors:** Soil moisture sensors are vital tools in optimizing irrigation practices. These sensors measure the moisture content in the soil, providing real-time data on whether plants need water. By preventing overwatering, soil moisture sensors promote water-use efficiency in agriculture, landscaping, and urban green spaces. Farmers and landscapers can make informed decisions based on accurate soil moisture information, leading to resource savings and improved plant health.

Expanding the use of these water-saving technologies requires a multi-faceted approach, including education, incentives, and regulatory support. Public awareness campaigns can highlight the benefits of adopting such technologies, while governments and municipalities can provide financial incentives or rebates to encourage their widespread implementation. Building codes and regulations can also be updated to mandate the use of water-saving technologies in new constructions, contributing to a more water-efficient future.

Rainwater Harvesting

1. **Cisterns and Storage Tanks:** Cisterns and storage tanks are fundamental components of rainwater harvesting systems. These structures collect rainwater from rooftops and store it for later use. The harvested rainwater can be utilized for various purposes, including landscape irrigation, flushing toilets, and even as a supplementary source for domestic water needs. The versatility of cisterns and storage tanks makes them adaptable to both residential and commercial applications, providing a reliable alternative to traditional water sources.

2. **Rain Gardens:** Rain gardens exemplify a sustainable landscaping approach that incorporates rainwater harvesting. These gardens are designed to capture and absorb rainwater runoff, preventing it from entering stormwater drains. Typically planted with native vegetation, rain gardens use plants and soil to filter pollutants from the captured rainwater. Besides mitigating runoff, rain gardens enhance urban biodiversity, promote soil health, and create aesthetically pleasing green spaces.

3. **Permeable Pavements:** Permeable pavements represent an innovative solution to combat surface runoff in urban areas. These pavements allow rainwater to infiltrate through the surface, reducing runoff and minimizing the strain on stormwater drainage systems. Permeable concrete, pavers, and asphalt are designed to facilitate water penetration while maintaining structural integrity. The use of permeable pavements contributes to groundwater recharge, lowers the risk of flooding, and enhances overall urban water resilience.

4. **Fog Nets:** Fog nets offer a unique approach to rainwater harvesting in regions with limited traditional rainfall. Deployed in arid or fog-prone coastal areas, these nets capture moisture from fog, converting it into liquid water for collection. The mesh nets capture tiny water droplets, and the collected water is directed into storage containers. While not a universal solution, fog nets provide an additional source of water in areas where conventional rainfall is scarce, showcasing the adaptability of rainwater harvesting methods.

5. **Rainwater Harvesting in Agriculture:** Agricultural practices can benefit significantly from rainwater harvesting techniques. Contour trenches, check dams, and other methods can be employed to capture and store rainwater for crop irrigation. Implementing rainwater harvesting in agriculture helps farmers adapt to variable rainfall patterns, reduces dependence on traditional irrigation sources, and enhances overall water-use efficiency. This approach is particularly critical in regions prone to drought, where rainwater harvesting contributes to agricultural resilience.

Expanding the adoption of rainwater harvesting requires a combination of policy support, infrastructure development, and community engagement. Governments can incentivize rainwater harvesting through subsidies or tax benefits, while educational programs can raise awareness about the benefits of these systems. Integrating rainwater harvesting into urban planning and landscaping guidelines ensures its incorporation into new developments, fostering a culture of sustainable water use.

Drought Mitigation Strategies

1. **Drought-Resistant Crops:** Developing and cultivating drought-resistant crops is a proactive strategy to mitigate the impact of water scarcity on agriculture. Plant breeding programs focus on selecting and enhancing traits that enable crops to withstand periods of reduced water availability. Drought-resistant crops not only ensure food security but also contribute to the sustainability of agricultural systems in regions prone to water stress.

2. **Water Recycling and Reuse:** Water recycling and reuse play a crucial role in drought mitigation by maximizing the use of available water resources. Advanced water treatment technologies can purify wastewater, making it suitable for various non-potable purposes. By recycling and reusing treated wastewater for irrigation, industrial processes, and other applications, communities can reduce their dependence on freshwater sources during periods of drought.

3. **Aquifer Recharge:** Managed aquifer recharge is a strategic approach to enhance groundwater reserves, especially in regions where aquifers are a primary source of water. This involves intentionally directing water, often from excess runoff or treated wastewater, into underground aquifers. By replenishing aquifers during times of abundance, communities create a natural buffer against the impacts of drought, ensuring a more reliable and sustained water supply.

4. **Xeriscaping:** Xeriscaping, or water-wise landscaping, is a landscaping approach that conserves water by using drought-resistant plants and efficient irrigation techniques. This strategy minimizes the need for supplemental irrigation, reduces water consumption, and creates visually appealing and environmentally friendly landscapes. Xeriscaping is particularly effective in regions facing water restrictions or experiencing prolonged drought conditions.

5. **Water-Efficient Industrial Practices:** Industries can contribute to drought mitigation by adopting water-efficient practices. Closed-loop systems, water recycling, and process optimization reduce the overall water footprint of industrial operations. By prioritizing water efficiency, industries not only

lower their environmental impact but also enhance their resilience to water shortages, ensuring continued productivity during drought periods.

Implementing these drought mitigation strategies requires a collaborative effort involving governments, businesses, communities, and individuals. Robust water management policies, including regulations on water use and incentives for water-efficient practices, create a supportive framework. Public awareness campaigns educate communities about the importance of water conservation and the role of drought mitigation strategies in building water resilience.

7. Crop Productivity and Water Security

Crop productivity and water security are intricately linked facets of global agriculture, wielding profound implications for food availability, economic stability, and environmental sustainability. As the world contends with burgeoning population growth, the specter of climate change, and the looming threat of water scarcity, delving deeper into the complex interplay between crop productivity and water security becomes imperative.

Factors Influencing Crop Productivity

1. Water Availability

- **Description:** Water stands as the lifeblood of crop productivity. Inadequate water, whether stemming from erratic rainfall or insufficient irrigation, induces water stress, impeding crucial plant processes like photosynthesis and nutrient uptake. Conversely, optimal water availability becomes a linchpin for supporting the physiological underpinnings of plant growth.

2. Climate Conditions

- **Description:** Climate intricacies, encompassing temperature fluctuations and precipitation patterns, exert a formidable influence on crop growth trajectories. Extreme temperatures, drought episodes, or excessive rainfall can conspire to undermine crop productivity. The development and deployment of climate-resilient crop varieties and adaptive agricultural practices emerge as linchpins in mitigating the adverse impacts of evolving climatic dynamics.

3. Soil Quality

- **Description:** The vitality and fertility of the soil constitute pivotal determinants of crop productivity. Soil composition, nutrient richness, and organic matter content wield a profound influence on a plant's capacity to absorb water and nutrients. The adoption of sustainable soil management practices, including crop rotation and organic farming, emerges as a linchpin for fostering improved soil quality and subsequently augmenting crop yields.

4. Crop Genetics

- **Description:** The genetic makeup of crops assumes a decisive role in shaping their resilience, adaptability, and productivity. Rigorous breeding programs, spotlighting the development of drought-resistant, pest-resistant, and high-yielding crop varieties, contribute indomitably to sustained agricultural productivity.

5. Farm Management Practices

- **Description:** Effective farm management, embracing astute irrigation methodologies, judicious pest control, and prudent crop rotation, materializes as a harbinger of crop productivity. The institutionalization of sustainable farming practices, engineered to optimize water utilization, minimize soil erosion, and enhance nutrient cycling, emerges as a bedrock for the enduring viability of agricultural systems.

6. Technology Adoption

- **Description:** The assimilation of technology, spanning precision agriculture, sensor-driven irrigation, and state-of-the-art machinery, assumes a transformative role in amplifying crop productivity. These technological interventions empower farmers to fine-tune resource utilization, monitor crop conditions with granularity, and make informed, data-driven decisions conducive to augmented yields.

Expanding on these factors necessitates a multi-pronged strategy encompassing educational initiatives, regulatory frameworks, and research endeavors to synergistically bolster agricultural resilience in the face of a dynamic and challenging environment.

Impact of Water Availability on Agricultural Systems

1. Water Scarcity and Stress

- **Description:** Regions grappling with water scarcity often find themselves ensnared in the throes of water stress, an affliction that reverberates through agricultural systems. Inadequate water availability precipitates diminished crop yields, a contraction in agricultural output, and heightened susceptibility to climatic vicissitudes. The deployment of efficient water management practices and cutting-edge technologies emerges as an imperative in mitigating the ramifications of water scarcity.

2. Irrigation Challenges

- **Description:** Irrigated agriculture, an indispensable facet for ensuring consistent water access, encounters its own set of challenges. Inefficient irrigation systems, profligate water usage, and over-extraction from water reservoirs can amplify water scarcity woes. Adherence to sustainable irrigation practices, encompassing precision methods like drip irrigation and astute water resource management, assumes pivotal significance in upholding water security.

3. Crop Water Use Efficiency

- **Description:** The concept of crop water use efficiency emerges as a barometer gauging the effectiveness with which crops convert water into biomass. Water-efficient crops and judicious irrigation techniques that maximize the utilization of available water stand as bulwarks for sustained agricultural productivity. A nuanced comprehension of the water requirements of crops becomes instrumental in fine-tuning water use efficiency.

4. Impact on Livelihoods

- **Description:** Water availability's direct imprint on the livelihoods of farmers and communities enmeshed in agriculture is profound. Crop failures precipitated by water scarcity can unleash economic downturns, induce food insecurity, and elevate vulnerability to the scourge of poverty. Strengthening water security in agriculture emerges as a linchpin for fortifying the resilience of rural communities against adversities.

5. Climate Change Effects

- **Description:** The specter of climate change introduces an additional layer of complexity to water availability in agriculture. Altered precipitation patterns, escalating temperatures, and a surge in extreme weather events all conspire to influence crop growth dynamics and water availability. Strategic adaptations, ranging from the cultivation of climate-resilient crops to the implementation of agile water management practices, emerge as pivotal measures to counterbalance these effects.

Delving into the nuances of these impacts necessitates a holistic strategy wherein community involvement, policy advocacy, and technological innovation coalesce to fortify the foundations of agricultural sustainability.

Strategies to Enhance Water Security in Agriculture

1. Efficient Irrigation Practices

- **Description:** The adoption of efficient irrigation practices, epitomized by methodologies such as drip irrigation and state-of-the-art sprinkler systems, stands as a vanguard in optimizing water use in agriculture. These precision irrigation methods abate water wastage, ensure uniform water distribution, and contribute substantively to augmenting crop water use efficiency.

2. Water Harvesting and Storage

- **Description:** The strategic harvesting of rainwater coupled with judicious storage emerges as a sustainable strategy for enhancing water security. Constructing reservoirs, deploying check dams, and instituting rainwater harvesting systems collectively contribute to bolstering water availability, particularly during protracted dry spells.

3. Drought-Resistant Crop Varieties

- **Description:** The development and cultivation of drought-resistant crop varieties stands as a linchpin in fortifying agricultural resilience against water scarcity. Research and breeding programs directed at identifying genetic traits that confer drought resilience pave the way for a robust and adaptive agricultural landscape.

4. Precision Agriculture

- **Description:** The ethos of precision agriculture, leveraging a suite of technologies including sensors, drones, and sophisticated data analytics, emerges as a transformative force. By minutely monitoring soil and crop conditions, farmers are empowered to make informed decisions, curtail resource utilization, and amplify overall water efficiency.

5. Integrated Water Resource Management

- **Description:** The institutionalization of integrated water resource management transcends the silos of sourcing, utilization, and disposal. This holistic approach not only advocates for judicious water usage but also minimizes water pollution, fostering agricultural practices that seamlessly align with broader water management objectives.

6. Capacity Building and Education

- **Description:** The dissemination of knowledge and the elevation of awareness regarding water-efficient practices, sustainable farming methodologies, and the imperative of water conservation resonate as foundational pillars. Capacity-building initiatives, both at the individual and community levels, empower stakeholders to assimilate water-saving technologies and embrace practices that fundamentally contribute to water security.

7. Policy Support and Governance

- **Description:** Governments, wielding a pivotal role in the orchestration of water security, can shape the narrative through the formulation and implementation of policies. Effective governance structures, regulations governing water use, and incentives designed to propagate water-saving practices collectively contribute to the overarching goal of bolstering water security in agriculture.

8. Principles of Agricultural Drainage

Irrigation and Drainage Relations: Sources of Excess Water and Salts in Soil

Irrigation and Drainage Relations

Irrigation and drainage are intricately connected components of agricultural water management, playing a pivotal role in determining soil conditions and influencing crop productivity. The interaction between these two processes is fundamental for maintaining optimal soil moisture levels and preventing issues such as waterlogging and salinity. A deeper understanding of the relations between irrigation and drainage is crucial for fostering sustainable agriculture.

1. **Irrigation:** Irrigation serves as a lifeline for agriculture, particularly in regions where natural rainfall is insufficient to meet the water requirements of crops. It involves the deliberate application of water to the soil, ensuring that plants receive the necessary moisture for growth and development. The judicious use of irrigation enhances crop yields, allows for multiple cropping seasons, and contributes to food security.

2. **Drainage:** Drainage is the process of removing excess water from the soil, preventing waterlogging and creating favorable conditions for plant roots to thrive. While water is essential for plant growth, excess water in the root zone can lead to reduced oxygen availability, hindering root function and nutrient uptake. Effective drainage systems are designed to mitigate these risks, maintaining soil structure and supporting optimal crop performance.

3. **Relation:** The relationship between irrigation and drainage is symbiotic, forming a delicate balance crucial for agricultural success. When crops receive adequate irrigation, they thrive, utilizing water for photosynthesis, nutrient absorption, and overall growth. However, without proper drainage, excess water can accumulate, leading to waterlogging. This, in turn, compromises the aeration of the soil, nutrient availability, and ultimately, the health of the plants. Therefore, a well-coordinated approach that integrates both irrigation and drainage is necessary to create an environment where crops can flourish.

Expanding on the relation between irrigation and drainage involves considering advanced technologies and precision agriculture methods that enable farmers to optimize water use. Implementing smart irrigation systems, soil moisture sensors, and drainage solutions tailored to specific soil types contribute to a more efficient and sustainable water management approach.

Sources of Excess Water in Soil

1. **Excessive Rainfall:** Excessive rainfall, especially during monsoon seasons, can lead to a surplus of water in the soil. While rainfall is essential for natural irrigation, inadequate drainage systems may struggle to cope with heavy downpours, resulting in waterlogging. This excess water can linger in the soil, impeding agricultural activities and causing harm to crops.

2. **Poorly Designed Irrigation Systems:** Inefficient or poorly designed irrigation systems can contribute to waterlogging issues. Over-irrigation or uneven water distribution can saturate the soil, leading to standing water and creating unfavorable conditions for plant roots. Implementing modern irrigation technologies that allow for precise control over water application can mitigate this challenge.

3. **High Water Table:** A high-water table, often exacerbated by factors like poor drainage or proximity to water bodies, can contribute to excess water in the soil. During periods of increased rainfall or irrigation, the rising water table may lead to waterlogging. Strategic planning and drainage solutions are essential in managing areas with high water tables.

4. **Compacted Soils:** Soil compaction reduces the permeability of the soil, limiting water infiltration and drainage. Compacted soils are more prone to surface runoff and water accumulation, particularly in areas with heavy machinery or foot traffic. Implementing soil conservation practices, such as cover cropping and reduced tillage, can improve soil structure and mitigate compaction-related drainage issues.

Expanding on these sources of excess water in the soil involves delving into regional variations and specific soil characteristics that influence drainage dynamics. Tailoring drainage solutions to the unique challenges of different landscapes ensures a more effective and sustainable approach to managing excess water.

Sources of Excess Salts in Soil

1. **Poor-Quality Water:** The quality of water used for irrigation significantly influences soil salinity. Poor-quality water containing high levels of salts, especially in arid regions, can contribute to salinity problems in the soil. As water evaporates from the soil surface, salts accumulate, posing a threat to plant health. The use of brackish or saline water for irrigation requires careful management to prevent salt buildup.

2. **Natural Soil Salinity:** Some soils inherently possess high salt content, a condition often exacerbated by factors like arid climates or proximity to coastal areas. When these soils are irrigated, salts can become concentrated in the root zone, hindering water uptake by plants. Implementing soil

amendments and choosing salt-tolerant crops are strategies to address natural soil salinity.

3. **Fertilizer Application:** The application of fertilizers, particularly those containing soluble salts, can contribute to soil salinity. Overuse or improper application of fertilizers leads to salt accumulation in the soil. Adopting precision agriculture practices, including controlled-release fertilizers and soil testing, helps optimize nutrient application and minimize the risk of salinity.

4. **Insufficient Drainage:** Inadequate drainage exacerbates salt accumulation in the soil. When water is not efficiently removed, salts can concentrate in the root zone as water evaporates, leading to salinity issues. Improving drainage systems, incorporating organic matter into the soil, and adopting leaching practices are strategies to manage soil salinity effectively.

Expanding on these sources of excess salts in the soil involves exploring the intricate balance between nutrient management and soil health. Implementing sustainable agricultural practices that prioritize soil conservation and careful water management is essential for mitigating salinity issues.

Strategies for Managing Irrigation and Drainage

1. **Precision Irrigation:** Precision irrigation involves the application of water with a high degree of accuracy, matching the water needs of crops precisely. Technologies such as drip irrigation, soil moisture sensors, and weather forecasting contribute to efficient water use. Precision irrigation minimizes water wastage and helps prevent both waterlogging and salinity.

2. **Proper Drainage Systems:** Investing in well-designed drainage systems is crucial for preventing waterlogging and maintaining soil health. Subsurface drainage, contouring, and the use of drainage ditches are effective measures to remove excess water. Tailoring drainage solutions to the specific needs of the landscape ensures optimal performance.

3. **Soil Testing:** Regular soil testing provides valuable insights into soil fertility and salinity levels. By understanding the composition of the soil, farmers can make informed decisions about nutrient management and irrigation practices. Soil testing guides the implementation of targeted amendments to address specific soil challenges.

4. **Crop Selection:** Choosing crop varieties that are well-adapted to local soil conditions and water quality is essential for mitigating the impact of excess water and salts. Salt-tolerant crops can thrive in saline soils, while strategic crop rotations help break the cycle of nutrient depletion and salinity buildup.

5. **Cover Crops:** Integrating cover crops into agricultural practices enhances soil structure, reduces erosion, and improves water infiltration. Cover crops also contribute organic matter to the soil, enhancing its water-holding capacity and promoting nutrient cycling. This approach supports effective drainage and mitigates excess water issues.
6. **Managed Aquifer Recharge:** In regions facing declining water tables, managed aquifer recharge involves intentional replenishment of aquifers during periods of excess water. This strategy helps maintain sustainable groundwater levels, providing a valuable water resource during dry periods. Managed aquifer recharge contributes to both water conservation and effective drainage.
7. **Leaching:** Controlled leaching is a practice where excess water is applied to the soil to flush out accumulated salts. This helps prevent salt buildup in the root zone and mitigates soil salinity. Leaching is particularly effective in regions with poor-quality irrigation water.
8. **Integrated Water Management:** Adopting an integrated approach to water management considers the interconnected nature of irrigation, drainage, and soil health. This holistic strategy involves coordinating irrigation schedules with drainage practices, optimizing water use efficiency, and promoting overall sustainability in agricultural water management.

Expanding on these strategies involves exploring innovative technologies, community engagement, and policy support to foster a comprehensive approach to irrigation and drainage management. Implementing these measures collectively contributes to a resilient and sustainable agricultural ecosystem.

9. Development of Drainage Problems in Different Crops and Soils

Effective drainage is paramount for maintaining soil health and optimizing crop productivity. While water is essential for plant growth, excessive water or poor drainage can lead to a myriad of problems, adversely affecting different crops and soil types. Understanding the nuanced development of drainage issues in various crops and soils is crucial for implementing targeted solutions.

Types of Drainage Problems

1. Waterlogging

- **Description:** Waterlogging occurs when the soil is saturated with water, leading to a water table rise and restricting the movement of air within the soil. This condition is detrimental to crops as it creates an oxygen-deficient environment in the root zone.

- **Development in Different Crops**
  - *Rice:* While rice is adapted to semi-aquatic conditions, prolonged waterlogging during the wrong growth stages can lead to reduced oxygen availability, affecting root function and nutrient uptake.
  - *Wheat:* Waterlogging in wheat fields hampers root growth and nutrient absorption, resulting in stunted plants with diminished grain yield.
  - *Maize:* Maize is sensitive to waterlogging, especially during the early growth stages. It can lead to poor germination, reduced stand establishment, and overall yield loss.

2. Poor Drainage

- **Description:** Inadequate drainage leads to the accumulation of excess water in the soil, impeding the movement of water away from the root zone. This can result in a range of problems, including nutrient imbalances and root diseases.

- **Development in Different Crops**
  - *Potatoes:* Poorly drained soils contribute to the development of diseases like potato scab. Additionally, waterlogged conditions can lead to reduced tuber quality and increased vulnerability to pests.
  - *Soybeans:* Soybeans are susceptible to poor drainage, impacting nodulation and nitrogen fixation. This can result in reduced yields and nutrient deficiencies.
  - *Citrus fruits:* Citrus trees can suffer from root rot in poorly drained soils, affecting overall tree health, fruit quality, and yield.

3. Surface Drainage

Definition and Principles: Surface drainage refers to the removal of excess water from the soil surface to prevent waterlogging and improve aeration. It involves the shaping of the land to create slopes or channels that guide water away from the fields.

Methods of Surface Drainage

1. **Graded Channels:** Creating gentle slopes or graded channels directs water flow away from fields, preventing water accumulation.
2. **Open Ditches:** Ditches alongside fields collect and channel excess water away, promoting surface runoff.
3. **Contour Plowing:** Plowing along the contour lines of the land helps slow down water runoff, reducing soil erosion and promoting infiltration.

Benefits of Surface Drainage

- **Prevents Waterlogging:** Surface drainage prevents the saturation of the topsoil, reducing the risk of waterlogging.
- **Erosion Control:** By guiding water away, surface drainage minimizes soil erosion, preserving valuable topsoil.
- **Improved Aeration:** Proper surface drainage enhances soil aeration, creating favorable conditions for root growth.

Applications

Surface drainage is particularly effective in areas with gently sloping topography. It is widely used in agricultural fields, orchards, and pasture lands to prevent water stagnation.

4. Subsurface Drainage

Definition and Principles: Subsurface drainage involves the removal of excess water from the root zone of the soil. This method utilizes underground systems, such as pipes or tiles, to efficiently drain water from the soil profile.

Methods of Subsurface Drainage

1. **Tile Drains:** Perforated pipes, known as tile drains, are installed below the surface to collect and transport excess water away from the root zone.
2. **French Drains:** These are trenches filled with gravel or rock that allow water to percolate into a perforated pipe, which then carries the water away.
3. **Mole Drains:** Mole drains are created by a specialized plow that forms channels beneath the surface to facilitate water movement.

Benefits of Subsurface Drainage

- **Lower Water Table:** Subsurface drainage helps maintain a lower water table, preventing waterlogging in the root zone.
- **Enhanced Root Growth:** Improved soil aeration and reduced water stress contribute to healthier root systems and better nutrient uptake.
- **Extended Growing Season:** Subsurface drainage allows for earlier planting and extended growing seasons, reducing the risk of crop damage.

Applications

Subsurface drainage is particularly effective in areas with poorly draining soils, high water tables, or heavy clay compositions. It is commonly employed in row-crop agriculture, horticulture, and areas prone to seasonal waterlogging.

Surface vs. Subsurface Drainage

1. Complementary Nature

- *Surface Drainage:* Primarily addresses issues related to excess water at the soil surface.
- *Subsurface Drainage:* Targets waterlogged conditions within the soil profile.

2. Installation and Visibility

- *Surface Drainage:* Visible structures like channels and ditches are created on the land surface.
- *Subsurface Drainage:* Components are installed underground, making the system less visible.

3. Applicability

- *Surface Drainage:* Well-suited for areas with gentle slopes and surface water runoff.
- *Subsurface Drainage:* Ideal for areas with poor soil drainage, high water tables, and heavier soils.

4. Maintenance

- *Surface Drainage:* Regular maintenance is required to prevent clogging and ensure proper water flow.
- *Subsurface Drainage:* Maintenance is typically lower as components are underground, but occasional inspections are essential.

5. Cost Considerations

- *Surface Drainage:* Initial installation costs can be lower, but ongoing maintenance expenses may accrue.
- *Subsurface Drainage:* Initial investment can be higher, but maintenance costs are generally lower over the long term.

5. Salinity

- **Description:** Salinity arises when soluble salts accumulate in the soil, affecting water availability to plants. High salt concentrations can disrupt osmotic balance, leading to water stress in crops.

• Development in Different Crops

- *Tomatoes:* Tomatoes are sensitive to salinity, and excess salts can lead to physiological disorders, reduced fruit quality, and diminished yields.

- *Carrots:* Carrots can experience stunted growth and deformities in the presence of high soil salinity, affecting marketable yield and quality.
- *Lettuce:* Salinity stress in lettuce can result in reduced leaf size, increased bitterness, and overall poor crop quality.

Impact of Drainage Problems on Different Soils

1. Clay Soils

- **Description:** Clay soils have fine particles that allow for water retention but impede water movement. Inadequate drainage in clay soils can lead to waterlogging and compaction.

• Impact on Different Crops

- *Barley:* Barley is susceptible to waterlogging in clay soils, impacting root development and nutrient uptake. This can result in reduced grain quality and yield.
- *Cabbage:* Poor drainage in clay soils can lead to increased susceptibility to diseases like clubroot in cabbage crops, affecting plant health and marketable yield.
- *Apples:* Apple trees in clay soils may experience root suffocation and nutrient deficiencies due to waterlogging, leading to reduced fruit production.

2. Sandy Soils

- **Description:** Sandy soils have larger particles that allow for rapid drainage but offer lower water retention capacity. Poor drainage in sandy soils can lead to nutrient leaching and water stress.

• Impact on Different Crops

- *Peanuts:* Peanuts grown in sandy soils with poor drainage may experience water stress, impacting pod development and reducing overall yield.
- *Watermelon:* Watermelon plants in sandy soils may suffer from nutrient leaching and inadequate water supply, leading to reduced fruit quality and size.
- *Cotton:* Poor drainage in sandy soils can result in nutrient deficiencies in cotton plants, affecting fiber quality and yield.

3. Loamy Soils

- **Description:** Loamy soils, with a balanced mixture of sand, silt, and clay, offer good drainage and water retention. However, poor drainage can still occur, impacting crop health.

- **Impact on Different Crops**
  - *Corn:* Corn crops in poorly drained loamy soils may experience shallow root development, reducing resilience to adverse weather conditions and compromising overall yield.
  - *Strawberries:* Strawberries are sensitive to waterlogged conditions in loamy soils, leading to root diseases and diminishing fruit quality.
  - *Peppers:* Poor drainage in loamy soils can lead to nutrient imbalances in pepper plants, affecting both fruit quality and yield.

Strategies for Mitigating Drainage Problems

1. Subsurface Drainage

- **Description:** Subsurface drainage involves the installation of pipes or tiles beneath the soil surface to facilitate the removal of excess water. It is effective in addressing both waterlogging and poor drainage issues.

- **Applicability in Different Crops and Soils**
  - Subsurface drainage is beneficial for crops like rice, wheat, and maize, especially in clay soils prone to waterlogging.
  - In sandy soils, subsurface drainage helps prevent nutrient leaching and ensures adequate water supply for crops like peanuts and watermelon.
  - Loamy soils benefit from subsurface drainage to maintain optimal moisture levels for crops like corn and strawberries.

2. Cover Crops

- **Description:** Cover crops are grown between main crops to improve soil structure, reduce erosion, and enhance water infiltration. They contribute to mitigating drainage problems by promoting healthy soil conditions.

- **Applicability in Different Crops and Soils**
  - Cover crops are effective in clay soils prone to compaction, benefiting crops like barley and cabbage.
  - In sandy soils, cover crops prevent nutrient leaching and water stress, supporting crops like peanuts and cotton.
  - Loamy soils benefit from cover crops to maintain a balanced soil structure for crops like corn and peppers.

3. Raised Bed Farming

- **Description:** Raised bed farming involves creating elevated planting beds, improving drainage and preventing waterlogging. It is particularly effective in areas with heavy or poorly drained soils.

- **Applicability in Different Crops and Soils**
  - Raised bed farming is beneficial for crops like rice and tomatoes in clay soils prone to waterlogging.
  - In sandy soils, raised beds help retain water and nutrients for crops like watermelon and peanuts.
  - Loamy soils benefit from raised bed farming to enhance drainage and root development for crops like strawberries and corn.

4. Improved Irrigation Practices

- **Description:** Precision irrigation methods, such as drip irrigation and soil moisture sensors, help optimize water application, preventing waterlogging and conserving water in sandy soils.

- **Applicability in Different Crops and Soils**
  - Precision irrigation is crucial for water-intensive crops like rice in clay soils to prevent waterlogging.
  - In sandy soils, drip irrigation ensures efficient water use for crops like watermelon and peanuts.
  - Loamy soils benefit from precision irrigation for crops like corn and strawberries to maintain optimal moisture levels.

10. Drainage Requirements of Crops: Enhancing Agricultural Productivity

In the realm of agriculture, the drainage requirements of crops play a pivotal role in determining the success and productivity of farming endeavors. Efficient drainage systems are essential for maintaining a balanced soil environment, preventing waterlogging, and fostering optimal conditions for crop growth. This extended discussion delves deeper into the multifaceted aspects of drainage requirements, exploring additional factors influencing these needs, diverse drainage methods for specific crops, and the broader importance of effective drainage in sustainable agriculture.

Drainage Requirements of Crops: Enhancing Agricultural Productivity

Understanding the drainage requirements of crops is crucial for promoting optimal growth and mitigating water-related challenges in agricultural settings. Proper drainage ensures that excess water is efficiently removed from the root zone, preventing waterlogging and creating favorable conditions for crop development. In this discussion, we explore the varying drainage needs of different crops, factors influencing these requirements, and the importance of effective drainage in sustainable agriculture.

Factors Influencing Drainage Requirements

1. Soil Type

- *Clay Soils:* These soils have lower drainage capacity, and crops grown in clayey areas may require enhanced drainage to prevent waterlogging.
- *Sandy Soils:* Sandy soils generally have better drainage, but excessive drainage can result in rapid water percolation, leading to water stress for some crops.

2. Topography

- *Sloping Terrain:* Areas with slopes facilitate natural drainage, preventing water accumulation. Crops in sloping regions may have lower drainage requirements.
- *Flat Terrain:* Flat landscapes may necessitate artificial drainage systems to prevent waterlogging.

3. Rainfall Patterns

- *High Rainfall Areas:* Regions with abundant rainfall may require efficient drainage systems to prevent waterlogging and soil erosion.
- *Low Rainfall Areas:* While drainage is essential, careful water conservation practices are also crucial in regions with low rainfall.

4. Crop Types

- *Water-Intensive Crops:* Crops with high water requirements, such as rice, may need specialized drainage systems to maintain an optimal water balance.
- *Drought-Tolerant Crops:* Crops adapted to arid conditions may benefit from well-managed drainage to prevent water stress.

5. Rooting Depth

- *Shallow-Rooted Crops:* Crops with shallow root systems may be more susceptible to waterlogging, necessitating effective surface drainage.
- *Deep-Rooted Crops:* Crops with deeper roots may require subsurface drainage to maintain aeration and prevent waterlogging at lower soil depths.

6. Climate Variability

- *Seasonal Rainfall Fluctuations:* Regions experiencing seasonal variations in rainfall may need adaptive drainage strategies to manage excess water during wet periods and conserve moisture during dry spells.

- *Extreme Weather Events:* Unpredictable weather events, such as heavy storms or prolonged droughts, underscore the need for flexible drainage systems capable of handling diverse conditions.

7. Soil Structure and Composition

- *Compacted Soils:* Compacted soils impede water movement and drainage. Crops in such areas may require interventions like deep tillage or aeration to enhance drainage pathways.
- *Organic Matter Content:* Soils rich in organic matter tend to have better drainage due to improved structure. Management practices that enhance organic matter can indirectly influence drainage capabilities.

8. Crop Growth Stages

- *Germination and Seedling Stage:* Crops are often sensitive to waterlogging during germination. Adequate surface drainage ensures that excess water does not hinder the emergence of seedlings.
- *Flowering and Fruit Development:* Efficient drainage becomes critical during flowering and fruiting stages, as water stress or excess moisture can impact pollination and fruit set.

9. Land Use and Management Practices

- *Intensive Agriculture:* Intensively cultivated fields may experience higher runoff and soil compaction, demanding advanced drainage solutions to counteract these effects.
- *Cover Cropping:* Incorporating cover crops can positively influence drainage by improving soil structure and reducing erosion, benefiting subsequent crops.

Future Trends in Drainage Management

1. Smart Drainage Systems

- Advancements in sensor technology and data analytics enable the development of smart drainage systems. These systems can dynamically adjust drainage parameters based on real-time weather forecasts, soil conditions, and crop requirements.

2. Climate-Resilient Drainage Practices

- The integration of climate resilience into drainage planning involves anticipating and adapting to changing climate conditions. This may include the implementation of flexible drainage infrastructure and the incorporation of climate-smart agriculture principles.

3. Regenerative Drainage

- Regenerative drainage practices aim to not only manage water but also improve soil health and ecosystem services. This holistic approach considers the entire agroecosystem, emphasizing sustainable land management.

11. Biodrainage

Traditional drainage systems often involve the use of pipes, ditches, and other infrastructure to manage excess water. However, there is a growing recognition of the environmental impact and limitations of these conventional approaches. Biodrainage emerges as an innovative and sustainable alternative, leveraging natural processes and vegetation to achieve effective water management.

Understanding Biodrainage

1. Definition

Biodrainage refers to the use of living organisms, particularly plants and trees, in managing water levels and improving soil conditions. Instead of relying solely on engineering solutions, biodrainage harnesses the ecological functions of vegetation to enhance water movement and mitigate water-related issues.

2. Principles

Biodrainage operates on fundamental ecological principles, recognizing the symbiotic relationship between vegetation and water management. The key principles include:

- **Vegetation as Allies:** Plants play a crucial role in biodrainage. Their root systems enhance soil structure, creating channels for water movement and improving infiltration.
- **Biochemical Processes:** The root activities of plants contribute to biochemical processes in the soil, influencing water retention, nutrient cycling, and the breakdown of organic matter.
- **Biodiversity Boost:** Biodrainage encourages the cultivation of diverse plant species, promoting biodiversity and creating resilient ecosystems that can adapt to changing environmental conditions.

Biodrainage Techniques

1. Agroforestry Systems: Agroforestry involves the intentional integration of trees and shrubs into agricultural landscapes. In the context of biodrainage:

- **Tree Plantations:** Integrating strategically placed trees within agricultural fields can enhance drainage. Trees with deep root systems, such as willows and poplars, are particularly effective in lowering the water table.

- **Windbreaks:** Planting rows of trees as windbreaks not only helps in preventing soil erosion but also contributes to enhanced drainage by breaking up the force of winds and allowing for better water infiltration.

2. Cover Crops: Cover crops play a crucial role in preventing soil erosion, improving soil structure, and contributing to biodrainage. This technique involves:

- **Winter Cover Crops:** Planting cover crops, like clover or rye, during the off-season helps improve soil structure, prevent erosion, and contribute to enhanced drainage when the main crops are planted. The cover crop's root system creates pathways for water movement and improves overall soil health.

- **Green Manure:** Some cover crops, when plowed back into the soil, act as green manure, enhancing organic matter content and soil structure, thereby aiding in better drainage.

3. Constructed Wetlands: Constructed wetlands represent a biodrainage technique that mimics natural wetland ecosystems:

- **Designed Wetland Areas:** Implementing constructed wetlands involves creating specific areas with water-loving plants that naturally assist in water purification and drainage. The plants help in nutrient uptake, and the wetland serves as a natural filter, reducing nutrient runoff into water bodies.

4. Riparian Buffer Strips:

Riparian buffer strips involve maintaining natural vegetation along water bodies to protect water quality and contribute to effective drainage:

- **Vegetated Buffer Zones:** Maintaining strips of natural vegetation along water bodies, known as riparian buffer strips, can help prevent nutrient runoff, reduce soil erosion, and contribute to better drainage. The roots of riparian plants stabilize the soil, reducing the risk of erosion and enhancing water infiltration.

Advantages of Biodrainage

1. Eco-friendly Solution: Biodrainage is inherently environmentally friendly, aligning with sustainable and regenerative agricultural practices:

- **Reduced Environmental Impact:** Unlike traditional drainage methods that may involve the use of materials like concrete and plastic, biodrainage relies on natural processes, minimizing the environmental footprint associated with construction and maintenance.

- **Enhanced Ecosystem Services:** Biodrainage systems provide additional ecosystem services, such as habitat creation, carbon sequestration, and improved water quality, contributing to overall environmental health.

2. Soil Health Improvement: Biodrainage contributes to the improvement of soil health through various mechanisms:

- **Microbial Activity:** The presence of plants in biodrainage systems enhances soil microbial activity. Microorganisms play a crucial role in nutrient cycling, organic matter decomposition, and the creation of a healthy soil ecosystem.
- **Enhanced Nutrient Availability:** Biodrainage helps in nutrient cycling, making essential nutrients more available to plants. The organic matter contributed by cover crops and other vegetation serves as a nutrient reservoir for subsequent crops.
- **Reduced Soil Compaction:** The root systems of plants in biodrainage systems improve soil structure, reducing soil compaction and promoting better water infiltration.

3. Biodiversity Conservation: Biodrainage actively promotes biodiversity, contributing to the conservation of plant and animal species:

- **Diverse Plant Species:** Biodrainage encourages the cultivation of diverse plant species, creating a mosaic of habitats that supports a wide range of organisms. This diversity enhances ecosystem resilience and adaptability to environmental changes.
- **Wildlife Habitat:** The presence of natural vegetation in biodrainage systems provides habitats for various wildlife species. Riparian buffer strips, for example, can serve as corridors for wildlife movement.
- **Pollinator Support:** Biodrainage systems with diverse flowering plants attract pollinators, supporting the health of pollinator populations crucial for crop production.

4. Low Maintenance Requirements: Biodrainage systems generally require lower maintenance compared to traditional drainage infrastructure:

- **Self-regulating Systems:** Once established, biodrainage systems often become self-regulating. Natural processes and vegetation play a significant role in maintaining the system without the need for extensive human intervention.
- **Cost Savings:** The reduced need for maintenance translates into cost savings over the long term. While initial establishment costs may vary, the ongoing operational costs are typically lower.

Index

A

Acid Rain 21, 172
Aeroponics 132
Afforestation 7, 55, 60, 64, 65, 75, 107, 119, 120, 124, 131, 133, 134, 135, 136, 140, 166, 175, 325
Agroforestry 34, 58, 59, 63, 70, 71, 74, 75, 78, 113, 115, 118, 124, 131, 135, 138, 140, 144, 148, 149, 150, 151, 155, 160, 163, 16164, 168, 169, 172, 174, 176, 177, 180, 204, 205, 231, 232, 239, 241, 242, 243, 244, 245, 246, 247, 248-262, 298, 354
AI 106, 108
Anthropogenic 8, 20, 30, 33, 34, 112, 114, 117, 171

B

Biodiversity 5, 35, 40, 59, 60, 61, 67, 70, 71, 73, 75, 91, 106, 113, 115, 118, 119, 123, 124, 125, 126, 131, 133, 134, 136, 137, 140, 141, 145, 149, 150, 151, 157, 158, 159, 160, 163, 166, 167, 169, 171, 175, 182, 192, 201, 202, 203, 209, 213, 216, 227, 229, 230, 231, 236, 237, 239, 243, 244, 246, 249, 250, 256, 258, 260, 262, 265, 272, 319, 320, 323324, 331, 336, 354, 356
Biological Nitrogen Fixation 244
Bioremediation 22, 24, 103, 106135, 141

C

Capillary 17, 18, 286, 287, 289, 292, 293, 296, 297
Carbon Sequestration 4, 6, 140, 157, 180, 221, 227, 232, 243, 244, 262, 355
Cation Exchange Capacity (CEC) 23, 153
Chronological 10
Climate Change 6, 15, 23, 24, 30, 34, 78, 88, 96, 97, 98, 107, 108, 119, 120, 122, 123, 127, 128, 131, 132, 139, 146, 155, 157, 161, 163, 167, 170, 203, 204, 206, 221, 243, 244, 256, 258, 262, 264, 271, 273, 281, 285, 286, 288, 290, 291, 316, 322, 328, 334, 338, 340
Compost 24, 72, 149, 154, 160, 173, 174, 179, 185, 189, 194, 199, 206, 210, 224, 227, 228, 229, 238
Contour Ploughing 7, 47, 52, 53, 55, 58, 63, 66, 70, 91, 131, 135, 149, 150, 165, 168, 172, 174, 177, 185, 203, 232, 239, 270, 271, 297, 299
Cover crops 16, 34, 39, 40, 41, 44, 50, 52, 59, 61, 69, 70, 92, 131, 143, 144, 145, 148, 149, 150, 153, 160, 162, 168, 173, 175, 176, 199, 206, 207, 208, 214, 215, 222, 226, 232, 234, 274, 345, 350, 353, 355, 356

D

Decomposition 1, 2, 5, 6, 8, 9, 10, 11, 15, 20, 21, 27, 28, 29, 31, 32, 34, 41, 130, 131, 143, 153, 154, 172, 176, 179, 183, 184, 186, 187, 188, 191, 192, 193, 197, 205, 206, 220, 221, 222, 224, 227, 253, 356
Desalination 85, 108, 132, 139
Drainage 2, 9, 12, 13, 14, 15, 17, 18, 53, 61, 129, 131, 133, 153, 166, 168, 170, 171, 172, 173, 174, 175, 178, 180, 190, 194, 195, 198, 205, 285, 286, 287, 288, 294, 297, 298, 313, 330, 332, 336, 342, 343, 344, 345, 346, 347, 348, 349, 350, 351, 352, 353, 354, 355, 356
Drought 5, 18, 37, 40, 83, 96, 98, 108, 112, 128, 129, 132, 139, 148, 150, 151, 155, 156, 158, 169, 175, 176, 177, 205, 215, 221, 245, 254, 287, 288, 294, 295, 297, 311, 315, 316, 319, 321, 322, 323, 324, 325, 334, 336, 337, 338, 341, 352
Dryland 149, 150

E

Ecosystems 4, 21, 52, 53, 115, 126, 209, 235, 236
Erosion 15, 26, 38, 39, 41, 42, 43, 44, 45, 46, 47, 49, 52, 43, 54, 56, 57, 58, 59, 61, 62, 64, 65, 66, 67, 72, 74, 75, 78, 92, 96, 125, 126, 127, 129, 130, 135, 145, 149, 159, 161 162, 164, 165, 168, 172, 176, 180, 185, 203, 205,

209, 214, 219, 223, 232, 233, 234, 244, 245, 324, 333, 347
Evaporation 99, 130, 223, 288, 323, 325, 326

F
FAO 122
Fertility 4, 11, 25, 26, 125, 152, 153, 169, 202, 225, 244
Filtration 65, 94, 103, 211, 269, 272
Fire 123
Furrow 92, 266

G
Germination 83, 187, 192, 353
GIS 7, 63, 68, 72, 105, 116, 120, 139, 278, 279
Granular 14
Green Manure 175, 207, 208, 226

H
Halophyte 133
Humid 151
Hydroponics 132

I
Industrialization 123
Integrated Pest Management (IPM) 144, 151, 159, 216, 23,
Integrated Water Resource Management (IWRM) 107
Irrigation 19, 91, 92, 93, 94, 97, 98, 99, 100, 101, 104, 105, 106, 107, 132, 138. 146, 147, 148, 151, 152, 153, 155, 164, 210, 211, 235, 263, 264, 265, 266, 267, 268, 269, 271, 272, 274, 275, 276, 277, 278, 279, 280, 284, 293, 298, 305, 306, 307, 309, 311, 325, 330, 334, 325, 339, 340, 342, 343, 344, 351

L
Landscapes 230
Leaching 130, 186, 189, 191, 194, 307, 345
Livestock 164, 248

M
Mangrove 119
Manure 175, 176, 206, 207, 208, 224, 226, 355
Metagenomics 30
MoEFCC 121
Mulching 151, 177, 210, 221, 222, 223, 232, 238, 271, 274, 299
Mycorrhizal Associations 197, 300

N
NLRMP 121
NRSC 121

P
Pasture 156, 158
Pedogenesis 1, 8
Phytoremediation 103, 116, 133, 136, 139, 181
Pollution 113, 118, 168, 209
Purification

R
Rainfall 44, 45, 96, 211, 315, 318, 319, 343, 352
Rainfed 150
Rainwater 85, 90, 91, 98, 164, 175, 210, 211, 212, 274, 313, 329, 330, 331, 332, 333, 334, 336
Ravine 133
Reforestation 62, 63, 65, 115, 11, 181
Remote sensing 7, 26, 105, 126, 127, 288, 311
Reservoirs 85, 86, 87, 88, 89, 268
Rhizosphere 33, 229, 290
Robotics 106

S
Saline 118, , 133, 171, 173, 186, 191, 192
Satellite 68, 105, 107, 162, 279, 311, 317
Sedimentation 48, 54, 55, 58, 61, 65, 125, 126, 130
Shelterbelts 51, 182, 247, 248
Silvopasture 163, 231, 247, 248
Sprinklers 91, 92, 93, 269
Subterranean 89
Symbiosis 33

T
Tensiometer 293, 301, 308, 309
Tillage 37, 38, 42, 69, 148, 149, 151, 161, 179, 185, 219, 234, 274
Topography 9, 153, 171, 327, 352
Transpiration 83, 288, 289, 300, 323, 325, 326, 327,

U
Urbanization 26, 65, 166
USDA 3, 7, 35, 36, 37

V
Variable Rate Technology (VRT) 104
Vegetation 17, 44, 45, 46, 48, 50, 60, 148, 171, 234, 328, 354

W
Wasteland 111112, 113, 114, 115, 116, 117, 118, 119, 120, 121, 122, 132, 133, 134, 136
Waterlogging 125, 129, 187, 192, 345, 346, 347
Water retention 18, 262
Water Use Efficiency (WUE) 152
Weathering 20, 66, 153, 183
Wetlands 6, 103, 109, 151, 355
Wildlife 231, 356
Windbreak 50, 51, 132, 149, 163, 182, 233, 244, 247, 248, 249, 355